Power System Small Signal Stability Analysis and Control

Power System Small Signal Stability Analysis and Control

Debasish Mondal

HALDIA INSTITUTE OF TECHNOLOGY, P.O. –HATIBERIA,
HIT, PURBA MEDINIPUR, HALDIA–721657

Abhijit Chakrabarti

BENGAL ENGINEERING & SCIENCE
UNIVERSITY, P.O. –B. GARDEN,
SHIBPUR, HOWRAH–711103

Aparajita Sengupta

BENGAL ENGINEERING AND SCIENCE
UNIVERSITY, P.O. –B. GARDEN,
SHIBPUR, HOWRAH–711103

AMSTERDAM • BOSTON • HEIDELBERG • LONDON
NEW YORK • OXFORD • PARIS • SAN DIEGO
SAN FRANCISCO • SINGAPORE • SYDNEY • TOKYO

Academic Press is an imprint of Elsevier

Academic Press is an imprint of Elsevier
32 Jamestown Road, London NW1 7BY, UK
The Boulevard, Langford Lane, Kidlington, Oxford OX5 1GB, UK
Radarweg 29, PO Box 211, 1000 AE Amsterdam, The Netherlands
225 Wyman Street, Waltham, MA 02451, USA
525 B Street, Suite 1800, San Diego, CA 92101-4495, USA

First edition **2014**

Notice
Knowledge and best practice in this field are constantly changing. As new research and experience
broaden our understanding, changes in research methods, professional practices, or medical treatment
may become necessary.

Practitioners and researchers must always rely on their own experience and knowledge in evaluating
and using any information, methods, compounds, or experiments described herein. In using such
information or methods they should be mindful of their own safety and the safety of others, including
parties for whom they have a professional responsibility.

To the fullest extent of the law, neither the Publisher nor the authors, contributors, or editors, assume
any liability for any injury and/or damage to persons or property as a matter of products liability,
negligence or otherwise, or from any use or operation of any methods, products, instructions, or ideas
contained in the material herein.

Library of Congress Cataloging-in-Publication Data
Application submitted

British Library Cataloguing in Publication Data
A catalogue record for this book is available from the British Library

For information on all **Academic Press** publications
visit our web site at store.elsevier.com

Printed and bound in USA
14 15 16 17 18   10 9 8 7 6 5 4 3 2 1

ISBN: 978-0-12-800572-9

To my all Teachers

—Debasish Mondal

Contents

Acknowledgments

Authors express their sincere thanks and deepest sense of gratitude to all the reviewers of this book for their valuable suggestions and comments to enrich and upgrade the contents of this book.

Authors would also wish to convey their gratitude to all faculty members of the Department of Electrical Engineering, Bengal Engineering and Science University (BESU), West Bengal, India, for their wholehearted cooperation to make this work turn into a reality. Thanks to the faculty members of the Department of Applied Electronics and Instrumentation Engineering, Haldia Institute of Technology (HIT), West Bengal, India, for their cooperation.

The authors also acknowledge the interest and effort of the entire editorial and management teams of Elsevier Inc. Science and Technology Books and Energy & Power Group.

Last but not the least, we feel proud for the respect and encouragement we had received from our family members to carry out this work.

The authors cordially invite any constructive criticism or comment from the reader about the book.

Author Biography

Debasish Mondal received his B.Tech. and Master of Engineering degrees in the years 1998 and 2000 form Calcutta University, Kolkata, India, and Bengal Engineering and Science University, Shibpur, Howrah, India, respectively. He did Ph.D. in Engineering in 2012 from the Department of Electrical Engineering, Bengal Engineering and Science University. He has 14 years of industrial and teaching experience and has 20 publications in different international journals and conferences. He holds a permanent post of Associate Professor at the Haldia Institute of Technology, Haldia, India. His active research interests are in the areas of power systems stability, FACTS device, soft computing, robust control, and nonlinear control system. Dr. Mondal is a life member of Institution of Engineers (IE) India.

Abhijit Chakrabarti received B.E., M.Tech., and Ph.D. (Tech) degrees in 1978, 1987, and 1991 from NIT, Durgapur; IIT, Delhi and Calcutta University, respectively. He is a professor at the Department of Electrical Engineering, Bengal Engineering and Science University, Shibpur, Howrah, India. He has 26 years of research and teaching experience and has around 121 research papers in national and international journal and conferences. He is an author of 12 important books in different fields of electrical engineering. Professor Chakrabarti is a recipient of Pandit Madan Mohan Malviya award, and the power medal, merit, and best paper awards (twice) from Central Electricity Authority, India. He holds several responsible administrative positions in different academic bodies and institutions. He also acted as expert in different selection committees of Universities in India. He has an active research interest in the areas of power systems stability, power electronics, circuit theory, congestion management, and FACTS Devices. He is a fellow of the Institution of Engineers (India).

Aparajita Sengupta (nee Rai Chaudhuri) completed her B.E. in Electrical Engineering from Jadavpur University, Kolkata, in 1992. She did her M.Tech. in Control Systems in 1994 and Ph.D. in Engineering in 1997 from the Department of Electrical Engineering, IIT, Kharagpur. In 1997, she joined as a Lecturer in Electrical Engineering in the present Bengal Engineering and Science University, Shibpur, Howrah, India. Now she is working as an Associate Professor there. She has 40 publications to her credit in theoretical control and its applications in power systems, electrical machines, power electronics, as well as aerospace systems. Her research topics include nonlinear uncertain systems, real-time systems, robust control, H-infinity optimal control, and its applications in various fields of electrical engineering. She has also been involved in sponsored research to the tune of nearly Rs. 1 crore.

Preface

In the wide and longitudinal transmission circuits, long transmission lines are common features for transporting electric power from generating stations to the load centers, and there, 220 and 400 kV lines are most common. Stability of these transmission systems is thus a major concern, and it has been observed that there were blackouts because of small-signal oscillations in the power systems. The utility companies encounter the problem of voltage stability and small signal stability, particularly during heavy-loading periods. If this problem is not damped properly, they may cause instability. Damping of these oscillations can be enhanced by eliminating the nonlinear loads in the distribution system. Installation of HVDC line between two regional EHV grids in addition to EHV AC link can eliminate the problem of low-frequency oscillations provided that the line power flows in AC and DC links are carefully monitored during heavy-loading periods. This book aims at exploring the issue of this small-signal stability problem in single-machine infinite bus (SMIB) power systems and in multimachine power systems and its mitigation applying power system controllers.

Traditional and efficient ways to mitigate these problems are to derive additional signals for the generator excitation systems and to compensate the fluctuations of power flow through the transmission networks. Power system stabilizer (PSS) and flexible alternating current transmission system (FACTS) devices have remarkable capability to perform these tasks. The objective of this book is to present a detailed study, simulation, and analysis of the small signal stability problem and its mitigation applying PSS and FACTS (SVC, STATCOM, and TCSC) controllers. The simulation results have been presented employing eigenvalue and time domain analysis in MATLAB.

The text starts with fundamental discussions on small-signal stability problems and their common features. Basic models of synchronous machine are overviewed. The small-signal models of different power system components relevant to the text and their installation in power system are described. The modeling of FACTS controllers and their installation in a SMIB power system and in multimachine systems are the major thrust area of this book. The effectiveness and performance of the FACTS controllers (SVC, TCSC, and STATCOM) are investigated and compared; in particular, the superiority of these FACTS controllers over traditional PSS controllers has been shown in face of commonly occurring power system disturbances. The methods of design of optimal FACTS controller and finding their best location in a multimachine power system are also important aspects of this book.

This book does not claim to be a detailed study of general electric power system, power system dynamics, synchronous machine, and all types of problems related to the small signal oscillations.

The salient features of this book are as follows: (i) Easy understanding of power system small signal stability problem and its different ways of mitigation. (ii) Detail derivation of multimachine two-axis model for the evaluation of system matrix and eigenvalue. (iii) Model of supplementary damping controllers for PSS and advance FACTS devices. (iv) Multimachine model with PSS in order to mitigate small signal stability problem. (v) A new and simple method of finding optimal location of PSS in a multimachine power system. (vi) Multimachine small signal model with FACTS controllers: SVC, TCSC, and VSC-based FACTS device (STATCOM). (vii) Knowledge of design and application of optimal FACTS controller applying soft computation techniques (Genetic Algorithm, Particle Swarm Optimization) and H_∞ optimization method. (viii) Chapterwise illustrations/solved problems and exercise.

The book has been written utilizing the long experience of the authors in teaching control system and electrical engineering. The contents of this book will be useful for all levels of students (UG and PG) and professors of electrical engineering for the study and analysis of power system small signal stability problem and its control in particular; it will be a ready reference for students of postgraduate and doctoral degrees in electrical and allied branches of engineering. This book will also be a good resource for practicing engineers and research scholars/fellows of power system.

Debasish Mondal
Abhijit Chakrabarti
Aparajita Sengupta

Concepts of Small-Signal Stability

1.1 INTRODUCTION

Small-signal (or small disturbance) stability is the ability of the power system to maintain synchronism under small disturbances such as small variations in loads and generations. Physically power system stability can be broadly classified into two main categories – *angle stability* or rotor angle stability and *voltage stability* [1].

Angle stability or rotor angle stability can be defined as "the ability of interconnected synchronous machines of a power system to remain in synchronism." This stability problem involves the study of electromechanical oscillations inherent in power systems. A fundamental issue here is the manner in which the power output of synchronous machines varies as their rotors oscillate.

Voltage stability can be broadly defined as "the ability of a system to maintain steady acceptable voltages at all buses following a system contingency or disturbance." A system enters into a state of voltage instability when a disturbance, increase in load demand, or change in system condition causes a progressive and uncontrollable drop in voltage. The main factor causing instability is the inability of the power system to meet the demand for active and reactive power flow through the transmission network. A power system, at a given operating state, is *small disturbance voltage stable* if, following any small disturbance, voltages near the loads do not change or remain close to the predisturbance values. The concept of small disturbance voltage stability is related to the steady-state stability and can be analyzed using the small-signal (linearized) model of the system.

The category of angle stability can be considered in terms of two main subcategories:

1. *Steady-state/dynamic*: This form of instability results from the inability to maintain synchronism and/or dampen out system transients and oscillations caused by small system changes, such as continual changes in load and/or generation.
2. *Transient*: This form of instability results from the inability to maintain synchronism after large disturbances such as system faults and/or equipment outages.

The aim of transient stability studies being to determine if the machines in a system will return to a steady synchronized state following a large disturbance.

The literature of this book will focus in particular on the steady-state/dynamic stability subcategory and on the techniques that can be used to analyze and control the dynamic stability of a power system following a small disturbance.

1.2 SWING EQUATION

This equation bears the dynamics of oscillations of rotor of a synchronous generator. Consider a generating unit consisting of a three-phase synchronous generator and prime mover, as shown in Figure 1.1.

The motion of the synchronous generator's rotor is determined by Newton's second law, which is given as [2]

$$J\alpha_m(t) = T_m(t) - T_e(t) = T_a(t) \tag{1.1}$$

where J is the total moment of inertia of the rotating masses (prime mover and generator) (kg m^2), α_m is the rotor angular acceleration (rad/s^2), T_m is the mechanical torque supplied by the prime mover minus the retarding torque due to mechanical losses (e.g., friction) (N m), T_e is the electrical torque, accounting for the total three-phase power output and losses (N m), and T_a is the net accelerating torque (N m).

The machine and electrical torques, T_m and T_e, are positive for generator operation.

The rotor angular acceleration is given by

$$\alpha_m(t) = \frac{d\omega_m}{dt} = \frac{d^2\theta_m(t)}{dt^2} \tag{1.2}$$

$$\omega_m(t) = \frac{d\theta_m}{dt} \tag{1.3}$$

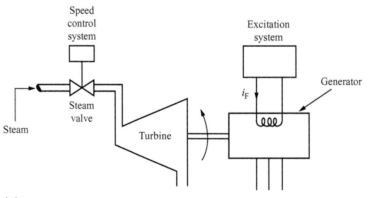

FIGURE 1.1

Generating unit.

where ω_m is the rotor angular velocity (rad/s) and θ_m is the rotor angular position with respect to a stationary axis (rad).

In steady-state conditions, the mechanical torque equals the electrical torque and the accelerating torque is zero. There is no acceleration and the rotor speed is constant at the synchronous velocity. When the mechanical torque is more than the electrical torque, then the acceleration torque is positive and the speed of the rotor increases. When the mechanical torque is less than the electrical torque, then the acceleration torque is negative and the speed of the rotor decreases. Since we are interested in the rotor speed relative to the synchronous speed, it is convenient to measure the rotor angular position with respect to a synchronously rotating axis instead of a stationary one.

We therefore define

$$\theta_m(t) = \omega_{msyn} t + \delta_m(t) \tag{1.4}$$

where ω_{msyn} is the synchronous angular velocity of the rotor (rad/s) and δ_m is the rotor angular position with respect to a synchronously rotating reference.

To understand the concept of the synchronously rotating reference axis, consider the diagram in Figure 1.2. In this example, the rotor is rotating at half the synchronous speed, $\omega_{msyn}/2$, such that in the time it takes for the reference axis to rotate 45°, the rotor only rotates 22.5° and the rotor angular position with reference to the rotating axis changes from −45° to −67.5°.

Using Equations (1.2) and (1.4) in Equation (1.1), we have

$$J\alpha_m(t) = J\frac{d^2\theta_m(t)}{dt^2} = J\frac{d^2\delta_m(t)}{dt^2} = T_m(t) - T_e(t) = T_a(t) \tag{1.5}$$

Being that we are analyzing a power system, we are interested in values of power more than we are in values of torque. It is therefore more convenient to work with expressions of power. Furthermore, it is convenient to consider this power in per unit rather than actual units.

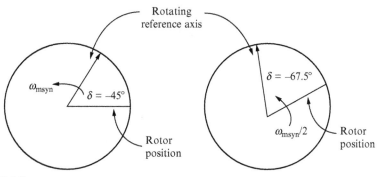

FIGURE 1.2

Synchronously rotating reference axis.

Power is equal to the angular velocity times the torque and per-unit power can be obtained by dividing by S_{rated}, so that

$$\frac{J\omega_m}{S_{rated}}\frac{d^2\delta_m(t)}{dt^2} = \frac{\omega_m T_m(t) - \omega_m T_e(t)}{S_{rated}} = \frac{P_m(t) - P_e(t)}{S_{rated}} = P_{mpu}(t) - P_{epu}(t) \quad (1.6)$$

P_{mpu} is the mechanical power supplied by the prime mover minus mechanical losses (per unit), P_{epu} is the electrical power output of generator plus electrical losses (per unit), and S_{rated} is the generator volt-ampere rating.

We here define a constant value known as the normalized inertia constant, or "H" constant:

$$H = \frac{\text{stored kinetic energy at synchronous speed}}{\text{generator volt-ampere rating}}$$

$$= \frac{\frac{1}{2}J\omega_{msyn}^2}{S_{rated}} \quad \text{(J/VA or per-unit seconds)}$$

Equation (1.6) becomes

$$2H\frac{\omega_m(t)}{\omega_{msyn}^2}\frac{d^2\delta_m(t)}{dt^2} = P_{mpu}(t) - P_{epu}(t) = P_{apu}(t) \quad (1.7)$$

where P_{apu} is the accelerating power.

We define per-unit rotor angular velocity as

$$\omega_{pu}(t) = \frac{\omega_m(t)}{\omega_{msyn}} \quad (1.8)$$

Equation (1.7) becomes

$$\frac{2H\omega_{pu}(t)}{\omega_{msyn}}\frac{d^2\delta_m(t)}{dt^2} = P_{mpu}(t) - P_{epu}(t) = P_{apu}(t) \quad (1.9)$$

When a synchronous generator has P poles, the synchronous electrical angular velocity, ω_{syn}, known more correctly as the synchronous electrical radian frequency, can be related to the synchronous mechanical angular velocity by the following relationship:

$$\omega_{syn} = \frac{P}{2}\omega_{msyn} \quad (1.10)$$

To understand how this relationship arises, consider that the number of mechanical radians in one full revolution of the rotor is 2π. If, for instance, a generator has four poles (two pairs) and there are 2π electrical radians between poles in a pair, then the electrical waveform will go through $2 \times 2\pi = 4\pi$ electrical radians within the same revolution of the rotor. In general, the number of electrical radians in one revolution is the number of mechanical radians times the number of pole pairs (the number of poles divided by two).

The relationship shown in Equation (1.10) also holds for the electrical angular acceleration $\alpha(t)$, the electrical radian frequency $\omega_r(t)$, and the electrical power angle $\delta(t)$ values:

$$\alpha(t) = \frac{P}{2}\alpha_m(t)$$

$$\omega_r(t) = \frac{P}{2}\omega_m(t)$$

$$\delta(t) = \frac{P}{2}\delta_m(t) \tag{1.11}$$

From Equation (1.8), we have

$$\omega_{pu}(t) = \frac{\omega_m(t)}{\omega_{msyn}} = \frac{\frac{2}{P}\omega_r(t)}{\frac{2}{P}\omega_{syn}} = \frac{\omega_r(t)}{\omega_{syn}} \tag{1.12}$$

Therefore, Equation (1.9) can be written in electrical terms rather than mechanical:

$$\frac{2H}{\omega_{syn}}\omega_{pu}(t)\frac{d^2\delta(t)}{dt^2} = P_{mpu}(t) - P_{epu}(t) = P_{apu}(t) \tag{1.13}$$

Equation (1.13) represents the equation of motion of synchronous machine. It is commonly referred to as the "swing equation" because it represents swing in rotor angle δ during disturbances and it is the fundamental equation in determining rotor dynamics in transient stability studies.

The swing equation is nonlinear because $P_{epu}(t)$ is a nonlinear function of rotor angle δ and because of the $\omega_{pu}(t)$ term. The rotor speed, however, does not vary a great deal from the synchronous speed during transients, and a value of $\omega_{pu}(t) \approx 1.0$ is often used in hand calculations. Defining $M = \frac{2H}{\omega_{syn}}$, the equation in the preceding text becomes

$$M\frac{d^2\delta(t)}{dt^2} = T_m - T_e \tag{1.14}$$

It is often desirable to include a component of damping torque, not accounted for in the calculation of T_e, separately. This is accomplished by introducing a term proportional to speed deviation in the preceding equation. The equation of motion considering damping torque has been shown later in Equation (1.22).

1.3 NATURE OF OSCILLATIONS

Oscillations in the power system have the following properties:

1. Oscillations are due to natural modes of the system and therefore cannot be completely eliminated.

2. With increase in complexity of the power system, the frequency and damping of oscillations may increase and new ones may be added.
3. Automatic voltage regulator (AVR) control is the primary source of introducing negative damping torque in the power system. With increase in the number of controls, negative damping may further increase.
4. Inter-area oscillations are associated with weak transmission lines and larger line loadings.
5. Inter-area oscillations may involve more than one utility.
6. Damping of the system is to be enhanced to control these tie-line oscillations.

1.4 MODES OF OSCILLATIONS AND ITS STUDY PROCEDURE

The disturbance is considered to be small, and therefore, the equations that describe the resulting response of the system can be linearized. The electromechanical oscillations are of two types [3]:

(i) *Local mode oscillations*, which are associated with the swing of units at a generating station with respect to the rest of the power systems. Typical range of frequency of oscillations is 1-3 Hz. The term local is used because the oscillations are localized at one station or a small part of the power system.
(ii) *Inter-area mode oscillations*, which are associated with the swing of many machines in one part of the system against the machines in other parts or areas. Typical range of frequency of these types of oscillations is less than 1 Hz. They are caused by two or more groups of closely coupled machines being interconnected by weak ties.

There are two methods of analysis that are available in literature to study the aforementioned electromechanical oscillations:

(a) A linearized single-machine infinite-bus system case that investigates only local oscillations.
(b) A multimachine linearized analysis that computes the eigenvalues and also finds those machines that contribute to a particular eigenvalue; both local and inter-area modes can be studied in such framework.

The modes of oscillation referred to earlier are electromechanical in nature. There are another two types of oscillation modes that are also analyzed for a synchronous machine – (i) *control mode or exciter mode* and (ii) *torsional oscillation mode*.

Control modes are associated with generator or the exciter units and other control equipments. Poorly tuned exciters, speed governors, HVDC converters, and static Var compensators are the usual causes of instability of these modes. The frequency of the control mode is close to 3 Hz.

Torsional oscillation modes are associated with the turbine-generator shaft rotational system. Instability of these types of modes is generally caused by interaction among control equipments and interaction between exciter control, speed governors,

HVDC controls, and series-compensated line, etc. The frequency of this mode of oscillation is usually in the range of 10-50 Hz.

Voltage stability or dynamic voltage stability is analyzed by monitoring the eigenvalues of the linearized power system with progressive loading. Instability occurs when a pair of complex conjugate eigenvalue crosses the right half of *s*-plane. This is referred to as dynamic voltage instability. Mathematically, this phenomenon is called *Hopf bifurcation*. As real power is related to rotor angle instability, similarly reactive power is central to voltage instability analyses. Deficit or excess reactive power leads to voltage instability either locally or globally, and any increase in loading may lead to voltage collapse. The analysis of voltage stability normally requires simulation of the power system modeled by nonlinear differential algebraic equations.

Since small-signal stability is based on a linearized model of the system around its equilibrium operating points, formulation of the problem is very important. The formulation of the state equations for small-signal stability analysis involves the development of linearized equations about an operating point and elimination of all variables other than the state variables. The small-signal models of different power system components are described in Chapter 3, and they are used in successive chapters for small-signal stability analysis.

1.5 SYNCHRONIZING TORQUE AND DAMPING TORQUE

The nature of the system response to small disturbances depends on a number of factors such as the initial conditions, the transmission system strength, and the type of generator excitation control and largely on the value of electrical torque. In electrical power systems, the change in electrical torque of a synchronous machine following a perturbation due to disturbance can be resolved into two components [4]:

$$\Delta T_e = T_s \Delta \delta + T_D \Delta \omega \qquad (1.15)$$

Here, $T_s \Delta \delta$ and $T_D \Delta \omega$ are, respectively, the component of electrical torque change in phase with the rotor angle deviation ($\Delta \delta$) and speed deviation ($\Delta \omega$). $T_s \Delta \delta$ and $T_D \Delta \omega$ are referred to as the *synchronizing torque* component and *damping torque* component; T_s and T_D are, respectively, denoted as the synchronizing torque coefficient and damping torque coefficient.

System stability depends on the existence of both components of torque for the synchronous machines. A lack of sufficient synchronizing torque results in instability through an *aperiodic drift* in rotor angle. On the other hand, lack of sufficient damping torque results in *oscillatory instability*.

For a generator connected radially to a large power system and continuously acting as AVR, the instability happens because of insufficient damping torque, whereas in the absence of AVR, the instability is due to lack of sufficient synchronizing torque. Figure 1.3a shows the synchronous machine stable operation in the presence or in the absence of AVR. Instability is normally through oscillations of increasing amplitude. Figure 1.3b and c illustrates the nature of small-signal instability response of generators acting with and without an AVR, respectively.

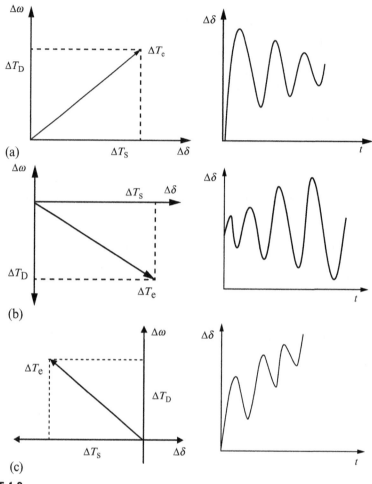

FIGURE 1.3

(a) Stable operation with or without AVR: positive synchronizing torque (T_s) and positive damping torque (T_D). (b) Oscillatory instability with AVR: positive synchronizing torque (T_s) and negative damping torque (T_D). (c) Aperiodic drift without AVR: negative synchronizing torque (T_s) and positive torque (T_D).

1.6 SMALL-SIGNAL OSCILLATIONS IN A SYNCHRONOUS GENERATOR CONNECTED TO AN INFINITE BUS

A synchronous generator connected to an infinite-bus bar of voltage $V_b \angle 0°$ through a reactance X_e is given in Figure 1.4. Here, armature and line resistances are neglected for the purpose of analysis. The generator is represented by the classical model, in which the voltage E' behind the transient reactance X'_d remains constant at

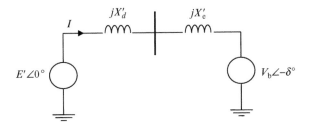

FIGURE 1.4

Single-machine infinite-bus system.

the predisturbance value. Let E' lead the bus voltage by angle $\angle \delta°$, which keeps changing as the machine undergoes small oscillations following the disturbance. If we take E' as the reference phasor, then [5]

$$I = \frac{E'\angle 0° - V_b\angle - \delta°}{j(X_d' + X_e)} = \frac{E'\angle 0° - V_b\angle - \delta°}{jX_T} \tag{1.16}$$

where $X_T = X_d' + X_e$. The complex power behind X_d' is given by

$$S = E'I^* = \frac{E'V_b\sin\delta}{X_T} + j\frac{E'(E' - V_b\cos\delta)}{X_T} \tag{1.17}$$

Since all resistances are neglected, the air gap power is equal to the terminal power and, in per unit, equal to the air gap torque. We have

$$T_e = P = \frac{E'V_b\sin\delta}{X_T} \tag{1.18}$$

For small increments, linearizing equation (1.18) around the operating condition represented by $\delta = \delta_o$ results in

$$\Delta T_e = \Delta P = \frac{E'V_b\cos\delta_o}{X_T}\Delta\delta \tag{1.19}$$

Writing $\omega_{sys} = \omega_o$, the rotor equation of motion (1.14) is given by

$$\frac{2H}{\omega_o}\frac{d^2\delta}{dt^2} = T_m - T_e \tag{1.20}$$

Equation (1.20) is linearized to represent small-signal oscillations and is given by

$$\frac{2H}{\omega_o}\frac{d^2\Delta\delta}{dt^2} = \Delta T_m - \Delta T_e \tag{1.21}$$

If we add now damping torque ($\Delta T_D = D\Delta\omega$) and assume that governor action T_m (mechanical torque) remains constant and $\Delta T_m = 0$, Equation (1.21) becomes

$$\frac{2H}{\omega_o}\frac{d^2\Delta\delta}{dt^2} = -\frac{E'V_b\cos\delta_o}{X_T}\Delta\delta - D\Delta\omega \tag{1.22}$$

If δ is the angular position of the rotor in electrical radians with respect to a synchronously rotating reference and δ_o is its value at $t=0$,

$$\delta = \omega_r t - \omega_{syn} t + \delta_o \qquad (1.23)$$

Taking the time derivative, we have

$$\frac{d\delta}{dt} = \omega_r - \omega_{syn} \qquad (1.24)$$

Linearizing equation (1.24), we have

$$\frac{d\Delta\delta}{dt} = \Delta\omega_r \qquad (1.25)$$

Again, $\Delta\omega = \dfrac{\Delta\omega_r}{\omega_o} = \dfrac{1}{\omega_o}\dfrac{d\Delta\delta}{dt}$.

Substituting in Equation (1.22) gives

$$\frac{2H}{\omega_o}\frac{d^2\Delta\delta}{dt^2} + \frac{D}{\omega_o}\frac{d\Delta\delta}{dt} + \frac{E'V_b\cos\delta_o}{X_T}\Delta\delta = 0 \qquad (1.26)$$

The term $\dfrac{E'V_b\cos\delta_o}{X_T}$ is the synchronizing torque coefficient T_s. If the steady-state air gap power P_o prior to the disturbance is

$$P_o = \frac{E'V_b\sin\delta_o}{X_T} \quad \text{and} \quad T_s = \frac{E'V_b\cos\delta_o}{X_T}$$

then it follows that $\dfrac{P_o}{T_s} = \tan\delta_o$ or

$$T_s = P\cot\delta_o \qquad (1.27)$$

Equation (1.27) provides an approximate value of the synchronizing torque coefficient in per unit. If the steady-state power (P_o) of a synchronous generator connected to an infinite-bus bar is 0.8 pu and $E' \angle 0°$ leads the bus voltage by $\delta_o \angle 30°$, then the per-unit value of synchronizing torque coefficient is $T_s = 1.3856$ pu.

If we replace the $\dfrac{d}{dt}$ operator by s in Equation (1.26), then for undamped $(D=0)$ equation of motion is

$$\frac{2H}{\omega_o}s^2\Delta\delta + T_s\Delta\delta = 0$$

or

$$\frac{2H}{\omega_o}s^2\Delta\delta + T_s\Delta\delta = 0 \qquad (1.28)$$

$$s = \pm j\sqrt{\frac{T_s}{2H/\omega_o}} = \pm j\sqrt{\frac{\omega_o E'V_b\cos\delta_o}{2HX_T}}\,\text{rad/s} \qquad (1.29)$$

For a case when $P_o = 0.8$, $\delta_o = 30°$, $T_s = 1.3856$ pu, and $H = 6$ s, the frequency of oscillations of a 50 Hz system is

$$s = \pm j \sqrt{\frac{1.3856 \times 314}{2 \times 6}}; \quad \omega_n = 2\pi f_o = 6.02\,\text{rad/s}$$

Again from Equation (1.25), we have

$$\frac{d^2 \Delta \delta}{dt^2} + \frac{D}{2H}\frac{d\Delta\delta}{dt} + \frac{\omega_o T_s}{2H}\Delta\delta = 0 \tag{1.30}$$

Therefore, the characteristic equation is given by

$$s^2 + \frac{D}{2H}s + \frac{\omega_o T_s}{2H} = 0 \tag{1.31}$$

that can be written in general form

$$s^2 + 2\zeta\omega_n s + \omega_n^2 = 0 \tag{1.32}$$

Therefore, the undamped natural frequency is given by $\omega_n = \sqrt{\frac{\omega_o T_s}{2H}}$ rad/s and the damping ratio (ζ) is given by

$$\zeta = \frac{1}{2}\left(\frac{D}{2H\omega_n}\right) = \frac{1}{2}\left(\frac{D}{\sqrt{2H\omega_o T_s}}\right)$$

It is very clear that ΔT_e will not yield any imaginary term since all resistances and control actions have been neglected and damping will not be represented by this simplified synchronous machine model. In fact, it was suspected that the damping of oscillations in synchronous machine originates in winding resistances due to copper loss by oscillating current and in the damping controllers (such as power system stabilizers) that introduce phase shift. The introduction of damper windings was based on this perception, and the damper bars were made of brass in order to offer relatively large resistance to cause large dissipations of power.

1.7 AN ILLUSTRATION

A generator supplies power in steady state to an infinite-bus 50 Hz system shown in Figure 1.5. Assuming there are some contingencies and transmission line #2 gets outage, find out the following: (i) undamped natural frequency of oscillations, (ii) damping ratios, and (iii) damped frequency of oscillations for three different damping coefficients 0, 5, and −5. The pu values of the system parameters on a 1500 MVA, 25 kV base, are given as follows [6]:

$$P = 1\,\text{pu}; \quad Q = 0.3\,\text{pu}; \quad V = 1\angle 17°, \quad V_b = 0.99\angle 0°; \quad |X_{L_1}| = |X_{L_2}| = 0.5; \quad |X_{tr}| = 0.2; \quad |X_d'| = 0.25; \quad H = 3\,\text{MW s/MVA}$$

All voltage magnitudes and reactances are expressed in pu.

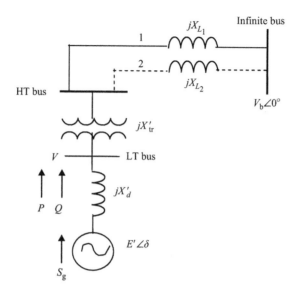

FIGURE 1.5

Schematic diagram of a generator connected to infinite-bus system.

Solution

The generator bus voltage is taken as reference phasor. Generator output current (I_g) is obtained as

$$I_g = \frac{(P+jQ)^*}{V^*} = \frac{1-j0.3}{1.0} = (1-j0.3)\,\text{pu}$$

$$= 1.044\angle -16.7°\,\text{pu}$$

Voltage behind transient reactance (X'_d) is obtained as

$$E' = V + jX'_d I_g = (1+j0) + j0.25(1-j0.3)$$

$$= (1.075 + j0.25)\,\text{pu} = 1.104\angle 13.09°\,\text{pu}$$

Hence, the angle by which E' leads V_b is $(17° + 13.09°) = 30.09°(\delta_o)$.

The total reactance of the system is given by $|X| = |X_L| + |X_{tr}| + |X'_d|$, taking one line in the system. Therefore, $|X| = 0.5 + 0.2 + 0.25 = 0.95\,\text{pu}$.

∴ The synchronizing torque coefficient (T_s) is given by

$$T_s = \frac{E'V_b\cos\delta_o}{X_T} = \frac{1.104 \times 0.99 \times \cos 30.09}{0.95} = 0.995\,\text{pu}$$

(i) Undamped natural frequency of oscillation is obtained as

$$\omega_n = \sqrt{\frac{\omega_o T_s}{2H}} \ \text{rad/s}$$

$$= \sqrt{0.995 \times \frac{2\pi \times 50}{2 \times 3}} = 7.22 \ \text{rad/s}$$

$$f_n = 1.15 \text{Hz}$$

∴ The undamped natural frequency ω_n (=7.22 rad/s) is independent of damping coefficient and hence will remain the same for all the three given damping coefficients in the question.

(ii) $$\xi = \frac{1}{2}\left(\frac{D}{\sqrt{2H\omega_o T_s}}\right)$$

For $D = 0$; $\xi = 0$

For $D = 5$; $\xi = \frac{1}{2}\left(\frac{5}{\sqrt{2 \times 3 \times 2\pi \times 50 \times 0.995}}\right) = 0.058$

For $D = -5$; $\xi = \frac{1}{2}\left(\frac{5}{\sqrt{2 \times 3 \times 2\pi \times 50 \times 0.995}}\right) = -0.058$

(iii) The damped frequency (ω_d) can be obtained from the following formula:

$$\omega_d = \omega_n\sqrt{1 - \xi^2}$$

For $D = 0$, $\omega_d = \omega_n\sqrt{1 - \xi^2} = 7.22\sqrt{1 - 0} = 7.22 \ \text{rad/s} = 1.15 \text{Hz}$

$D = 5$, $\omega_d = 7.22\sqrt{1 - (0.058)^2} = 7.208 \ \text{rad/s} = 1.148 \text{Hz}$

$D = -5$, $\omega_d = 7.22\sqrt{1 - (-0.058)^2} \ \text{rad/s} = 1.148 \text{Hz}$

Thus, the magnitude of damped frequency (ω_d) is the same for positive and negative values of damping coefficient.

EXERCISES

1.1   What is small-signal stability problem? Explain different categories of small-signal stability problem. Explain the effect of synchronizing torque and the damping torque on dynamic stability of a synchronous machine with and without AVR.

1.2   For a three-phase synchronous generator and prime mover unit, derive the analytic expression of swing equation

$$\frac{2H}{\omega_{syn}}\omega_{pu}(t)\frac{d^2\delta(t)}{dt^2} = P_{mpu}(t) - P_{epu}(t) = P_{apu}(t)$$

where H is the normalized inertia constant (s), ω_{syn} is the synchronous electrical angular velocity of the rotor (rad/s), ω_{pu} is the per-unit rotor angular velocity, P_{mpu} is the mechanical power supplied by the prime mover minus mechanical losses (pu), P_{epu} is the electrical power output of generator plus electrical losses (pu), and P_{apu} is the accelerating power (pu).

1.3   Write down different properties of small-signal oscillations. What do you mean by "local oscillations" and "inter-area oscillations" of synchronous machine? What are the different procedures of analysis of these oscillations?

1.4   A two-pole, 50 Hz, 11 kV turbo-alternator has a rating 100 MW, power factor 0.85 lagging. The rotor has a moment of inertia of a 10,000 kg m². Calculate H and M.

1.5   Find the expression for undamped natural frequency of power system oscillation and the damping ratio for a synchronous generator connected to an infinite-bus system.

___

References

[1] D.P. Kothari, I.J. Nagrath, Modern Power System Analysis, McGraw-Hill, Singapore, 2003.
[2] P.M. Anderson, A.A. Fouad, Power System Control and Stability, Iowa State University Press, Ames, IA, 1977.
[3] P.W. Sauer, M.A. Pai, Power System Dynamics and Stability, Pearson Education Pte. Ltd., Singapore, 1998.
[4] P. Kundur, Power System Stability and Control, McGraw-Hill, New York, 1994.
[5] M.A. Pai, D.P. Sengupta, K.R. Padiyar, Small Signal Analysis of Power Systems, Narosa Publishing House, India, 2004.
[6] A. Chakrabarti, S. Halder, Power System Analysis Operation and Control, PHI learning Pvt. Ltd., India, New Delhi, 2010.

Fundamental Models of Synchronous Machine

2.1 INTRODUCTION

Synchronous machine is the heart of the power system network. Its models have different variations in the literature. These variations are mainly in sign conventions and in the representation of damping windings. Among these versions, the model based on *Park's* reference frame is most popular. The Park's transformation is originated from Blondel's two-reaction theorem [1]. The main problem in synchronous machine modeling is the analysis of transient voltage and current as inductances of the coils are functions of rotor positions.

In an alternator (Figure 2.1), the transient voltage and current relationship of the stator coils 'a', 'b', and 'c' and the field coil "f" can be represented by the following equations:

$$V_a = R_a i_a + \frac{d}{dt}(L_a i_a) + \frac{d}{dt}(M_{ab} i_b) + \frac{d}{dt}(M_{ac} i_c) + \frac{d}{dt}(M_{bf} i_f) \qquad (2.1)$$

$$V_b = R_b i_b + \frac{d}{dt}(L_b i_b) + \frac{d}{dt}(M_{ba} i_a) + \frac{d}{dt}(M_{bc} i_c) + \frac{d}{dt}(M_{bf} i_f) \qquad (2.2)$$

$$V_c = R_c i_c + \frac{d}{dt}(L_c i_c) + \frac{d}{dt}(M_{ca} i_a) + \frac{d}{dt}(M_{cb} i_b) + \frac{d}{dt}(M_{cf} i_f) \qquad (2.3)$$

$$V_f = R_f i_f + \frac{d}{dt}(L_f i_f) + \frac{d}{dt}(M_{fa} i_a) + \frac{d}{dt}(M_{fb} i_b) + \frac{d}{dt}(M_{fc} i_c) \qquad (2.4)$$

Here, L stands for self-inductance of a coil and M stands for the mutual inductance. M_{ab} is the mutual inductance between the a-phase and the b-phase windings in the stator. M_{ca} is the mutual inductance between the c-phase and the a-phase windings in the stator, and similarly, other symbols have usual significances.

Equations (2.1)–(2.4) are extremely difficult to solve since the stator and the rotor coils are in relative motion. These equations are derived on the basis of measuring the voltages and currents in each coil at its respective terminals. In other words, the "observer" is stationary with respect to the coil. The observer shifts from the stationary stator coils to the rotating field coil "f" in order to measure the applied and induced voltages in the same coil. Instead of fixing the observer's reference to each

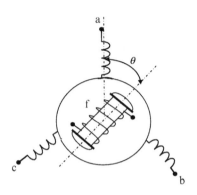

FIGURE 2.1

Schematic of a three-phase synchronous machine.

of the coils in the machines and then computing the voltages, inductances, and currents, the net combined effects of these coils in terms of the mmf and the fluxes that they produced may be viewed. An advantage of this is that the three-phase currents circulating through the three-phase winding of the armature produce a sinusoidal mmf wave form of constant amplitude that rotates at synchronous speed. The field winding carrying a direct current and rotating at synchronous speed also gives rise to a sinusoidal mmf wave. The two mmf waves rotating synchronously are obviously stationary with respect to each other.

For a cylindrical rotor machine, the fluxes produced by these mmf's are almost similar in shape. In the case of salient pole machines, however, the flux waves get distorted, especially along the interpolar gaps, as shown in Figure 2.2. The air gap in

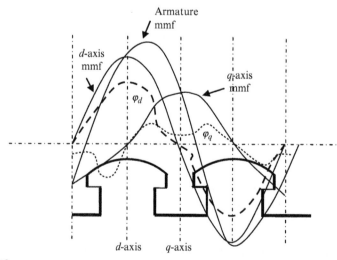

FIGURE 2.2

Air gap flux wave forms.

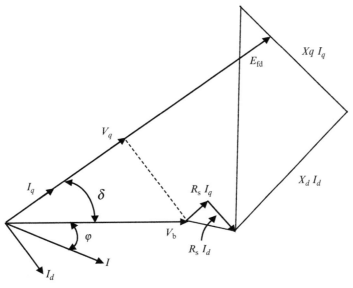

FIGURE 2.3

Phasor diagram of a synchronous generator.

this space is larger and the reluctance offered to magnetic fluxes along this path is also larger. Self-inductance, which is (number of turns)^2/reluctance, is, therefore, less when the flux wave along the interpolar axis, that is, the quadrature axis or q-axis, was considered. The vector diagram for a salient pole synchronous generator, connected to the infinite bus, based on well-known "*Blondel's two-reaction theory*," is given in Figure 2.3, which represents the steady-state performance of synchronous generator.

The steady-state equations that may be derived from the vector diagram are represented along two axes, the direct axis (d-axis) that coincides with the pole axis and the quadrature axis (q-axis) that is the interpolar axis.

The steady-state voltage equations are

$$V_d = V_b \sin\delta = -R_s I_d + X_q I_q \tag{2.5}$$

$$V_q = V_b \cos\delta = -R_s I_q - X_d I_d + E_{fd} \tag{2.6}$$

where E_{fd} is the steady-state-induced emf or open-circuit voltage, V_b is the generator terminal voltage or the infinite bus voltage (line impedance $=0$), δ is the load voltage, I_d and I_q are the steady-state current components along the d-axis and q-axis, respectively, V_d and V_q are the steady-state applied voltage components along the d-axis and q-axis, respectively, X_d is the synchronous reactance along the direct axis (d-axis), X_q is the synchronous reactance along the quadrature axis (q-axis), and R_s is the armature resistance.

All voltages, currents, and impedances are expressed in per unit (pu).

2.2 SYNCHRONOUS MACHINE DYNAMIC MODEL IN THE a–b–c REFERENCE FRAME

The two-axis (d-axis and q-axis) model of the synchronous machine is widely used in the literatures [2,4]. It has three-phase windings on the rotor, a field winding on the direct axis of the rotor, and a damper winding in the quadrature axis of the rotor. Consider a two-pole machine shown in Figure 2.4. It has been assumed that stator has three coils in a balanced symmetrical configuration centered 120 electrical degree apart. Rotor has four coils in a balanced symmetrical configuration located in pairs 90 electrical degree apart. The relationship between the flux linkages and the currents must reflect a conservative coupling field. It is a multiport device with five ports on the electric side and one port on the mechanical side (Figure 2.4c). The fundamental KVL equations for the electric side using the motor convention and the Newton's laws for mechanical equations at the shaft are obtained as follows [2]:

- **Electric equations**

$$v_a = i_a r_s + \frac{d\lambda_a}{dt} \tag{2.7}$$

$$v_b = i_b r_s + \frac{d\lambda_b}{dt} \tag{2.8}$$

$$v_c = i_c r_s + \frac{d\lambda_c}{dt} \tag{2.9}$$

$$v_{fd} = i_{fd} r_{fd} + \frac{d\lambda_{fd}}{dt} \tag{2.10}$$

$$v_{1q} = i_{1q} r_{1q} + \frac{d\lambda_{1q}}{dt} \tag{2.11}$$

where λ is the flux linkage and r is the winding resistance. All other symbols have their usual significances.

- **Mechanical equations**

$$\frac{d\theta}{dt} = \omega \tag{2.12}$$

$$J\frac{d\omega}{dt} = T_m - T_e - T_D \tag{2.13}$$

where J is the inertia constant, T_m is the mechanical power, T_e is the torque of electric origin, and T_D is the damping or friction windage torque.

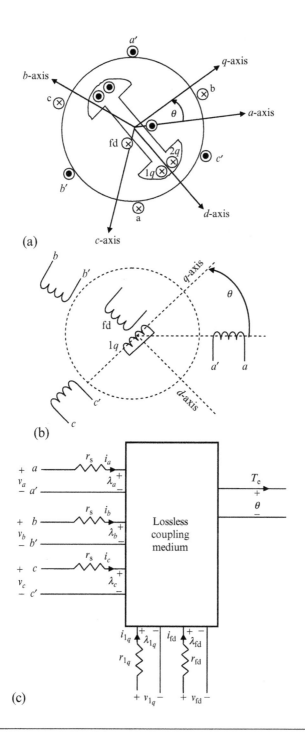

FIGURE 2.4

A two-pole synchronous machine (a) physical structure, (b) coupled circuit view, and (c) multiport view.

Assuming machine flux linkages are linear function of currents, the relationships between the flux linkages and the currents are given by

$$\lambda_{abc} = L_{ss}(\theta_{shaft})i_{abc} + L_{sr}(\theta_{shaft})i_{rotor} \tag{2.14}$$

$$\lambda_{rotor} = L_{rs}(\theta_{shaft})i_{abc} + L_{rr}(\theta_{shaft})i_{rotor} \tag{2.15}$$

or

$$\begin{bmatrix} \lambda_a \\ \lambda_b \\ \lambda_c \\ \lambda_{fd} \\ \lambda_{1q} \end{bmatrix} = \begin{bmatrix} L_{ss}(\theta) & L_{sr}(\theta) \\ L_{sr}^T(\theta) & L_{rr}(\theta) \end{bmatrix} \begin{bmatrix} i_a \\ i_b \\ i_c \\ i_{fd} \\ i_{1q} \end{bmatrix} \tag{2.16}$$

where $L_{ss}(\theta)$ is given by the following expressions:

$$\begin{bmatrix} L_{ls} + L_A - L_B \cos 2\theta & -\frac{1}{2}L_A - L_B \cos\left(2\theta - \frac{2\pi}{3}\right) & -\frac{1}{2}L_A - L_B \cos\left(2\theta + \frac{2\pi}{3}\right) \\ -\frac{1}{2}L_A - L_B \cos\left(2\theta - \frac{2\pi}{3}\right) & L_{ls} + L_A - L_B\left(\cos 2\theta + \frac{2\pi}{3}\right) & -\frac{1}{2}L_A - L_B \cos 2\theta \\ -\frac{1}{2}L_A - L_B \cos\left(2\theta + \frac{2\pi}{3}\right) & -\frac{1}{2}L_A - L_B \cos 2\theta & L_{ls} + L_A - L_B\left(\cos 2\theta - \frac{2\pi}{3}\right) \end{bmatrix} \tag{2.17}$$

$$L_{sr}(\theta) = L_{sr}^T = \begin{bmatrix} L_{sfd} \sin\theta & L_{sq} \cos\theta \\ L_{sfd} \sin\left(\theta - \frac{2\pi}{3}\right) & L_{sq} \cos\left(\theta - \frac{2\pi}{3}\right) \\ L_{sfd} \sin\left(\theta + \frac{2\pi}{3}\right) & L_{sq} \cos\left(\theta + \frac{2\pi}{3}\right) \end{bmatrix} \tag{2.18}$$

$$L_{rr}(\theta) = \begin{bmatrix} L_{fd} & 0 \\ 0 & L_{qq} \end{bmatrix} \tag{2.19}$$

Equations (2.7)–(2.13) are the differential equations together with the $\lambda - i$ constitutive algebraic (2.16). A straightforward substitution of Equation (2.16) in Equations (2.7)–(2.11) would give rise to time-varying differential equations, since θ is a function of t. Such equations are not amenable to analysis. This difficulty is avoided through a transformation called "Park's transformation."

2.3 PARK'S TRANSFORMATION AND DYNAMIC MODEL IN THE d–q–o REFERENCE FRAME

In Park's transformation, the time-varying differential equations (2.7)–(2.13) are converted into time-invariant differential equations. The transformation converts the a–b–c variables to a new set of variables called the d–q–o variables, and the transformation is given by

$$\begin{bmatrix} v_d \\ v_q \\ v_o \end{bmatrix} = [P] \begin{bmatrix} v_a \\ v_b \\ v_c \end{bmatrix} \tag{2.20}$$

$$\begin{bmatrix} \lambda_d \\ \lambda_q \\ \lambda_o \end{bmatrix} = [P] \begin{bmatrix} \lambda_a \\ \lambda_b \\ \lambda_c \end{bmatrix} \tag{2.21}$$

$$\begin{bmatrix} i_d \\ i_q \\ i_o \end{bmatrix} = [P] \begin{bmatrix} i_a \\ i_b \\ i_c \end{bmatrix} \tag{2.22}$$

where

$$P \triangleq \frac{2}{3} \begin{bmatrix} \sin\theta & \sin\left(\theta - \dfrac{2\pi}{3}\right) & \sin\left(\theta + \dfrac{2\pi}{3}\right) \\ \cos\theta & \cos\left(\theta - \dfrac{2\pi}{3}\right) & \cos\left(\theta + \dfrac{2\pi}{3}\right) \\ \dfrac{1}{2} & \dfrac{1}{2} & \dfrac{1}{2} \end{bmatrix} \tag{2.23}$$

and

$$P^{-1} = \begin{bmatrix} \sin\theta & \sin\left(\theta - \dfrac{2\pi}{3}\right) & \sin\left(\theta + \dfrac{2\pi}{3}\right) \\ \cos\theta & \cos\left(\theta - \dfrac{2\pi}{3}\right) & \cos\left(\theta + \dfrac{2\pi}{3}\right) \\ \dfrac{1}{2} & \dfrac{1}{2} & \dfrac{1}{2} \end{bmatrix} \tag{2.24}$$

Though this transformation is not a power-invariant transformation, it is widely used in the industry. The stator equations (2.7)–(2.9) can be rewritten as

$$\begin{bmatrix} v_a \\ v_b \\ v_c \end{bmatrix} = \begin{bmatrix} r_s & 0 & 0 \\ 0 & r_s & 0 \\ 0 & 0 & r_s \end{bmatrix} \begin{bmatrix} i_a \\ i_b \\ i_c \end{bmatrix} + \frac{d}{dt} \begin{bmatrix} \lambda_a & 0 & 0 \\ 0 & \lambda_b & 0 \\ 0 & 0 & \lambda_c \end{bmatrix} \tag{2.25}$$

Multiplying both sides of (2.25) by P, we get

$$P\begin{bmatrix} v_a \\ v_b \\ v_c \end{bmatrix} = P\begin{bmatrix} r_s & 0 & 0 \\ 0 & r_s & 0 \\ 0 & 0 & r_s \end{bmatrix}P^{-1}P\begin{bmatrix} i_a \\ i_b \\ i_c \end{bmatrix} + P\frac{d}{dt}(P^{-1}P)\begin{bmatrix} \lambda_a \\ \lambda_b \\ \lambda_c \end{bmatrix} \qquad (2.26)$$

Thus, using the transformation in Equation (2.22), Equation (2.26) becomes

$$\begin{bmatrix} v_d \\ v_q \\ v_o \end{bmatrix} = \begin{bmatrix} r_s & 0 & 0 \\ 0 & r_s & 0 \\ 0 & 0 & r_s \end{bmatrix}\begin{bmatrix} i_d \\ i_q \\ i_o \end{bmatrix} + P\frac{d}{dt}\left[P^{-1}\begin{bmatrix} \lambda_d \\ \lambda_q \\ \lambda_o \end{bmatrix}\right] \qquad (2.27)$$

$$= \begin{bmatrix} r_s & 0 & 0 \\ 0 & r_s & 0 \\ 0 & 0 & r_s \end{bmatrix}\begin{bmatrix} i_d \\ i_q \\ i_o \end{bmatrix} + \begin{bmatrix} 0 & -\omega & 0 \\ \omega & 0 & 0 \\ 0 & 0 & 0 \end{bmatrix}\begin{bmatrix} \lambda_d \\ \lambda_q \\ \lambda_o \end{bmatrix} + \begin{bmatrix} \frac{d\lambda_d}{dt} \\ \frac{d\lambda_q}{dt} \\ \frac{d\lambda_o}{dt} \end{bmatrix} \qquad (2.28)$$

It is to be noted that the second term in Equation (2.28) is a speed-dependent term.
The rotor equations (2.10) and (2.11) are

$$\begin{bmatrix} v_{fd} \\ v_{1q} \end{bmatrix} = \begin{bmatrix} r_{fd} & 0 \\ 0 & r_{1q} \end{bmatrix}\begin{bmatrix} i_{fd} \\ i_{1q} \end{bmatrix} + \begin{bmatrix} \frac{d\lambda_{fd}}{dt} \\ \frac{d\lambda_{1q}}{dt} \end{bmatrix} \qquad (2.29)$$

The mechanical equation is given by

$$J\frac{d^2\delta}{dt^2} = T_m - T_e - T_D \qquad (2.30)$$

where T_m is the mechanical torque, $T_e = -\frac{3}{2}(\lambda_d i_q - \lambda_q i_d)$ is the electrical torque, and T_D is the damping or torque. Defining the rotor angle of the machine from the synchronous frame of reference as $\delta = \theta - \omega_s t$, the mechanical equations in the state space form with δ and $\omega = \frac{d\delta}{dt}$ as state variables are

$$\frac{d\delta}{dt} = \omega - \omega_s \qquad (2.31)$$

$$J\frac{d\omega}{dt} = T_m - T_e - T_D \qquad (2.32)$$

With this, the complete set of state space equations for the machine in the d–q–o coordinates combining (2.28), (2.29), (2.31), and (2.32) is

$$\frac{d\lambda_d}{dt} = -r_s i_d + \omega\lambda_q + v_d \qquad (2.33)$$

$$\frac{d\lambda_q}{dt} = -r_s i_q - \omega\lambda_d + v_q \qquad (2.34)$$

$$\frac{d\lambda_o}{dt} = -r_s i_o + v_o \tag{2.35}$$

$$\frac{d\lambda_{\text{fd}}}{dt} = -r_{\text{fd}} i_{\text{fd}} + v_{\text{fd}} \tag{2.36}$$

$$\frac{d\lambda_{1q}}{dt} = -r_{1q} i_{1q} + v_{1q} \tag{2.37}$$

$$\frac{d\delta}{dt} = \omega - \omega_s \tag{2.38}$$

$$J\frac{d\omega}{dt} = T_{\text{m}} + \frac{3}{2}\left(\lambda_d i_q - \lambda_q i_d\right) - T_{\text{D}} \tag{2.39}$$

The fictitious positions of the two coils d and q after the transformation are shown in Figure 2.5. Applying Park's transformation to the a–b–c variables, the flux linkages in d–q–o variables are derived as follows:

Consider Equation (2.16)

$$\begin{bmatrix} \lambda_a \\ \lambda_b \\ \lambda_c \\ \lambda_{\text{fd}} \\ \lambda_{1q} \end{bmatrix} = \begin{bmatrix} L_{\text{ss}}(\theta) & L_{\text{sr}}(\theta) \\ L_{\text{sr}}^{\text{T}}(\theta) & L_{\text{rr}}(\theta) \end{bmatrix} \begin{bmatrix} i_a \\ i_b \\ i_c \\ i_{\text{fd}} \\ i_{1q} \end{bmatrix} \tag{2.40}$$

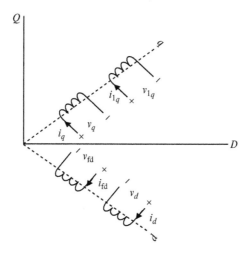

FIGURE 2.5

The stator and rotor coils after Park's transformation.

The transformation is only for the a–b–c variables, and, hence, it is written as

$$
\begin{bmatrix} \lambda_d \\ \lambda_q \\ \lambda_o \\ \lambda_{fd} \\ \lambda_{1q} \end{bmatrix} = \begin{bmatrix} P & 0 \\ 0 & I \end{bmatrix} \begin{bmatrix} \lambda_a \\ \lambda_b \\ \lambda_c \\ \lambda_{fd} \\ \lambda_{1q} \end{bmatrix} \tag{2.41}
$$

where I is the 2×2 matrix. Therefore, Equation (2.40) becomes

$$
\begin{bmatrix} \lambda_d \\ \lambda_q \\ \lambda_o \\ \lambda_{fd} \\ \lambda_{1q} \end{bmatrix} = \begin{bmatrix} P & 0 \\ 0 & I \end{bmatrix} \begin{bmatrix} L_{ss}(\theta) & L_{sr}(\theta) \\ L_{sr}^T(\theta) & L_{rr}(\theta) \end{bmatrix} \begin{bmatrix} P^{-1} & 0 \\ 0 & I \end{bmatrix} \begin{bmatrix} P & 0 \\ 0 & I \end{bmatrix} \begin{bmatrix} i_a \\ i_b \\ i_c \\ i_{fd} \\ i_{1q} \end{bmatrix}
$$

$$
= \begin{bmatrix} P & 0 \\ 0 & I \end{bmatrix} \begin{bmatrix} L_{ss}(\theta) & L_{sr}(\theta) \\ L_{sr}^T(\theta) & L_{rr}(\theta) \end{bmatrix} \begin{bmatrix} P^{-1} & 0 \\ 0 & I \end{bmatrix} \begin{bmatrix} i_d \\ i_q \\ i_o \\ i_{fd} \\ i_{1q} \end{bmatrix}
$$

$$
= \begin{bmatrix} L_{dqo} & L_{dqo-r} \\ L_{r-dqo} & L_{rr} \end{bmatrix} \begin{bmatrix} i_d \\ i_q \\ i_o \\ i_{fd} \\ i_{1q} \end{bmatrix} \tag{2.42}
$$

In Equation (2.31), the various submatrices are given by

$$
L_{dqo} = \begin{bmatrix} L_{ls} + L_{md} & 0 & 0 \\ 0 & L_{ls} + L_{mq} & 0 \\ 0 & 0 & L_{ls} \end{bmatrix}, \quad L_{dqo-r} = \begin{bmatrix} L_{sfd} & 0 & 0 \\ 0 & L_{sq} & 0 \\ 0 & 0 & 0 \end{bmatrix},
$$

$$
L_{r-dqo} = \begin{bmatrix} \frac{3}{2}L_{sfd} & 0 & 0 \\ 0 & \frac{3}{2}L_{sq} & 0 \\ 0 & 0 & 0 \end{bmatrix}, \quad L_{rr} = \begin{bmatrix} L_{fd} & 0 \\ 0 & L_{qq} \end{bmatrix},
$$

where $L_{md} = \dfrac{3}{2}(L_A + L_B)$ and $L_{mq} = \dfrac{3}{2}(L_A - L_B)$.

2.4 PER UNIT (PU) REPRESENTATION AND SCALING [2]

It is customary to scale the synchronous machine equations using the traditional concept of pu. This scaling process is presented here as a change of variables and a change of parameters and also converts model to "generator" notation. To conform to the usual generator convention, we reverse the signs of i_a, i_b, and i_c as $-i_a$, $-i_b$, and $-i_c$, which makes the d–q–o currents as $-i_d$, $-i_q$, and $-i_o$, in Equation (2.22). The new per unitized variables are as follows:

(i) ψ_d, ψ_q, and ψ_o replacing λ_d, λ_q, and λ_o

(ii) V_d, V_q, and V_o replacing v_d, v_q, and v_o

(iii) $-I_d$, $-I_q$, and $-I_o$ replacing $-i_d$, $-i_q$, and $-i_o$

(iv) V_{fd}, V_{1q}, I_{fd}, I_{1q}, ψ_{fd}, ψ_{1q}, R_s, R_{fd}, and R_{1q} replacing v_{fd}, v_{1q}, i_{fd}, i_{1q}, λ_{fd}, λ_{1q}, r_s, r_{fd}, and r_{1q}, respectively

(v) The inertia constant $H = \dfrac{1}{2}\dfrac{J\omega_s^2}{\text{Base volt-amps}}$

With this change of new variables, Equations (2.33)–(2.39) become as follows:

- **Electric equations**

$$\frac{1}{\omega_s}\frac{d\psi_d}{dt} = R_s I_d + \frac{\omega}{\omega_s}\psi_q + V_d \tag{2.43}$$

$$\frac{1}{\omega_s}\frac{d\psi_q}{dt} = R_s I_q - \frac{\omega}{\omega_s}\psi_d + V_q \tag{2.44}$$

$$\frac{1}{\omega_s}\frac{d\psi_o}{dt} = R_s I_o + V_o \tag{2.45}$$

$$\frac{1}{\omega_s}\frac{d\psi_{fd}}{dt} = -R_{fd}I_{fd} + V_{fd} \tag{2.46}$$

$$\frac{1}{\omega_s}\frac{d\psi_{1q}}{dt} = -R_q I_{1q} + V_{1q} \tag{2.47}$$

- **Rotor mechanical equations**

$$\frac{d\delta}{dt} = \omega - \omega_s \tag{2.48}$$

$$\frac{2H}{\omega_s}\frac{d\omega}{dt} = T_m - \left(\psi_d I_q - \psi_q I_d\right) - T_D \tag{2.49}$$

The flux, current, and voltage variables as well as T_m and T_D are all in pu. δ is in radians and ω is in radians/s. $M = \frac{2H}{\omega_s}$ is also used commonly in the literature. The unit of H is s, while M has the units of $(s)^2$. The pu speed can also be expressed as $v = \frac{\omega}{\omega_s}$. Therefore, Equations (2.48) and (2.49) become

$$\frac{d\delta}{dt} = \omega_s(v - 1)$$

$$2H\frac{dv}{dt} = T_M - \left(\psi_d I_q - \psi_q I_d\right) - T_D \tag{2.50}$$

The flux–current relations (2.42) become, after per unitization and scaling,

$$
\begin{bmatrix} \psi_d \\ \psi_q \\ \psi_o \\ \psi_{fd} \\ \psi_{1q} \end{bmatrix}
=
\begin{bmatrix}
X_d & 0 & 0 & X_{md} & 0 \\
0 & X_q & 0 & 0 & X_{mq} \\
0 & 0 & X_{ls} & 0 & 0 \\
X_{md} & 0 & 0 & X_{fd} & 0 \\
0 & X_{mq} & 0 & 0 & X_{1q}
\end{bmatrix}
\begin{bmatrix} -I_d \\ -I_q \\ -I_o \\ I_{fd} \\ I_{1q} \end{bmatrix}
\tag{2.51}
$$

where the reactances are defined as $X_d = X_{ls} + X_{md}$ and $X_q = X_{ls} + X_{mq}$. It is convenient to define leakage reactance on the rotor windings as $X_{lfd} = X_{fd} - X_{md}$ and $X_{l1q} = X_{1q} - X_{mq}$.

2.5 PHYSICAL SIGNIFICANCE OF PU SYSTEM

Equation (2.51) can be interpreted as follows. The zero sequence quantities are completely decoupled and, therefore, can be ignored. The direct and quadrature axis flux–current relationships given the by first two equations in (2.51) can be represented by the circuit diagram as shown in Figure 2.6.

- **Direct axis (*d*-axis) transient reactance (X_d')**

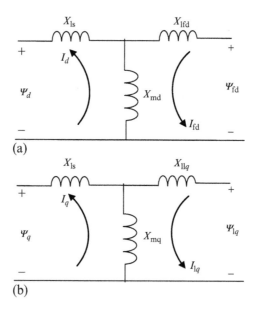

(a)

(b)

FIGURE 2.6

Equivalent circuit of flux–current relationships. (a) Direct axis flux-current relationship, (b) Quadrature axis flux-current relationship.

This reactance is seen from the direct axis terminals with field winding short-circuited in Figure 2.6:

$$X'_d = X_{ls} + \frac{X_{md}X_{lfd}}{X_{md} + X_{lfd}}$$

$$= X_{ls} + \frac{X_{md}X_{lfd}}{X_{fd}}$$

$$= X_d + \frac{X_{md}X_{lfd}}{X_{fd}} - X_{md}$$

$$= X_d + \frac{X_{md}X_{lfd}}{X_{fd}} - \frac{X_{md}(X_{md} + X_{lfd})}{X_{fd}} \tag{2.52}$$

$$X'_d = X_d - \frac{X^2_{md}}{X_{fd}} \tag{2.53}$$

- **Quadrature axis (q-axis) transient reactance (X'_d)**

This is the transient reactance seen from the quadrate (q) axis terminals with damper windings short-circuited in Figure 2.6. Similarly,

$$X'_q = X_q - \frac{X^2_{mq}}{X_{1q}} \tag{2.54}$$

- **Time constants**

The following time constants are defined for the field and damper windings, respectively:

$$T'_{do} = \frac{X_{fd}}{\omega_s R_{fd}}, \quad T'_{qo} = \frac{X_{1q}}{\omega_s R_{1q}} \tag{2.55}$$

We define new variables as

$$T'_{do} = \frac{X_{md}}{X_{fd}}\psi_{fd}, \quad E_{fd} = \frac{X_{md}}{R_{fd}}V_{fd},$$

$$E'_q = \frac{X_{md}}{X_{fd}}\psi_{fd}, \quad E'_d = -\frac{X_{mq}}{X_{1q}}\psi_{1q} \tag{2.56}$$

where E'_q is proportional to the rotor field flux in the direct axis, whereas E'_d is proportional to the damper winding flux in the quadrature axis. With these definitions, we first transform the flux–current relationships (2.51) and then the rotor equations (2.48) and (2.49) as follows:

2.6 STATOR FLUX–CURRENT RELATIONSHIPS

From Equations (2.51) and (2.53), we obtain

$$\psi_d = -X_d I_d + X_{md} I_{fd}$$

$$= -\left(X'_d + \frac{X^2_{md}}{X_{fd}}\right) I_d + X_{md} I_{fd}$$

$$= -X'_d I_d + \frac{X_{md}}{X_{fd}}(-X_{md} I_d + X_{fd} I_{fd}) \tag{2.57}$$

$$\psi_d = -X'_d I_d + \frac{X_{md}}{X_{fd}} \psi_{fd} \tag{2.58}$$

Using Equation (2.56), we get

$$\psi_d = -X'_d I_d + E'_q \tag{2.59}$$

Similarly, from Equations (2.51) and (2.54), we have

$$\psi_q = -X_q I_q + X_{mq} I_{1q}$$

$$= -\left(X'_q + \frac{X^2_{mq}}{X_{1q}}\right) I_q + X_{mq} I_{1q}$$

$$= -X'_q I_q + \frac{X_{mq}}{X_{1q}}(-X_{mq} I_q + X_{1q} I_{1q})$$

$$= -X'_q I_q + \frac{X_{mq}}{X_{1q}} \psi_{1q} \tag{2.60}$$

Using Equation (2.56), we get

$$\psi_q = -X'_q I_q - E'_d \tag{2.61}$$

2.7 ROTOR DYNAMIC EQUATIONS

Considering Equations (2.46) and (2.47),

$$\frac{1}{\omega_s} \frac{d\psi_{fd}}{dt} = -R_{fd} I_{fd} + V_{fd} \tag{2.62}$$

$$\frac{1}{\omega_s} \frac{d\psi_{1q}}{dt} = -R_{1q} I_{1q} + V_{1q} \tag{2.63}$$

Multiplying Equation (2.62) by $\frac{X_{md}}{R_{fd}}$ and using Equations (2.55) and (2.56),

$$\frac{X_{fd}}{\omega_s R_{fd}} \frac{X_{md}}{X_{fd}} \frac{d\psi_{fd}}{dt} = -R_{fd} \frac{X_{md}}{R_{fd}} I_{fd} + \frac{X_{md}}{R_{fd}} V_{fd}$$

$$T'_{do} \frac{dE'_q}{dt} = -X_{md} I_{fd} + E_{fd} \tag{2.64}$$

Now, from Equations (2.51), (2.53), and (2.56), we get

$$-X_{md} I_{fd} = -X_{md}(\psi_{fd} + X_{md} I_d)/X_{fd}$$

$$= -E'_q - \frac{X^2_{md}}{X_{fd}} I_d = -E'_q + (X'_d - X_d) I_d$$

$$= -E'_q - (X_d - X'_d) I_d \tag{2.65}$$

After substitution of Equation (2.65) in Equation (2.64), we get

$$T'_{do} \frac{dE'_q}{dt} = -E'_q - (X_d - X'_d) I_d + E_{fd} \tag{2.66}$$

Since the rotor q-axis winding is short-circuited, $V_{1q} = 0$. Therefore, Equation (2.63) becomes

$$\frac{1}{\omega_s} \frac{d\psi_{1q}}{dt} = -R_{1q} I_{1q} \tag{2.67}$$

Equation (2.67) is written by multiplying it by $\frac{-X_{mq}}{R_{1q}}$ as

$$\frac{X_{1q}}{\omega_s R_{1q}} \left(\frac{-X_{mq}}{X_{1q}} \right) \frac{d\psi_{1q}}{dt} = X_{mq} I_{1q} \tag{2.68}$$

Since $\frac{X_{1q}}{\omega_s R_{1q}} = T'_{qo}$, from Equation (2.55) and using the definition of E'_d in Equation (2.56), we get

$$T'_q \frac{dE'_d}{dt} = X_{mq} I_{1q}$$

From Equation (2.51), I_{1q} is expressed in terms of ψ_{1q} and I_q. Then, using the definition of E_d' in Equation (2.56) and X_q' in Equation (2.53), we get

$$T'_q \frac{dE'_d}{dt} = \frac{X_{mq}(\psi_{1q} + X_{mq} I_q)}{X_{1q}}$$

$$= -E'_d + \frac{X^2_{mq}}{X_{1q}} I_q$$

$$= -E'_d + \left(X_q - X'_q \right) I_q \tag{2.69}$$

This completes the conversion of the differential equation for ψ_{fd} and ψ_{1q} in terms of the new variables.

2.8 REDUCED ORDER MODEL

The reduced order model of synchronous machine can be obtained neglecting stator transients. Now, writing Equations (2.43), (2.44), (2.66), (2.69), (2.48), and (2.49) collectively, we have

$$\frac{1}{\omega_s}\frac{d\psi_d}{dt} = R_s I_d + \frac{\omega}{\omega_s}\psi_q + V_d \tag{2.70}$$

$$\frac{1}{\omega_s}\frac{d\psi_q}{dt} = R_s I_q - \frac{\omega}{\omega_s}\psi_d + V_q \tag{2.71}$$

$$T'_{do}\frac{dE'_q}{dt} = -E'_q - \left(X_d - X'_d\right)I_d + E_{fd} \tag{2.72}$$

$$T'_{qo}\frac{dE'_d}{dt} = -E'_d + \left(X_q - X'_q\right)I_q \tag{2.73}$$

$$\frac{d\delta}{dt} = \omega - \omega_s \tag{2.74}$$

$$\frac{2H}{\omega_s}\frac{d\omega}{dt} = T_m - \left(\psi_d I_q - \psi_q I_d\right) - T_D \tag{2.75}$$

In order to make dynamic analysis of a multimachine system, the following assumptions are made:

(i) The stator transients are neglected. It means the equivalent "transformer" voltages $\frac{1}{\omega_s}\frac{d\psi_d}{dt}$ and $\frac{1}{\omega_s}\frac{d\psi_q}{dt}$ in Equations (2.70) and (2.71) are neglected. This assumption is based on singular perturbation technique.

(ii) Speed deviations are small compared to ω_s, that is, $\omega \approx \omega_s$. This results in the pu electromagnetic torque being equal to the pu active power P_m.

(iii) The damping torque is assumed to be equal to $D(\omega - \omega_s)$. With these assumptions, Equations (2.70)–(2.75) mentioned earlier are rewritten as

$$0 = R_s I_d + \psi_q + V_d \tag{2.76}$$

$$0 = R_s I_q - \psi_d + V_q \tag{2.77}$$

$$T'_{do}\frac{dE'_q}{dt} = -E'_q - \left(X_d - X'_d\right)I_d + E_{fd} \tag{2.78}$$

$$T'_{qo}\frac{dE'_d}{dt} = -E'_d + \left(X_q - X'_q\right)I_q \tag{2.79}$$

$$\frac{d\delta}{dt} = \omega - \omega_s \tag{2.80}$$

$$\frac{2H}{\omega_s}\frac{d\omega}{dt} = T_m - \left(\psi_d I_q - \psi_q I_d\right) - T_D$$
$$= P_m - P_{ei} - T_D \tag{2.81}$$

Equations (2.76) and (2.77) are referred to as the *stator algebraic equations*. Using Equation (2.59) for ψ_d and Equation (2.61) for ψ_q, we can express Equations (2.76) and (2.77) in terms of voltages and currents only as follows:

$$0 = R_s I_d - X'_q I_q - E'_d + V_d$$

$$0 = R_s I_q + X'_d - E'_q + V_q$$

that is,

$$V_d = -R_s I_d + X'_q I_q + E'_d \tag{2.82}$$

$$V_q = -R_s I_q - X'_d I_d + E'_q \tag{2.83}$$

Equations (2.76) and (2.77) are the most popular version of stator algebraic equations.

Considering Equation (2.81), substituting ψ_d and ψ_q from Equations (2.76) and (2.77), and neglecting stator resistance R_s, the electric power is given by

$$P_{ei} = \psi_d I_q - \psi_q I_d = V_q I_q + V_d I_d$$
$$= \left(-X'_d I_d I_q + E'_q I_q\right) + \left(X'_q I_d I_q + E'_d I_d\right) \tag{2.84}$$
$$= \left(E'_d I_d + E'_q I_q\right) - \left(X'_d - X'_q\right)I_d I_q$$

Therefore, Equation (2.81) becomes

$$\frac{2H}{\omega_s}\frac{d\omega}{dt} = P_m - \left[\left(E'_d I_d + E'_q I_q\right) - \left(X'_d - X'_q\right)I_d I_q\right] - T_D \tag{2.85}$$

2.9 EQUIVALENT CIRCUIT OF THE STATOR ALGEBRAIC EQUATIONS

In a transmission network when different synchronous machines are interconnected together, all variables are expressed in a common reference frame, which is called synchronously rotating reference frame. This rotating reference frame is also known as D, Q, and O. Thus, for each machine, the transformation in the new variables is

$$\begin{bmatrix} V_D \\ V_Q \\ V_O \end{bmatrix} = P_s \begin{bmatrix} V_a \\ V_b \\ V_c \end{bmatrix} = P_s P^{-1} \begin{bmatrix} V_d \\ V_q \\ V_o \end{bmatrix} \tag{2.86}$$

where

$$P_s = \frac{2}{3} \begin{bmatrix} \cos \omega_s t & \cos\left(\omega_s t - \dfrac{2\pi}{3}\right) & \cos\left(\omega_s t + \dfrac{2\pi}{3}\right) \\ -\sin \omega_s t & -\sin\left(\omega_s t - \dfrac{2\pi}{3}\right) & -\sin\left(\omega_s t + \dfrac{2\pi}{3}\right) \\ \dfrac{1}{2} & \dfrac{1}{2} & \dfrac{1}{2} \end{bmatrix} \tag{2.87}$$

It can be proved that

$$P_s P^{-1} = \frac{2}{3} \begin{bmatrix} \sin \delta & \cos \delta & 0 \\ -\cos \delta & \sin \delta & 0 \\ 0 & 0 & 1 \end{bmatrix} \tag{2.88}$$

and

$$P P_s^{-1} = \begin{bmatrix} \sin \delta & -\cos \delta & 0 \\ \cos \delta & \sin \delta & 0 \\ 0 & 0 & 1 \end{bmatrix} \tag{2.89}$$

This transformation gives

$$\left(V_D + jV_Q\right) = \left(V_d + jV_q\right) e^{j(\delta - \pi/2)} \tag{2.90}$$

The same transformation applies to the current variables, so that

$$\left(I_D + jI_Q\right) = \left(I_d + jI_q\right) e^{j(\delta - \pi/2)} \tag{2.91}$$

The stator algebraic equations (2.82) and (2.83) can be combined into a single equation and also can be expressed in a single common reference frame. The synchronous frame of reference is denoted by the D-and Q-axes, which are orthogonal as shown in Figure 2.7. The d- and q-axes of the machine are also represented in Figure 2.7. The angle between D- and q-axes is denoted by δ and the angle between D and d-axes is denoted by $\left(\frac{\pi}{2} - \delta\right)$. This is consistent with Equations (2.90) and (2.91).

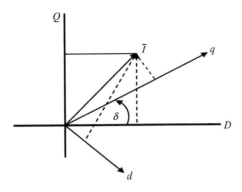

FIGURE 2.7

Relationship between synchronous frame of reference and synchronous machine.

The currents in these two reference frames are related by

$$I_D = I_d \sin\delta + I_q \cos\delta \qquad (2.92)$$

$$I_Q = -I_d \sin\delta + I_q \cos\delta \qquad (2.93)$$

Writing Equations (2.92) and (2.93) in matrix form,

$$\begin{bmatrix} I_D \\ I_Q \end{bmatrix} = \underbrace{\begin{bmatrix} \sin\delta & \cos\delta \\ -\cos\delta & \sin\delta \end{bmatrix}}_{T(\delta)} \begin{bmatrix} I_d \\ I_q \end{bmatrix} \qquad (2.94)$$

$T(\delta)$ is the rotation matrix, which is nonsingular and orthogonal. It is trivial to verify that

$$\begin{bmatrix} I_d \\ I_q \end{bmatrix} = [T^{-1}(\delta)] \begin{bmatrix} I_D \\ I_Q \end{bmatrix} \qquad (2.95)$$

$$\begin{bmatrix} I_d \\ I_q \end{bmatrix} = \begin{bmatrix} \sin\delta & -\cos\delta \\ \cos\delta & \sin\delta \end{bmatrix} \begin{bmatrix} I_D \\ I_Q \end{bmatrix} \qquad (2.96)$$

The voltage variables can also be expressed by the same relationship. The stator algebraic equations (2.82) and (2.83) are converted as follows:

$$\begin{bmatrix} V_D \\ V_Q \end{bmatrix} = \begin{bmatrix} -R_s & X'_q \\ -X'_d & -R_s \end{bmatrix} \begin{bmatrix} I_d \\ I_q \end{bmatrix} + \begin{bmatrix} E'_d \\ E'_q \end{bmatrix} \qquad (2.97)$$

$$\begin{bmatrix} V_D \\ V_Q \end{bmatrix} = [T] \begin{bmatrix} V_d \\ V_q \end{bmatrix} \qquad (2.98)$$

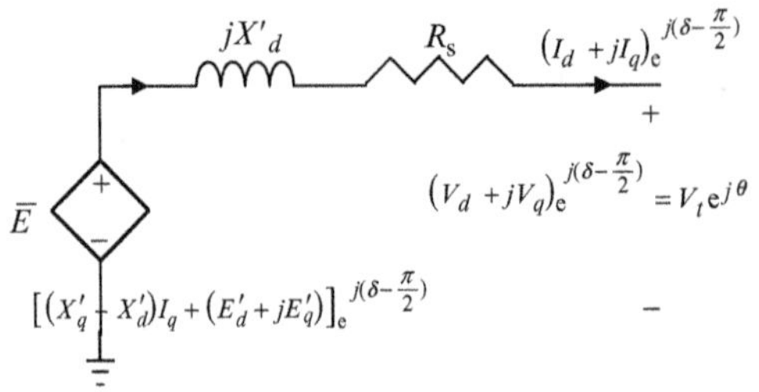

FIGURE 2.8

Synchronous machine stator equivalent circuit.

$$\begin{bmatrix} V_D \\ V_Q \end{bmatrix} = [T] \begin{bmatrix} -R_s & X'_q \\ -X'_d & -R_s \end{bmatrix} [T]^{-1} \begin{bmatrix} I_d \\ I_q \end{bmatrix} + [T] \begin{bmatrix} E'_d \\ E'_q \end{bmatrix} \quad (2.99)$$

Equation (2.99) can be expressed as

$$\left(V_D + jV_Q\right) + \left(R_s + jX'_d\right)\left(I_D + jI_Q\right) = \left[\left(E'_d + jE'_q\right) + \left(X'_q - X'_d\right)I_q\right]e^{j(\delta - \pi/2)} \quad (2.100)$$

Denoting $V_D + jV_Q = Ve^{j\theta_V}$ and $I_D + jI_Q = Ie^{j\theta_I}$, the relationships between the currents and the voltages in the two reference frames given by Equations (2.94) and (2.96), it can be verified that

$$Ve^{j\theta_V} = \left(V_d + jV_q\right)e^{j(\delta - \pi/2)} \quad (2.101)$$

and

$$Ie^{j\theta_I} = \left(I_d + jI_q\right)e^{j(\delta - \pi/2)}$$

$$\begin{aligned}\left(V_d + jV_q\right) &= Ve^{j(\theta_V - \delta + \pi/2)} \\ &= V\sin\left(\delta - \theta_V\right) + jV\cos\left(\delta - \theta_V\right)\end{aligned} \quad (2.102)$$

and

$$I_d + jI_q = I\sin\left(\delta - \theta_I\right) + jI\cos\left(\delta - \theta_I\right) \quad (2.103)$$

Thus, the synchronous machine equivalent circuit satisfying Equation (2.100) is presented in Figure 2.8.

2.10 SYNCHRONOUS MACHINE EXCITER

2.10.1 IEEE Type I exciter

The exciter provides the mechanism for controlling the synchronous machine terminal voltage magnitude. The basic elements that form different excitation systems are dc exciters (self-excited or separately excited), ac exciters, rectifiers (controlled or noncontrolled), magnetic or rotating or electronic amplifiers, excitation system stabilization feedback circuits, and signal sensing or processing circuits. We describe here models [3] for individual elements and finally present complete models for IEEE Type I excitation system. Figure 2.9 shows the basic excitation system with its stabilizing circuit. The models for individual components are described as follows:

• Self-excited dc circuit

The scaled model of a self-excited dc generator main exciter is

$$T_E \frac{dE_{fd}}{dt} = -(K_E + S_E(E_{fd}))E_{fd} + V_R \qquad (2.104)$$

where $K_E = \frac{r_f}{K_g}$ is the self-excited constant. r_f is the field resistance. K_g is the self-excited constant. $T_E = \frac{L_f}{R_g}$ is the exciter time constant. L_f is the unsaturated field inductance. $S_E(E_{fd})$ is the saturation function, dependent on E_{fd}, and can be expressed conveniently as $S_E(E_{fd}) = A_{ex}e^{B_{ex}E_{fd}}$. The typical values of the constants A_{ex} and B_{ex} are 0.09826 and 0.5527, respectively. V_R is the scaled output of the amplifier (pilot exciter), which is applied to the field of the main exciter. Figure 2.10 shows the block diagram representation of the main exciter using Equation (2.104).

• Amplifiers

In order to automatically control the terminal voltage of the synchronous machine, a transducer voltage must be compared to a reference voltage and amplified to produce the exciter input signal V_R. The amplifier may be characterized by a

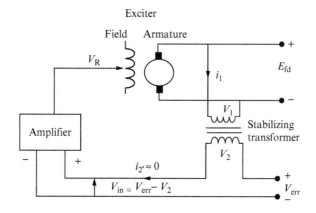

FIGURE 2.9

Excitation system.

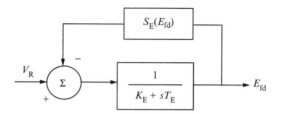

FIGURE 2.10

Main exciter model.

gain and with a time constant and can be modeled by the following first-order equation:

$$T_A \frac{dV_R}{dt} = -V_R + K_A V_{in} \tag{2.105}$$

$$V_R^{min} \leq V_R \leq V_R^{max}$$

where V_{in} is the amplifier input, T_A is the amplifier time constant, and K_A is the amplifier gain.

The equation mentioned earlier can be represented by the following block diagram as shown in Figure 2.11.

• *Stabilizer circuit*

In standard excitation systems, to achieve the desirable dynamic performance and to shape the regulator response, a stabilizing circuit is used. This may be accomplished by a series transformer whose input is connected to the output of the exciter and whose output voltage is subtracted from the amplifier input (Figure 2.12). As the secondary circuit of the transformer is connected to a high-impedance circuit, neglecting i_2, we have

$$V_1 = (R_1 + sL_1)i_1 \tag{2.106}$$

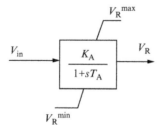

FIGURE 2.11

Amplifier model.

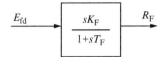

FIGURE 2.12

Stabilizer model.

$$V_2 = sMi_1 \tag{2.107}$$

Thus,

$$\frac{V_1}{V_2} = \frac{sM}{R_1 + sL_1} \tag{2.108}$$

Here, $V_1 = E_{fd}$ and $V_2 = R_F$ are the scaled outputs of the transformer (stabilizer feedback variables). Therefore, Equation (2.108) can further be expressed as

$$\frac{R_F}{E_{fd}} = \frac{sK_F}{1 + sT_F} \tag{2.109}$$

where $K_F = \dfrac{M}{R_1}$ and $T_F = \dfrac{L}{R_1}$; R, L, and M denote resistance, leakage inductance, and mutual inductance of the transformer, respectively. The block diagram representation of the stabilizer model is shown in Figure 2.11.

Combining models for individual components, the complete block diagram of the IEEE Type I exciter can be obtained as shown in Figure 2.13 [4].

Writing Equations (2.104), (2.15), and (2.109) together and replacing the expression for V_{in}, the complete dynamic equations for the IEEE Type I exciter are

$$T_E \frac{dE_{fd}}{dt} = -(K_E + S_E(E_{fd}))E_{fd} + V_R \tag{2.110}$$

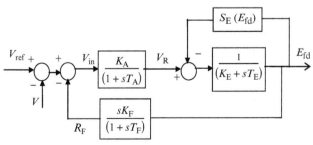

FIGURE 2.13

IEEE Type I exciter model.

$$T_A \frac{dV_R}{dt} = -V_R + K_A R_F - \frac{K_A K_F}{T_F} E_{fd} + K_A (V_{ref} - V + V_s) \tag{2.111}$$

$$T_F \frac{dR_F}{dt} = -R_F + \frac{K_F}{T_F} E_{fd} \tag{2.112}$$

where V_s is the supplementary signal derived from the power system stabilizers (PSSs) or flexible alternating current transmission system (FACTS) controllers. Linearizing equations (2.110)–(2.112), excluding supplementary signal (V_s), the state space representation of the exciter model for an m-machine system ($i=1$, 2, 3,...,m) can be obtained as

$$\begin{bmatrix} \Delta \dot{E}_{fd_i} \\ \Delta \dot{V}_{R_i} \\ \Delta \dot{R}_{F_i} \end{bmatrix} = \begin{bmatrix} f_{s_i}(E_{fd_i}) & \dfrac{1}{T_{E_i}} & 0 \\ -\dfrac{K_{A_i} K_{F_i}}{T_{A_i} T_{F_i}} & -\dfrac{1}{T_{A_i}} & \dfrac{K_{A_i}}{T_{A_i}} \\ \dfrac{K_{F_i}}{T_{F_i}^2} & 0 & -\dfrac{1}{T_{F_i}} \end{bmatrix} \begin{bmatrix} \Delta E_{fd_i} \\ \Delta V_{R_i} \\ \Delta R_{F_i} \end{bmatrix}$$

$$+ \begin{bmatrix} 0 \\ -\dfrac{K_{A_i}}{T_{A_i}} \\ 0 \end{bmatrix} \Delta V_i \tag{2.113}$$

$$\Delta E_{fd_i} = \begin{bmatrix} 1 & 0 & 0 \end{bmatrix} \begin{bmatrix} \Delta E_{fd_i} \\ \Delta V_{R_i} \\ \Delta R_{F_i} \end{bmatrix} \tag{2.114}$$

where

$$f_{s_i}(E_{fd_i}) = -\frac{K_{E_i} + E_{fd_i} \partial S_E(E_{fd_i}) + S_E(E_{fd_i})}{T_{E_i}}$$

with $S_{E_i}(E_{fd_i}) = 0.039 \exp(1.55 E_{fd_i})$.

The multimachine model with IEEE Type I exciter can be obtained including Equations (2.113) and (2.114) with synchronous machine differential algebraic equations given by Equations (2.78)–(2.81).

2.10.2 Static exciter

The block diagram representation of the static exciter is shown in Figure 2.14. The differential equations for the static exciter are given by

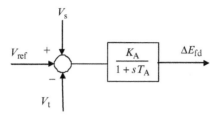

FIGURE 2.14

Static exciter or high-gain exciter.

$$T_A \frac{dE_{fd}}{dt} = -E_{fd} + K_A(V_{ref} - V + V_s) \qquad (2.115)$$

The supplementary signal V_s corresponding to the additional signal derived from the PSSs or FACTS controllers, which is introduced to damp rotor oscillations of the synchronous machine.

EXERCISES

2.1 Derive the synchronous machine dynamic model in the a–b–c reference frame. Using Park's transformation, convert this model in the d–q–o reference frame.

2.2 Construct the block diagram model of an IEEE Type I excitation system with inputs V_{ref} and V_t and output E_{fd}. Obtain the linearized state space model of the IEEE Type I excitation system.

2.3 For a WSCC type 3 machine, 9-bus system exciter parameters of the generators are given in Appendix A. Derive the transfer function of each exciter with V_t as input and E_{fd} as output.

2.4 Using the exciter model (2.110)–(2.112) with $K_E = 1.0$, $S_E = 0$, and $T_E = 0.5$ s, compute the response of E_{fd} for a constant input of $V_R = 1.0$. Use an initial value of $E_{fd} = 0$. All are in pu.

2.5 Express with per unitization and scaling the electric equations and rotor mechanical equations of the synchronous machine. What is the physical interpretation of per unitization?

2.6 Obtain the equivalent circuit representation of the stator algebraic equations of a synchronous machine in the synchronously rotating reference frame represented by D, Q, and O.

References

[1] R.H. Park, The two-reaction theory of synchronous machines-generalized method of analysis, part-I, AIEEE Trans. 48 (1929) 716–727.

[2] M.A. Pai, D.P. Sen Gupta, K.R. Padiyar, Small Signal Analysis of Power System, Narosa Publishing House, New Delhi, India, 2004.

[3] P. Kundur, Power System Stability and Control, McGraw-Hill, New York, 1994.

[4] P.W. Sauer, M.A. Pai, Power System Dynamics and Stability, Pearson Education Pte. Ltd., Singapore, 1998.

Models of Power Network and Relevant Power Equipments

3.1 INTRODUCTION

The stability of power systems is governed by its configuration and the characteristics of the different power system components having significant contribution in the power network. The synchronous machine, exciter, power system stabilizer (PSS), transmission lines, and the power electronic-based devices—flexible alternating current transmission system (FACTS)—are major power system components that play a large role in the generation, transmission, and distribution of power. These devices, under certain operating conditions, maintain synchronism with the rest of the interconnected power networks. Therefore, the examination of their characteristics and the accurate modeling of their dynamic performance are of fundamental importance to study power system stability. In this chapter, small-signal simulation models of the earlier mentioned devices have been described. These models are used in the successive chapters to study the dynamic behavior and investigate small-signal stability of power systems.

3.2 SIMPLE MODEL OF A SYNCHRONOUS GENERATOR

A synchronous generator can be modeled for the purpose of both steady state and dynamic state of operation. In steady state, at its basic form, a synchronous generator can be modeled as a voltage source delivering the required quantity of electrical power to the system. Since the excitation system of an alternator controls the magnitude of the terminal voltage, hence, it is customary to specify the magnitude of the terminal voltage and amount of real power required to be supplied to the system (i.e., V and P are specified) in the load flow studies. Since an alternator also supplies the reactive power to the system, hence, the reactive power limit also needs to be specified. A generator is required to supply power into a power network, and hence, positive P and Q quantities conventionally represent lagging current supplied to the system.

In transient stability studies, the synchronous machine can be represented by a constant $E' \angle \delta^\circ$ behind the d-axis transient reactance (x'_d). The machine is assumed to operate under balanced three-phase positive sequence condition and E' is held constant. E' is the excitation voltage of the alternator and V is the terminal voltage. In

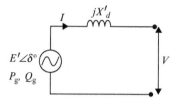

FIGURE 3.1

Classical representation of synchronous generator.

phasor equation, we then have $V = E' - j(x'_d I)$, where I is the current delivered by the machine during the transient condition (Figure 3.1):

$$V = |V|\angle 0°; \quad E' = |E'|\angle\delta°; \quad I = |I|\angle -\varphi \text{ (assuming lagging loads)}$$

Also, the total power is

$$S_g = VI^* = P_g + jQ_g \tag{3.1}$$

3.3 STEADY-STATE MODELING OF SYNCHRONOUS MACHINE (ANALYTICAL ASPECTS) [1]

A synchronous machine has two essential elements: the field and the armature. Normally, the field is on the rotor and the armature is on the stator. The field windings induce alternating voltages in the three-phase armature windings of the stator. The frequency of the stator electrical quantities is thus synchronized with the rotor mechanical speed, hence the designation "synchronous machine."

The synchronous machine under consideration is assumed to have three stator windings, one field winding, and two amortisseur or damper windings. These six windings are magnetically coupled. The magnetic coupling between the windings is a function of the rotor position.

In developing equations of a synchronous machine, the following assumptions are made:

(i) Three stator coils are in a balanced symmetrical configuration cantered 120 electrical degrees apart.
(ii) Rotor has four coils in a balanced symmetrical configuration located in pairs 90 electrical degrees apart.
(iii) The relationship between the flux linkages and the currents must be independent of θ_{shaft} when expressed in the *dqo* coordinate system.
(iv) Magnetic hysteresis and saturation effects are negligible.

Following these assumptions, the performance of the synchronous machine under balanced steady-state conditions may be analyzed by applying the per unit equations resulting from the dynamic model.

It is clear that under steady state, all time derivative terms drop out from the dynamic machine equations, and we must have constant speed ω and constant rotor angle δ, thus requiring $\omega = \omega_s$, and, therefore,

$$V_d = -R_s I_d - \psi_q \tag{3.2}$$

$$V_q = -R_s I_q + \psi_d \tag{3.3}$$

Assuming a balanced three-phase operation and damper winding currents are zero, the other algebraic equations can be solved in steady-state operation, which are as follows:

$$0 = -E_q' - (X_d - X_d')I_d + E_{\text{fd}} \tag{3.4}$$

$$0 = -\psi_{1d} + E_q' - (X_d' - X_{ls})I_d \tag{3.5}$$

$$0 = -E_d' - \left(X_q - X_q'\right)I_q \tag{3.6}$$

$$0 = -\psi_{2q} - E_d' - \left(X_q' - X_{ls}\right)I_q \tag{3.7}$$

$$0 = T_M - (\psi_d I_q - \psi_q I_d) - T_{\text{FW}} \tag{3.8}$$

$$\psi_d = E_q' - X_d' I_d \tag{3.9}$$

$$\psi_q = -E_d' - X_q' I_q \tag{3.10}$$

Except for Equations (3.9) and (3.10), these are all linear equations that can easily be solved for various steady-state representations. Substitution of Equations (3.9) and (3.10) in Equations (3.2) and (3.3) gives

$$V_d = -R_s I_d + E_d' + X_q' I_q \tag{3.11}$$

$$V_q = -R_s I_q + E_q' - X_d' I_d \tag{3.12}$$

These two real algebraic equations can be written as one complex equation of the form

$$\left(V_d + jV_q\right)e^{j(\delta - \pi/2)} = -\left(R_s + jX_q\right)\left(I_d + jI_q\right)e^{j(\delta - \pi/2)} + \overline{E} \tag{3.13}$$

Therefore, the relation between load voltage and internal voltage is

$$\overline{V} = -\left(R_s + jX_q\right)I^* + \overline{E} \tag{3.14}$$

The steady-state model of the synchronous machine mentioned earlier can be represented by the following circuit diagram.

$$\text{Again, } \overline{E} = \left[\left(E'_d - \left(X_q - X'_q \right) I_q + j \left(E'_q + \left(X_q - X'_d \right) I_d \right) \right) \right] e^{j(\delta - \pi/2)}$$

$$= j \left[\left(X_q - X'_d \right) I_d + E'_q \right] e^{(\delta - \pi/2)} \tag{3.15}$$

The internal voltage $\overline{E}$ can be further simplified, using Equation (3.4), as

$$\overline{E} = j \left[\left(X_q - X_d \right) I_d + E_{\text{fd}} \right] e^{j(\delta - \pi/2)}$$

$$\overline{E} = \left[\left(X_q - X_d \right) I_d + E_{\text{fd}} \right] e^{j\delta} \tag{3.16}$$

Here, $\delta =$ angle on $\overline{E}$. Also, the relation between the field current (I_{fd}) and the magnetizing reactance (X_{md}) is

$$I_{\text{fd}} = \frac{E_{\text{fd}}}{X_{\text{md}}} \tag{3.17}$$

The electrical torque is

$$T_{\text{Elec}} = \psi_d I_q - \psi_q I_d = V_d I_d + V_q I_q + R_s \left(I_d^2 + I_q^2 \right) \tag{3.18}$$

The torque is precisely the "real power" delivered by the controlled source of Figure 3.2. Therefore,

$$P_{\text{Out}} = \text{Real} \left(\overline{VI}^* \right) = \text{Real} \left(\left(V_d + jV_q \right) e^{-j(\delta - \pi/2)} \left(I_d - jI_q \right) e^{j(\delta - \pi/2)} \right)$$

$$= \text{Real} \left(\left(V_d + jV_q \right) \left(I_d - jI_q \right) \right)$$

$$= \psi_d I_q - \psi_q I_d \tag{3.19}$$

Also,

$$Q_{\text{Out}} = \text{Imag} \left(\overline{VI}^* \right) = \text{Imag} \left(\left(V_d + jV_q \right) e^{-j(\delta - \pi/2)} \left(I_d - jI_q \right) e^{j(\delta - \pi/2)} \right)$$

$$= \text{Imag} \left(\left(V_d + jV_q \right) \left(I_d - jI_q \right) \right)$$

$$= \psi_d I_q + \psi_q I_d \tag{3.20}$$

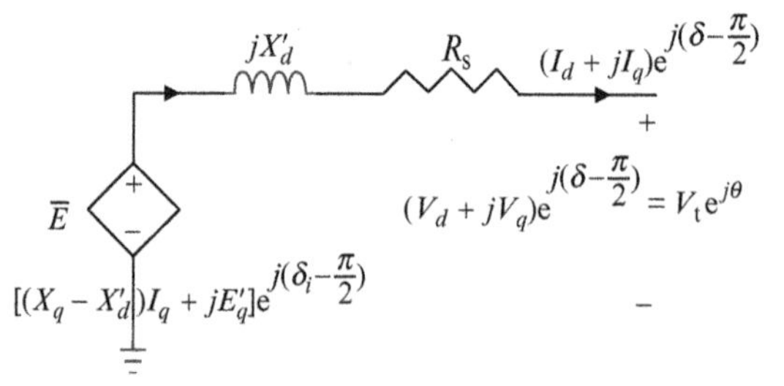

FIGURE 3.2

Synchronous machine steady-state circuit.

3.4 GOVERNOR MODEL [2]

If the load increases, the speed of the synchronous generator reduces slightly. The governor of any thermal unit reacts to this speed variation and permits the entry of some more steam from the boiler to the turbine that, in turn, increases the speed. The increased steam flow reduces the boiler pressure, which reinstates the increase of an adequate fuel, air, and water flow to release the steam pressure. Fortunately, the large *thermal inertia* of most boiler systems enables the load frequency performance of the turbine, generator, and load to be decoupled from that of the boiler, so that, for short duration of load change, the boiler pressure may be regarded as constant. The generator mainly determines the short-term response of the system to the load fluctuations.

Many forms of the governor system have been devised, all of which include, in some way or the other, the variation of the turbine-generator shaft speed as the basis on which the change of the position of the turbine working fluid control valve actuates. Typical speed droop characteristics for most governors range between 5% and 10%. The latest trend in the turbine governor design is to provide an electronic controller. A block diagram representation of the speed governor system is shown in Figure 3.3.

The speed-governing system of hydroturbine is more complicated. An additional feedback loop provides temporary droop compensation to *prevent* instability. This is necessitated by the large inertia of the *penstock gate*, which regulates the rate of water input to the turbine.

$$\text{Here,} \quad \Delta x_e = \frac{K_{SG}}{1 + sT_{SG}}\left(\Delta P_C - \frac{1}{R}\Delta\omega\right) \qquad (3.21)$$

Equation (3.21) plays an important role in modeling the governor operation. Let us consider a simple example. Assuming an increment $\Delta P_C = 1.0$ at $t=0$, for a

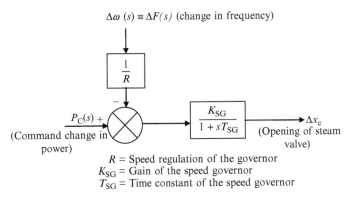

FIGURE 3.3

Block diagram representation of the speed governor system.

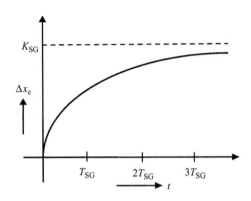

FIGURE 3.4

Speed governor response curve.

speed-governing system under test (i.e., operating on open loop resulting $\Delta\omega=0$), the increment in steam valve opening Δx_e is obtained from Equation (3.21) as

$$\Delta x_e = \frac{K_{SG}}{s(1+sT_{SG})}, \text{ using Laplace transformation of } \Delta P_C$$

$$= \frac{\dfrac{K_{SG}}{T_{SG}}}{s\left(s+\dfrac{1}{T_{SG}}\right)} \tag{3.22}$$

Mathematical manipulation yields

$$\Delta x_e = \frac{K_{SG}}{s} - \frac{K_{SG}}{s+\dfrac{1}{T_{SG}}} \tag{3.23}$$

which on inverse Laplace transform yields

$$\Delta x_e(t) = K_{SG}\left(1 - e^{-t/T_{SG}}\right) \text{ for } t \geq 0 \tag{3.24}$$

The response curve has been plotted in Figure 3.4. Thus, the governor action has been modeled utilizing the concept of transfer functions.

3.5 TURBINE MODEL [2]

Turbine dynamics are of prime importance as they also affect the overall response of the generating plant to load changes. The actual dynamics of course greatly depends on the type of turbine used. A nonreheat type of steam turbine has been shown in Figure 3.5.

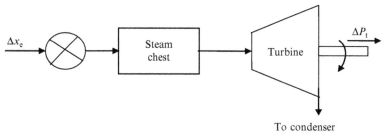

FIGURE 3.5

Block diagram of nonreheat-type turbine.

After passing the control valve, the high-pressure steam enters the turbine via the steam chest that introduces the delay T_T (in the order of 0.2–0.5 s) in the steam flow resulting in the transfer function:

$$G_T = \frac{\Delta P_t(s)}{\Delta x_e(s)} = \frac{1}{1+sT_T} \qquad (3.25)$$

The turbine governor block diagram has been shown in Figure 3.6.

Assuming the command increment to be ΔP_C, at steady state,

$$\Delta P_t = K_{SG}K_T\Delta P_C \qquad (3.26)$$

It insists to choose a scale factor so that $\Delta P_t = \Delta P_C$. This is equivalent to picking $K_{SG}K_T = 1$. This gives the model as shown in Figure 3.7.

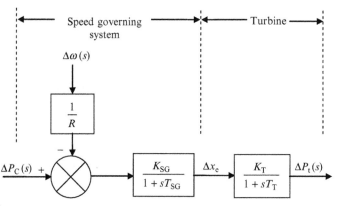

FIGURE 3.6

Turbine governor block diagram.

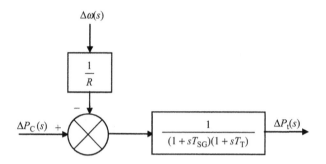

FIGURE 3.7

Block diagram for turbine governor modeling.

3.6 POWER NETWORK MODEL

The power networks interconnect the synchronous machine with transmission and distribution nodes along with rest of the power system components. The equivalent circuit of the power system networks together with stator networks of the synchronous machines and the real and reactive power loads is presented in Figure 3.8. The network equations for an n-bus system can be written in complex form [1].

The network equations for the *generator buses* are given by

$$V_i e^{j\theta_i}\left(I_{d_i} - I_{q_i}\right)e^{-j(\delta_i - \pi/2)} + P_{L_i(V_i)} + jQ_{L_i}(V_i) = \sum_{k=1}^{n} V_i V_k Y_{ik} e^{j(\theta_i - \theta_k - \alpha_{ik})} \qquad (3.27)$$

for $i = 1, 2, 3, \ldots, m$ (number of generator buses).

The network equation for the *load buses* is

$$P_{L_i(V_i)} + jQ_{L_i}(V_i) = \sum_{k=1}^{n} V_i V_k Y_{ik} e^{j(\theta_i - \theta_k - \alpha_{ik})} \qquad (3.28)$$

for $i = m+1, m+2, m+3, \ldots, n$ (number of load buses).

Here, $V_i e^{j\theta_i}\left(I_{d_i} - I_{q_i}\right)e^{-j(\delta_i - \pi/2)} = P_{G_i} + jQ_{G_i}$ is the complex power "injected" into bus i due to the generator, and $Y_{ik}e^{j\alpha_{ik}}$ is the ik th entry of the network bus admittance matrix. This matrix is formed using admittance of all of the branches of the form $\overline{Y}_{ik} = G_{ik} + jB_{ik}$.

The network equations for the generator buses (3.27) are separated into real and imaginary parts and are represented in power balance form as

$$I_{d_i} V_i \sin(\delta_i - \theta_i) + I_{q_i} V_i \cos(\delta_i - \theta_i) + P_{L_i}(V_i)$$
$$- \sum_{k=1}^{n} V_i V_k Y_{ik} \cos(\theta_i - \theta_k - \alpha_{ik}) = 0 \qquad (3.29)$$

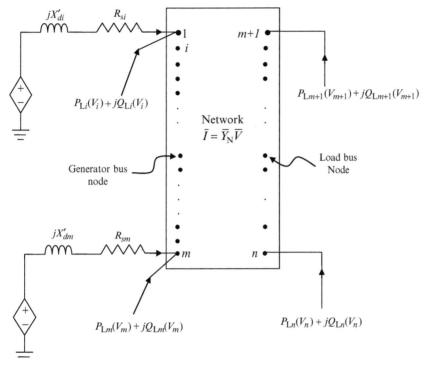

FIGURE 3.8

Synchronous machine dynamic circuit with power networks.

$$I_{d_i} V_i \cos(\delta_i - \theta_i) - I_{q_i} V_i \sin(\delta_i - \theta_i) + Q_{L_i}(V_i)$$
$$- \sum_{k=1}^{n} V_i V_k Y_{ik} \sin(\theta_i - \theta_k - \alpha_{ik}) = 0 \qquad (3.30)$$

where $i = 1, 2, 3, \ldots, m$ for generator buses.

Similarly, the power balance forms of the network equations for the load buses are

$$P_{L_i}(V_i) - \sum_{k=1}^{n} V_i V_k Y_{ik} \cos(\theta_i - \theta_k - \alpha_{ik}) = 0 \qquad (3.31)$$

$$Q_{L_i}(V_i) - \sum_{k=1}^{n} V_i V_k Y_{ik} \sin(\theta_i - \theta_k - \alpha_{ik}) = 0 \qquad (3.32)$$

where $i = m+1, m+2, m+3, \ldots, n$ for load buses.

These nonlinear network equations are linearized to include with the multimachine dynamic model for small-signal stability analysis.

Linearization of the network equations (3.29) and (3.30) pertaining to *generator buses* (PV buses) gives

$$0 = \left(I_{d_i} V_i \cos\left(\delta_i - \theta_i\right) - I_{q_i} V_i \sin\left(\delta_i - \theta_i\right)\right) \Delta\delta_i + V_i \sin\left(\delta_i - \theta_i\right) \Delta I_{d_i} + V_i \cos\left(\delta_i - \theta_i\right) \Delta I_{q_i}$$

$$+ \left(I_{d_i} \sin\left(\delta_i - \theta_i\right) + I_{q_i} \cos\left(\delta_i - \theta_i\right) - \left[\sum_{k=1}^{n} V_k Y_{ik} \cos\left(\theta_i - \theta_k - \alpha_{ik}\right)\right] + \frac{\partial P_{L_i}(V_i)}{\partial V_i}\right) \Delta V_i$$

$$+ \left(-I_{d_i} V_i \cos\left(\delta_i - \theta_i\right) + I_{q_i} V_i \sin\left(\delta_i - \theta_i\right) + \left[V_i \sum_{\substack{k=1 \\ \neq i}}^{n} V_k Y_{ik} \sin\left(\theta_i - \theta_k - \alpha_{ik}\right)\right]\right) \Delta\theta_i$$

$$- V_i \sum_{k=1}^{n} \left[Y_{ik} \cos\left(\theta_i - \theta_k - \alpha_{ik}\right)\right] \Delta V_k - V_i \sum_{\substack{k=1 \\ \neq i}}^{n} \left[V_k Y_{ik} \sin\left(\theta_i - \theta_k - \alpha_{ik}\right)\right] \Delta\theta_k$$

$$(3.33)$$

$$0 = \left(-I_{d_i} V_i \sin\left(\delta_i - \theta_i\right) - I_{q_i} V_i \cos\left(\delta_i - \theta_i\right)\right) \Delta\delta_i + V_i \cos\left(\delta_i - \theta_i\right) \Delta I_{d_i} - V_i \sin\left(\delta_i - \theta_i\right) \Delta I_{q_i}$$

$$+ \left(I_{d_i} \cos\left(\delta_i - \theta_i\right) - I_{q_i} \sin\left(\delta_i - \theta_i\right) - \left[\sum_{k=1}^{n} V_k Y_{ik} \sin\left(\theta_i - \theta_k - \alpha_{ik}\right)\right] + \frac{\partial Q_{L_i}(V_i)}{\partial V_i}\right) \Delta V_i$$

$$+ \left(I_{d_i} V_i \sin\left(\delta_i - \theta_i\right) + I_{q_i} V_i \cos\left(\delta_i - \theta_i\right) - \left[V_i \sum_{\substack{k=1 \\ \neq i}}^{n} V_k Y_{ik} \cos\left(\theta_i - \theta_k - \alpha_{ik}\right)\right]\right) \Delta\theta_i$$

$$- V_i \sum_{k=1}^{n} \left[Y_{ik} \sin\left(\theta_i - \theta_k - \alpha_{ik}\right)\right] \Delta V_k + V_i \sum_{\substack{k=1 \\ \neq i}}^{n} \left[V_k Y_{ik} \cos\left(\theta_i - \theta_k - \alpha_{ik}\right)\right] \Delta\theta_k$$

$$(3.34)$$

for $i = 1, 2, 3, \ldots, m$ (for the generator buses).

Linearization of the network equations (3.31) and (3.32) pertaining to *load buses* (PQ buses) results in

$$0 = \frac{\partial P_{L_i}(V_i)}{\partial V_i} \Delta V_i - \left[\sum_{k=1}^{n} V_k Y_{ik} \cos\left(\theta_i - \theta_k - \alpha_{ik}\right)\right] \Delta V_i$$

$$+ \left[\sum_{\substack{k=1 \\ \neq i}}^{n} V_i V_k Y_{ik} \sin\left(\theta_i - \theta_k - \alpha_{ik}\right)\right] \Delta\theta_i - V_i \sum_{k=1}^{n} \left[Y_{ik} \cos\left(\theta_i - \theta_k - \alpha_{ik}\right)\right] \Delta V_k$$

$$- V_i \sum_{\substack{k=1 \\ \neq i}}^{n} \left[V_k Y_{ik} \sin\left(\theta_i - \theta_k - \alpha_{ik}\right)\right] \Delta\theta_k \qquad (3.35)$$

$$0 = \frac{\partial Q_{L_i}(V_i)}{\partial V_i} \Delta V_i - \left[\sum_{k=1}^{n} V_k Y_{ik} \sin(\theta_i - \theta_k - \alpha_{ik}) \right] \Delta V_i$$

$$- \left[\sum_{\substack{k=1 \\ \neq i}}^{n} V_i V_k Y_{ik} \cos(\theta_i - \theta_k - \alpha_{ik}) \right] \Delta \theta_i$$

$$- V_i \sum_{k=1}^{n} [Y_{ik} \sin(\theta_i - \theta_k - \alpha_{ik})] \Delta V_k + V_i \sum_{\substack{k=1 \\ \neq i}}^{n} [V_k Y_{ik} \cos(\theta_i - \theta_k - \alpha_{ik})] \Delta \theta_k \quad (3.36)$$

for $i = m+1, m+2, \ldots, n$ (for the load buses).

These linearized network equations (3.33)–(3.36) are to be included with the synchronous machine differential algebraic equations (DAEs) to obtain the multimachine small-signal simulation model considering all network bus dynamics.

3.7 MODELING OF LOAD

A load model expresses the characteristics of the load at any instant of time as algebraic function of the bus voltage magnitude or the frequency at that instant. The active power component P_L and the reactive power component Q_L are considered separately. Appropriate voltage-dependent load can be incorporated into the network equations (3.35) and (3.36) of the dynamic model of the multimachine system by specifying the load functions. Traditionally, the voltage dependency of the load characteristics at any bus 'i' has been represented by the exponential model [3]:

$$P_{L_i} = P_{L_{io}} \left(\frac{V_i}{V_{io}} \right)^{n_{pi}} \quad (3.37)$$

$$Q_{L_i} = Q_{L_{io}} \left(\frac{V_i}{V_{io}} \right)^{n_{qi}} \quad (3.38)$$

where $i = m+1, m+2, \ldots, n$ (number of PQ buses). Here, $P_{L_{io}}$ and $Q_{L_{io}}$ are the nominal power and reactive power, respectively, at bus i, with corresponding nominal voltage magnitude V_{io}. P_{L_i} and Q_{L_i} are active and reactive power components of load when the bus voltage magnitude is V_i. n_{pi} and n_{qi} are the load indices. There are three types of static load models that can be configured based on the value of these indices:

- *Constant power type* ($n_p = n_q = 0$)
- *Constant current type* ($n_p = n_q = 1$)
- *Constant impedance type* ($n_p = n_q = 2$)

Therefore, for constant power-type load, Equations (3.37) and (3.38) become

$$P_{L_i} = P_{L_{io}} \quad (3.39)$$

$$Q_{L_i} = Q_{L_{io}} \tag{3.40}$$

For constant current-type load, we have

$$P_{L_i} = P_{L_{io}} \left(\frac{V_i}{V_{io}} \right) \tag{3.41}$$

$$Q_{L_i} = Q_{L_{io}} \left(\frac{V_i}{V_{io}} \right) \tag{3.42}$$

Again, for constant impedance-type characteristics, Equations (3.37) and (3.38) can be written as

$$P_{L_i} = P_{L_{io}} \left(\frac{V_i}{V_{io}} \right)^2 \tag{3.43}$$

$$Q_{L_i} = Q_{L_{io}} \left(\frac{V_i}{V_{io}} \right)^2 \tag{3.44}$$

For composite load, values of n_{pi} and n_{qi} depend on the aggregate characteristics of load components, and the exponential load model becomes then a polynomial load model. Therefore, the polynomial load model for real power is

$$P_{L_i} = P_{L_{io}} + P_{L_{io}} \left(\frac{V_i}{V_{io}} \right) + P_{L_{io}} \left(\frac{V_i}{V_{io}} \right)^2 \tag{3.45}$$

and for reactive power is

$$Q_{L_i} = Q_{L_{io}} + Q_{L_{io}} \left(\frac{V_i}{V_{io}} \right) + Q_{L_{io}} \left(\frac{V_i}{V_{io}} \right)^2 \tag{3.46}$$

for $i = m+1, m+2, \ldots, n$ (number of PQ buses).

Linearizing Equations (3.45) and (3.46), for constant power-type load, we have

$$\Delta P_{L_i} = 0 \tag{3.47}$$

$$\Delta Q_{L_i} = 0 \tag{3.48}$$

and for constant current-type characteristics,

$$\Delta P_{L_i} = \frac{P_{L_{io}}}{V_{io}} \Delta V_i \tag{3.49}$$

$$\Delta Q_{L_i} = \frac{Q_{L_{io}}}{V_{io}} \Delta V_i \tag{3.50}$$

and when load is considered constant impedance type,

$$\Delta P_{\mathrm{L}_i} = 2\frac{P_{\mathrm{L}_{io}}}{(V_{io})^2}V_i\Delta V_i \qquad (3.51)$$

$$\Delta Q_{\mathrm{L}_i} = 2\frac{Q_{\mathrm{L}_{io}}}{(V_{io})^2}V_i\Delta V_i \qquad (3.52)$$

In order to study the effect of different types of load on small-signal stability, these linearized load models given by Equations (3.47)–(3.52) are to be incorporated into the linearized network equations of the dynamic model of the multimachine power system.

3.8 POWER SYSTEM STABILIZER

The basic function of a PSS is to add damping to the generator rotor oscillations by controlling its excitation using auxiliary stabilizing signal such as machine speed, terminal frequency, or power. The stabilizing signals are processed through the PSS and its control acts on the power system through the exciter, i.e., without exciter, the PSSs have no effect of the power system. To improve the small-signal oscillations, the PSS must produce a component of electrical torque (damping torque) in phase with the rotor speed deviations. In Figure 3.9, the PSS with speed input (Δv) to the torque-angle loop has been shown through a block diagram.

The functional relationship between the generator speed and the torque (ΔT_{PSS}) is given by

$$\frac{\Delta T_{\mathrm{PSS}}}{\Delta v} = \mathrm{GEP}(s)G_{\mathrm{PSS}}(s) \qquad (3.53)$$

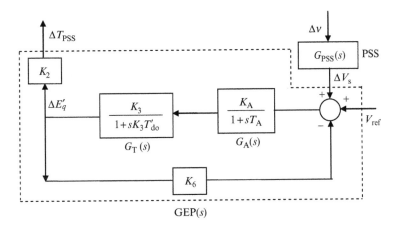

FIGURE 3.9

Power system stabilizer with speed as input.

where $\Delta v = \dfrac{\Delta \omega}{\omega_s}$ is the normalized speed deviation signal. The transfer function GEP(s) represents the characteristics of the generator $G_T(s)$, the excitation system $G_A(s)$, and the power system. Therefore, the transfer function of the PSS, utilizing shaft speed as input, is such that it must compensate for the phase lags introduced by the GEP(s) between the exciter input and the electrical torque to produce a component of torque in phase with speed changes so as to increase damping of the rotor oscillations. An ideal stabilizer characteristic would therefore be inversely proportional to GEP(s) [4,5], i.e.,

$$\text{Ideal } G_{PSS}(s) = \frac{D_{PSS}}{GEP(s)} \tag{3.54}$$

where D_{PSS} represents the desired damping contribution of the stabilizer. Such a stabilizer characteristic is impractical since perfect compensation for the lags of GEP(s) requires pure differentiation with its associated high gain at high frequencies. A practical speed stabilizer therefore must utilize lead-lag stages set to compensate for the phase lags in GEP(s), over the frequency range of interest. The gain must be attenuated at high frequencies to limit the impact of noise and minimize torsional interaction with PSS, and consequently, low-pass and possibly band-reject filters are required. A washout stage may be included to prevent steady-state voltage offset as system frequency changes. Therefore, a practical PSS is represented as

$$G(s) = K_{PSS} \frac{(1 + sT_1)}{(1 + sT_2)} \frac{sT_w}{1 + sT_w} \tag{3.55}$$

The time constants T_1 and T_2 should be set to provide damping over the range of frequencies at which oscillations are likely to occur. Typical values of these parameters are the following:

K_{PSS} (gain of PSS): 0.1-50 s
T_1 (lead time constant): 0.2-1.5 s
T_2 (lag time constant): 0.02-0.15 s
T_w (washout time): ≈ 10 s

Figure 3.10 represents a practical single-stage PSS connected with exciter. Neglecting washout stage block, the dynamic equation of the PSS can be obtained as

$$\left(\Delta \dot{V}_s = -\frac{1}{T_2} \Delta V_s + \frac{K_{PSS}}{T_2} \Delta v + \frac{K_{PSS}T_1}{T_2} \Delta \dot{v} \right) \tag{3.56}$$

In case of multimachine application, Equation (3.56) can be written as

$$\Delta \dot{V}_{si} = -\frac{1}{T_2} \Delta V_{si} + \frac{K_{PSS}}{T_2} \Delta v_i + \frac{K_{PSS}T_1}{T_2} \Delta \dot{v}_i \tag{3.57}$$

where $i =$ machine number where PSS is to be installed. As $\Delta v_i = \dfrac{\Delta \omega_i}{\omega_s}$, Equation (3.57) becomes

$$\Delta \dot{V}_{si} = -\frac{1}{T_2} \Delta V_{si} + \frac{K_{PSS}}{T_2} \frac{\Delta \omega_i}{\omega_s} + \frac{K_{PSS}T_1}{T_2} \frac{\Delta \dot{\omega}_i}{\omega_s} \tag{3.58}$$

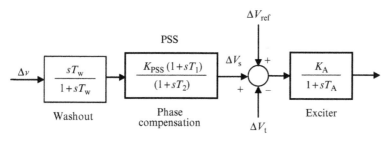

FIGURE 3.10

A practical power system stabilizer.

The expression of $\Delta\dot{\omega}_i$ has been described later in Chapter 5, Section 5.2.2, and can be substituted in Equation (3.58).

3.9 MODEL OF FACTS DEVICES

FACTS is an acronym for flexible alternating current transmission system. The philosophy of FACTS is to use power electronic-controlled devices to control power flows in a transmission network, thereby allowing transmission line plant to be loaded to its full capability. Power electronic-controlled devices, such as static VAR compensators, have been used in transmission networks for many years; however, the concept of FACTS as a total network control philosophy was introduced in 1988 by Dr. N. Hingorani from the Electric Power Research Institute in the United States.

> *FACTS is defined by the IEEE as "A power electronic based system and other static equipment that provide control of one or more ac transmission system parameters to enhance controllability and increase power transfer capability."*

FACTS controllers are capable of controlling the network conditions in a very fast manner, and this feature of FACTS can be exploited to solve many power system problems, to improve power system stability, to enhance system controllability, to increase power transfer capability, to mitigate subsynchronous resonance, etc. In this section, model and principle of operation of the most prominent FACTS controllers and their main steady-state characteristics relevant for power system stability analysis are briefly discussed.

3.9.1 Static Var compensator

Static Var compensator (SVC) is a type of FACTS device, used for shunt compensation to maintain bus voltage magnitude. SVC regulates bus voltage to compensate continuously the change of reactive power loading. The most popular configuration of this type of shunt-connected device is a combination of fixed capacitor C and a

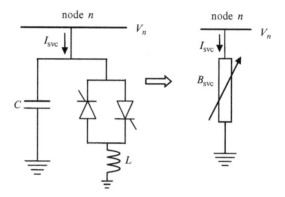

FIGURE 3.11

Advance SVC module.

thyristor-controlled reactor (TCR) as shown in Figure 3.11. The thyristor valves are fired symmetrically in an angle α, ranging from 90° to 180°.

The variable SVC equivalent susceptance B_{svc} at fundamental frequency can be obtained as follows [6]:

Let the source voltage (bus voltage) be expressed as $V_n(t) = V \sin \omega t$, where V is the peak value of the applied voltage and ω is the angular frequency of the supply voltage. The TCR current is then given by the following differential equation:

$$L\frac{di}{dt} - V_n(t) = 0 \tag{3.59}$$

where L is the inductance of the TCR. Integrating Equation (3.59), we get

$$i(t) = \frac{1}{L}\int V_n(t)\mathrm{d}t + C_n \tag{3.60}$$

where C_n is the constant.

Alternatively,

$$i(t) = -\frac{V}{\omega L}\cos\omega t + C_n \tag{3.61}$$

For the boundary condition $i(\omega t = \alpha) = 0$,

$$i(t) = \frac{V}{\omega L}(\cos\alpha - \cos\omega t) \tag{3.62}$$

where α is the the firing angle measured from positive to zero crossing of the applied voltage. To derive the fundamental component of the TCR current $I(\alpha)$, Fourier analysis is used, which in general is given as

$$I_1(\alpha) = a_1 \cos\omega t + b_1 \sin\omega t \tag{3.63}$$

where $b_1=0$ because of the odd-wave symmetry, that is, $f(t)=f(-t)$. Also, no even harmonics are generated because of the half-wave symmetry, i.e., $f(t+T/2)=-f(t)$. The coefficient a_1 is given by

$$a_1 = \frac{4}{T}\int_0^{T/2} f(t)\cos(2\pi t/T)\mathrm{d}t \tag{3.64}$$

Solving for $I(\alpha)$,

$$I_1(\alpha) = \frac{V}{\omega L}\left(1 - \frac{2\alpha}{\pi} - \frac{1}{\pi}\sin 2\alpha\right) \tag{3.65}$$

Expressing Equation (3.65) in terms of the *conduction angle* σ, the fundamental component of TCR current is given by

$$I_1(\alpha) = V\left(\frac{\sigma - \sin\sigma}{\pi X_L}\right) \tag{3.66}$$

where conduction angle σ is related by the equation, $\alpha + \frac{\sigma}{2} = \pi$, and X_L is the reactance of the linear inductor. Equation (3.66) can be written as

$$I_1(\alpha) = VB_{TCR}(\sigma) \tag{3.67}$$

where $B_{TCR}(\sigma)$ is the adjustable fundamental frequency susceptance controlled by the conduction angle to the law

$$B_{TCR}(\sigma) = \frac{\sigma - \sin\sigma}{\pi X_L} \tag{3.68}$$

The maximum value of $B_{TCR}(\sigma)$ is $\frac{1}{X_L}$, obtained with $\sigma = \pi$ or 180°, i.e., full conduction angle in the thyristor controller. The minimum value is zero, obtained with $\sigma = 0$ ($\alpha = 180°$).

From Equation (2.88), the TCR equivalent reactance X_{TCR} can be written as

$$X_{TCR} = \frac{\pi X_L}{\sigma - \sin\sigma} \tag{3.69}$$

The total effective reactance of the SVC, including the TCR and capacitive reactance, is determined by the parallel combination of both components:

$$X_{SVC} = \frac{X_C X_{TCR}}{X_C + X_{TCR}} \tag{3.70}$$

Expressing capacitive reactance and inductive reactance in complex notation, Equation (3.70) becomes

$$X_{SVC} = \frac{-jX_C\frac{\pi j X_L}{\sigma - \sin\sigma}}{-jX_C + \frac{\pi j X_L}{\sigma - \sin\sigma}} \tag{3.71}$$

$$X_{SVC} = \frac{\pi X_C X_L}{-jX_C(\sigma - \sin\sigma) + j\pi X_L} \tag{3.72}$$

$$X_{SVC} = \frac{X_C X_L}{jX_L - j\frac{X_C}{\pi}(\sigma - \sin\sigma)} \tag{3.73}$$

Writing conduction angle in more convenient form with $\sigma = 2(\pi - \alpha)$, the SVC equivalent susceptance B_{SVC} is given by

$$B_{svc} = \frac{-X_L - \frac{X_C}{\pi}(2(\pi - \alpha) + \sin(2\alpha))}{X_C X_L}$$

$$\left[\because X_{SVC} = -\frac{1}{jB_{svc}} \right] \tag{3.74}$$

while its profile, as a function of firing angle corresponding to a capacitive reactance of $X_C = 1.1708$ pu and a variable inductive reactance of $X_L = 0.4925$ pu, is shown in Figure 3.12. The linearized equivalent susceptance obtained from Equation (3.74) is given by

$$\Delta B_{svc} = \frac{2(\cos(2\alpha) - 1)}{X_L} \Delta\alpha \tag{3.75}$$

- **SVC controller**

The block diagram of a basic SVC with an auxiliary controller is shown in Figure 3.13. The voltage input, ΔV_{svc}, of the controller is measured from the SVC

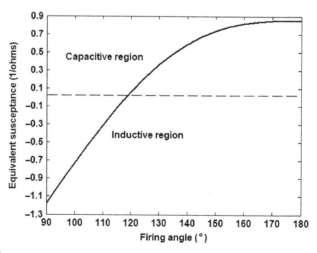

FIGURE 3.12

B_{svc} as function of firing angle α.

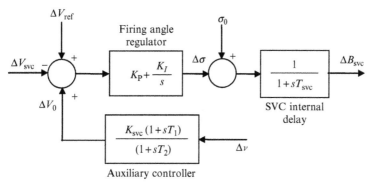

FIGURE 3.13

SVC block diagram with auxiliary controller.

bus. The firing angle of the thyristors that determines the value of susceptance is included in the network. This firing angle is regulated by a proportional-integral (PI) controller, which maintained the bus voltage at the reference (V_{ref}) value. The auxiliary controller with input signal, $\Delta v\,(=\Delta\omega/\omega_s)$, explicitly increases the system damping in addition to the SVC.

Setting K_P to zero, the linearized state-space equations of the SVC controller can be represented as

$$\Delta\dot{V}_0 = -\frac{1}{T_2}\Delta V_0 + \frac{K_{svc}}{\omega_s}\left(\frac{1}{T_2}\right)\Delta\omega + \frac{K_{svc}}{\omega_s}\left(\frac{T_1}{T_2}\right)\Delta\dot{\omega} \qquad (3.76)$$

$$\Delta\dot{\alpha} = -K_I\Delta V_0 + K_I\Delta V_{svc} - K_I\Delta V_{ref} \qquad (3.77)$$

$$\Delta\dot{B}_{svc} = -\frac{1}{T_{svc}}\Delta\alpha - \frac{1}{T_{svc}}\Delta B_{svc} \qquad (3.78)$$

where T_{svc} is the internal time delay of the SVC module. K_{svc}, T_1, and T_2 are the gain, lead, and lag time constants of the auxiliary controller, respectively. ΔB_{svc} is the linearized susceptance of the SVC.

The SVC linearized reactive power injection at the bus n can be derived from the following equation:

$$\Delta Q_n = \frac{\partial Q_n}{\partial\theta_n}\Delta\theta_n + \frac{\partial Q_n}{\partial V_n}\Delta V_n + \frac{\partial Q_n}{\partial\alpha}\Delta\alpha \qquad (3.79)$$

where $Q_n = -B_{svc}V_n^2$
which gives

$$\Delta Q_n = \begin{bmatrix} 0 & -2V_nB_{svc} & 2V_n^2(1-\cos2\alpha)/X_L \end{bmatrix}\begin{bmatrix} \Delta\theta_n \\ \Delta V_n \\ \Delta\alpha \end{bmatrix}$$

Therefore, SVC linearized power flow equation corresponding to the nth bus where SVC has been installed is

$$\begin{bmatrix} \Delta P_n \\ \Delta Q_n \end{bmatrix} = \begin{bmatrix} 0 & 0 & 0 \\ 0 & -2V_n B_{svc} & 2V_n^2(1-\cos 2\alpha)/X_L \end{bmatrix} \begin{bmatrix} \Delta \theta_n \\ \Delta V_n \\ \Delta \alpha \end{bmatrix} \qquad (3.80)$$

Installation of an SVC controller in a multimachine power system results in the addition of state variables, $\Delta x_{svc} = [\Delta V_0 \ \ \Delta \alpha \ \ \Delta B_{svc}]^T$, in the machine differential algebraic equations and the addition of SVC linearized power flow equations (3.80) in the nth bus network equations, respectively.

3.9.2 Static synchronous compensator

Static synchronous compensator (STATCOM) is a shunt-connected FACTS device. It is like the static counterpart of the rotating synchronous condenser, but it generates/absorbs reactive power at faster rate because no moving parts are involved. It is operated as a static Var compensator whose capacitive or inductive output currents are controlled to control the bus voltage with which it is connected. In principle, it performs the same voltage regulation as the SVC but in a more robust manner because unlike the SVC, its operation is not impaired by the presence of low voltages. It goes on well with advanced energy storage facilities, which opens the door for a number of new applications, such as energy deregulations and network security.

STATCOM operation is based on the principle of voltage source or current source converter [7]. The schematic of a STATCOM is shown in Figure 3.14a and b. When used with voltage source converter, its ac output voltage is controlled

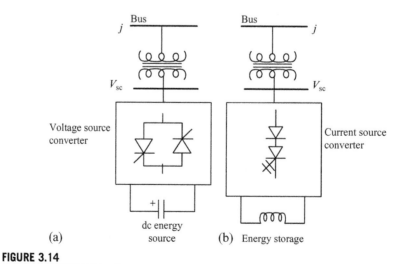

(a) (b) Energy storage

FIGURE 3.14

STATCOM schematic: (a) with voltage source converter (VSC) and (b) with current source converter (CSC).

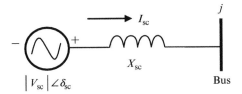

FIGURE 3.15

Equivalent circuit of shunt-operated STATCOM.

such that the required reactive power flow can be controlled at the generator/load bus with which it is connected. Due to the presence of dc voltage source in the capacitor, the voltage source converter converts its voltage to ac voltage source and controls the bus voltage. The exchange of reactive power between the converter and the ac system can be controlled by varying the amplitude of the three-phase output voltage, V_{sc}, of the converter, as illustrated in Figure 3.15. That is, if the amplitude of the output voltage is increased above that of the utility bus (j) voltage, then a current flows through the reactance from the converter to the ac system and the converter generates capacitive-reactive power for the ac system. If the amplitude of the output voltage is decreased below the utility bus voltage, then the current flows from the ac system to the converter and the converter absorbs inductive-reactive power from the ac system. If the output voltage equals the ac system voltage, the reactive power exchange becomes zero, in which case, the STATCOM is said to be in a floating state. Adjusting the phase shift between the converter-output voltage and the ac system voltage can similarly control real power exchange between the converter and the ac system. In other words, the converter can supply real power to the ac system from its dc energy storage if the converter-output voltage is made to lead the ac system voltage. On the other hand, it can absorb real power from the ac system for the dc system if its voltage lags behind the ac system voltage.

The reactive power of a STATCOM is produced by means of power electronic equipment of the voltage source converter (VSC). The VSC may be a two-level or three-level type, depending on the required output power and voltage. A number of VSCs are combined in a multipulse connection to form the STATCOM. In the steady state, the VSCs operate with fundamental frequency switching to minimize converter losses. However, during transient conditions caused by line faults, a pulse width-modulated mode is used to prevent the fault current from entering the VSCs. In this way, the STATCOM is able to withstand transients on the ac side without blocking. The STATCOM can also be designed to act as an active filter to reduce system harmonics and frequently includes the facility of having active power control.

- **Power flow model of STATCOM**

    Let $V_j \angle \theta_j$ be the utility bus voltage at bus j, $V_{sc} \angle \delta_{sc}$ be the inverted voltage (ac) at the output of STATCOM, referred to as jth bus side, X_{sc} be the reactance of the line between the jth bus and the STATCOM, and Q_{sc} be the reactive power exchange for the STATCOM with the bus.

Obviously,

$$Q_{sc} = \frac{|V_j|^2}{X_{sc}} - \frac{|V_j||V_{sc}|}{X_{sc}} \cos\left(\theta_j - \delta_{sc}\right)$$

$$= \frac{|V_j|^2 - |V_j||V_{sc}|}{X_{sc}}, \quad \text{if } \theta_j = \delta_{sc} \text{ (for a loss less STATCOM)} \tag{3.81}$$

Thus, if $|V_j| < |V_{sc}|$, Q_{sc} becomes negative and the STATCOM generates reactive power. On the other hand, if $|V_j| > |V_{sc}|$, Q_{sc} becomes positive and the STATCOM absorbs reactive power.

$$\text{Also,} \quad V_{sc} = |V_{sc}|\left(\cos\delta_{sc} + j\sin\delta_{sc}\right) \tag{3.82}$$

The maximum and minimum limits of $|V_{sc}|$ will be governed by the STATCOM capacitor rating. δ_{sc} may have any value between $0°$ and $180°$. Let us now draw the equivalent circuit of STATCOM in Figure 3.15.

Here, $I_{sc} = Y_{sc}(V_{sc} - V_j)$

where $Y_{sc} = \dfrac{1}{Z_{sc}} = G_{sc} + jB_{sc}$.

$$\therefore S_{sc}(\text{complex power flow}) = V_{sc}I_{sc}^* = V_{sc}Y_{sc}^*\left(V_{sc}^* - V_j^*\right) \tag{3.83}$$

However, $V_{sc} = |V_{sc}|(\cos\delta_{sc} + j\sin\delta_{sc})$ (Equation 3.82).

Substitution of V_{sc} in the expression of S_{sc} leads to the following equations:

$$P_{sc} = |V_{sc}|^2 G_{sc} - |V_{sc}||V_j|\left[G_{sc}\cos\left(\delta_{sc} - \theta_j\right) + B_{sc}\sin\left(\delta_{sc} - \theta_j\right)\right] \tag{3.84}$$

$$Q_{sc} = -|V_{sc}|^2 B_{sc} - |V_{sc}||V_j|\left[G_{sc}\sin\left(\delta_{sc} - \theta_j\right) - B_{sc}\cos\left(\delta_{sc} - \theta_j\right)\right] \tag{3.85}$$

To simplify these equations, let us assume the STATCOM is lossless (thus, $G_{sc} = 0$) and there is no capability of the STATCOM for active power flow (thus, $P_{sc} = 0$). Also, $\delta_{sc} \cong \theta_j$.

$$\therefore Q_{sc} = -|V_{sc}|^2 B_{sc} - |V_{sc}||V_j|B_{sc} \tag{3.86}$$

The power mismatch equation can be written now as

$$\begin{bmatrix} \Delta P_j \\ \Delta Q_{sc} \end{bmatrix} = \begin{bmatrix} \dfrac{\partial P_j}{\partial \delta_j} & \dfrac{\partial P_j}{\partial |V_{sc}|} \\ \dfrac{\partial Q_{sc}}{\partial \delta_j} & \dfrac{\partial Q_{sc}}{\partial |V_{sc}|} \end{bmatrix} \begin{bmatrix} \Delta\theta_j \\ \Delta|V_{sc}| \end{bmatrix} \tag{3.87}$$

At the end of iteration p, the variable voltage $|V_{sc}|$ can be corrected as

$$|V_{sc}|^{(p+1)} = |V_{sc}|^{(p)} + \Delta|V_{sc}|^{(p)} \tag{3.88}$$

- **STATCOM controller**

The STATCOM controller model used as a voltage controller is shown in Figure 3.16. Among the shunt controllers, the STATCOM performs better than

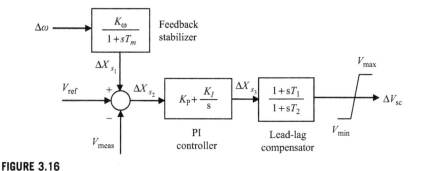

FIGURE 3.16

STATCOM voltage controller for damping small-signal oscillations.

SVC because of its better characteristics in the low-voltage region. The main function of an STATCOM, as with the conventional SVC, is to regulate the transmission line voltage at the point of connection. It has been observed for IEEE first bench mark system that the STATCOM equipped with only a voltage controller is not sufficient to damp all the oscillatory modes of the system. Thus, a need exists for an additional control signal along with the STATCOM voltage controller. The principal strategy in controlling small-signal oscillations using STATCOM damping controller is to use simple auxiliary stabilizing signals. It is known that the generator speed contains components of all the oscillatory modes; consequently, if the generator speed is used to control an STATCOM, all the oscillatory modes including swing modes and torsional modes will be affected. Therefore, the auxiliary signal employed is the generator speed deviation. The control system can be designed to maintain the magnitude of the bus voltage constant by controlling the magnitude and/or phase shift of the VSC output voltage.

The block diagram given in Figure 3.16 can be represented by the following state variable equations:

$$\Delta X_{s_1} = \left(\frac{K_\omega}{1+sT}\right)\Delta\omega \tag{3.89}$$

$$\Delta X_{s_3} = \left(K_P+\frac{K_I}{s}\right)\Delta X_{s_2} \tag{3.90}$$

$$\Delta V_{sc} = \left(\frac{1+sT_1}{1+sT_2}\right)\Delta X_{s_3} \tag{3.91}$$

Equating Equations (3.89)–(3.91), the linearized state-space model of a STATCOM controller is obtained as

$$\Delta\dot{X}_{s_2} = -\frac{1}{T_m}\Delta X_{s_2} + \frac{K_\omega}{T_m}\Delta\omega - \frac{1}{T_m}\Delta V_{meas} \tag{3.92}$$

$$\Delta \dot{X}_{s_3} = \left(-\frac{K_P}{T_m} + K_I\right)\Delta X_{s_2} + \frac{K_P K_\omega}{T_m}\Delta\omega - \frac{K_P}{T_m}\Delta V_{meas} \tag{3.93}$$

$$\Delta \dot{V}_{sc} = -\frac{1}{T_2}\Delta V_{sc} + \frac{1}{T_2}\Delta X_{s_3} + \frac{T_1}{T_2}\left(-\frac{K_P}{T_m} + K_I\right)\Delta X_{s_2}$$

$$+ \frac{T_1 K_P K_\omega}{T_2 T_m}\Delta\omega - \frac{T_1 K_P}{T_2 T_m}\Delta V_{meas} \tag{3.94}$$

The multimachine model with STATCOM controller can be formulated by adding the state variables $\Delta X_{STATCOM} = [\Delta X_{s_2} \ \Delta X_{s_3} \ \Delta V_{sc}]^T$ with the differential algebraic equations and the STATCOM linearized power flow equation (3.87) with the network equations of the multimachine system, respectively.

The STATCOM controller model used as a voltage controller (Figure 3.16) has three gains: the proportional gain, K_P; the integral gain, K_I; and the speed deviation feedback gain, K_ω. The objective is to damp all the swing modes at all series-compensation levels. Eigenvalue analysis is used to obtain the range of K_P, K_I, and K_ω for which the system is stable. It is found that the system is stable for typical values of STATCOM controller parameters as $6.0 \le K_P \le 0.1$, $132 \le K_I \le 0$, and $4 \le K_\omega \le 11$. These ranges of K_P, K_I, and K_ω, are used to select final gain parameters of the STATCOM controller by carrying out a step-response test on the system and ensuring the system settling time and a low generator speed overshoot.

3.9.3 Thyristor-controlled series compensator

The basic Thyristor-controlled series compensator (TCSC) configuration consists of a fixed series capacitor bank C in parallel with a TCR as shown in Figure 3.17. This simple model utilizes the concept of a variable series reactance. The series reactance is adjusted through appropriate variation of the firing angle (α), to allow specified amount of active power flow across the series-compensated line. The steady-state relationship between the firing angle α and the TCSC reactance X_{TCSC} at fundamental frequency can be derived as follows [8]:

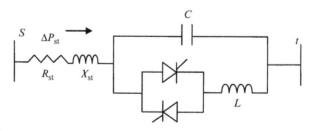

FIGURE 3.17

TCSC module.

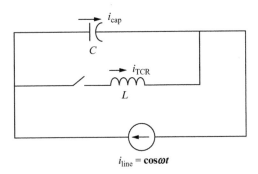

FIGURE 3.18

Equivalent circuit of a TCSC module.

The simplified TCSC equivalent circuit is shown in Figure 3.18. The transmission line current is assumed to be the independent input variable and is modeled as an external current source, $i_{line}(t)$.

It has been assumed that a loop current is trapped in the reactor-capacitor circuit and that the power system can be represented by an ideal, sinusoidal current source. Under these assumptions, the TCSC steady-state voltage and current equations that can be obtained from the analysis of a parallel LC circuit with a variable inductance are shown in Figure 3.18, and the asymmetrical current pulses through the TCSC thyristors are shown schematically in Figure 3.19. However, the analysis presented in the following text may be erroneous to the extent that the line current deviates from a purely sinusoidal nature. The original time reference (OR) is taken to be the positive going zero crossing of the voltage across the TCSC inductance. Also, an auxiliary time reference (AR) is taken at a time when the thyristor starts to conduct.

The line current is

$$i_{line} = \cos \omega t$$

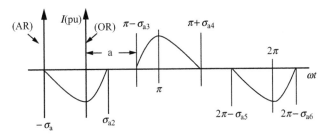

FIGURE 3.19

Asymmetrical thyristor current.

or, in the AR plane,

$$i_{line} = \cos(\omega t - \sigma_a) = \cos\omega t \, \cos\sigma_a + \sin\omega t \, \sin\sigma_a$$

Applying Kirchhoff current law (KCL) to the circuit shown in Figure 3.18,

$$i_{line} = i_{TCR} + i_{cap} \tag{3.95}$$

During the conduction period, the voltage across the TCSC inductive reactance and capacitive reactance coincides:

$$L\frac{di_{TCR}}{dt} = \frac{1}{C}\int i_{cap}dt + V_{cap}^+ \tag{3.96}$$

where V_{cap}^+ is the voltage across the capacitor when the thyristor turns on. Taking Laplace transformation of Equations (3.95) and (3.96),

$$I_{line} = \cos\sigma_a \frac{s}{s^2+\omega^2} + \sin\sigma_a \frac{\omega}{s^2+\omega^2} \tag{3.97}$$

$$I_{line} = I_{TCR} + I_{cap} \tag{3.98}$$

$$I_{cap} = s^2 LCI_{TCR} - CV_{cap}^+ \tag{3.99}$$

Substituting Equations (3.97) and (3.99) into Equation (3.98) and solving for I_{TCR},

$$I_{TCR} = \omega_0^2 \cos\sigma_a \frac{s}{(s^2+\omega_0^2)(s^2+\omega^2)} + \omega_0^2\omega\sin\sigma_a \frac{1}{(s^2+\omega_0^2)(s^2+\omega^2)} + \frac{\omega_0^2 CV_{cap}^+}{s^2+\omega_0^2} \tag{3.100}$$

where $\omega_0^2 = \frac{1}{LC}$. Expressing Equation (3.100) in the time domain leads to

$$i_{TCR} = A\cos(\omega t - \sigma_a) - A\cos\sigma_a\cos\omega_0 t$$
$$- B\sin\sigma_a\sin\omega_0 t + DV_{cap}^+\sin\omega_0 t \tag{3.101}$$

where $A = \frac{\omega_0^2}{\omega_0^2-\omega^2}$, $B = \frac{\omega_0\omega}{\omega_0^2-\omega^2}$, and $D = \omega_0 C$.

To express Equation (3.101) within the range $[-\sigma_a, \sigma_{a2}]$, it is only necessary to shift the equation to the original time reference by adding $\frac{\sigma_a}{\omega}$ to the time variable. Hence, substituting $t + \frac{\sigma_a}{\omega}$ in place of t,

$$i_{TCR} = A\cos\left(\omega\left(t+\frac{\sigma_a}{\omega}\right)-\sigma_a\right) - A\cos\sigma_a\cos\omega_0\left(t+\frac{\sigma_a}{\omega}\right)$$
$$- B\sin\sigma_a\sin\omega_0\left(t+\frac{\sigma_a}{\omega}\right) + DV_{cap}^+\sin\omega_0\left(t+\frac{\sigma_a}{\omega}\right)$$

Writing $\bar\omega = \frac{\omega_0}{\omega}$,

$$i_{TCR} = A\cos\omega t + \left(-A\cos\sigma_a\cos\bar\omega\sigma_a - B\sin\sigma_a\sin\bar\omega\sigma_a + DV_{cap}^+\sin\bar\omega\sigma_a\right)\cos\omega_0 t$$
$$+ \left(A\cos\sigma_a\sin\bar\omega\sigma_a - B\sin\sigma_a\cos\bar\omega\sigma_a + DV_{cap}^+\cos\bar\omega\sigma_a\right)\sin\omega_0 t \tag{3.102}$$

The TCSC capacitor voltage V_{cap}^+ is obtained, making the coefficient of $\sin \omega_0 t$ zero,

$$\text{i.e.,} \quad \left(A\cos\sigma_a \sin \varpi\sigma_a - B\sin\sigma_a \cos\varpi\sigma_a + DV_{cap}^+ \cos\varpi\sigma_a \right) = 0$$

$$\text{or} \quad V_{cap}^+ = \frac{B}{D}\sin\sigma_a - \frac{A}{D}\cos\sigma_a \tan(\varpi\sigma_a) \tag{3.103}$$

Substituting Equation (3.102) into Equation (3.103), the steady-state thyristor current is obtained as

$$i_{TCR} = A\cos\omega t + \left(-A\cos\sigma_a \cos\varpi\sigma_a - B\sin\sigma_a \sin\varpi\sigma_a \right.$$

$$\left. + D\left(\frac{B}{D}\sin\sigma_a - \frac{A}{D}\cos\sigma_a \tan(\varpi\sigma_a) \right)\sin\varpi\sigma_a \right)\cos\omega_0 t$$

$$= A\cos\omega t - A\cos\sigma_a \frac{\cos(\varpi\omega t)}{\cos(\varpi\sigma_a)} \tag{3.104}$$

where $\sigma_a = \pi - \alpha$. Equation (3.104) is symmetric and it is valid in the interval $\omega t \in [-\sigma_a, \sigma_a]$ and $\omega t \in [2\pi - \sigma_a, 2\pi + \sigma_a]$. Since the thyristor current has even and quarterly symmetry, its fundamental frequency component may be obtained by applying Fourier analysis of Equation (3.104):

$$I_{TCR} = \frac{4}{\pi}\int_0^{\sigma_a} \left(A\cos\omega t - A\cos\sigma_a \frac{\cos(\varpi\omega t)}{\cos(\varpi\sigma_a)} \right)\cos(\omega t)d(\omega t) \tag{3.105}$$

Let $\omega t = \theta$:

$$= \frac{2A}{\pi}\int_0^{\sigma_a}\left(2\cos^2\theta\right)d\theta - \frac{4A\cos\sigma_a}{\pi\cos(\varpi\sigma_a)}\int_0^{\sigma_a}\cos(\varpi\theta)(\cos\theta)d\theta \tag{3.106}$$

Solving Equation (3.106) results in

$$I_{TCR} = \frac{A}{\pi}(2\sigma_a + \sin(2\sigma_a))$$

$$- \frac{4A\cos^2\sigma_a}{(\varpi^2 - 1)}\left[\frac{\varpi\tan(\varpi\sigma_a) - \tan(\sigma_a)}{\pi} \right] \tag{3.107}$$

The TCSC fundamental impedance is

$$Z_{TCSC} = R_{TCSC} + jX_{TCSC} = \frac{V_{TCSC}}{I_{line}} \tag{3.108}$$

The voltage V_{TCSC} is equal to the voltage across the TCSC capacitor and Equation (3.108) can be written as

$$Z_{TCSC} = \frac{-jX_C I_{cap}}{I_{line}}$$

If the external power network is represented by an idealized current source, as seen from the TCSC terminals, this current source is equal to the sum of the currents flowing through the TCSC capacitor and inductor. The TCSC impedance can then be expressed as

$$Z_{TCSC} = \frac{-jX_C(I_{line} - I_{TCR})}{I_{line}}$$

Substituting the expression for I_{TCR} from Equation (3.107) and assuming $I_{line} = 1 \cos \omega t$,

$$Z_{TCSC} = -jX_C + \frac{-jX_C}{1 \cos \omega t} \left[\frac{A}{\pi}(2\sigma_a + \sin(2\sigma_a)) \right.$$
$$\left. - \frac{4A \cos^2 \sigma_a}{(\varpi^2 - 1)} \left[\frac{\varpi \tan(\varpi \sigma_a) - \tan(\sigma_a)}{\pi} \right] \right] \tag{3.109}$$

Let $\quad U_1 = \frac{-jX_C}{1 \cos \omega t} \left[\frac{A}{\pi}(2\sigma_a + \sin(2\sigma_a)) \right] \quad$ and $\quad U_2 = \frac{4AX_C \cos^2 \sigma_a}{(\varpi^2 - 1)1 \cos \omega t}$
$$\left[\frac{\varpi \tan(\varpi \sigma_a) - \tan(\sigma_a)}{\pi} \right]$$

Therefore,

$$Z_{TCSC} = -jX_C + U_1 + U_2 \tag{3.110}$$

Using the expression for $A = \dfrac{\omega_0^2}{\omega_0^2 - \omega^2}$, $\omega_0^2 = \dfrac{1}{LC}$, and $\sigma_a = \pi - \alpha$,

$$U_1 = \frac{X_C}{\pi \omega C \left(\dfrac{1}{\omega C - \omega L} \right)} [2(\pi - \alpha) + \sin(2(\pi - \alpha))], \quad [\because \cos \omega t|_{max} = 1]$$

$$= \frac{X_C + X_{LC}}{\pi} [2(\pi - \alpha) + \sin(2(\pi - \alpha))]$$

$$= C_1(2(\pi - \alpha) + \sin(2(\pi - \alpha))) \tag{3.111}$$

where $C_1 = \dfrac{X_C + X_{LC}}{\pi}$ and $X_{LC} = \dfrac{X_C X_L}{X_C - X_L}$.

Again,

$$U_2 = \frac{4AX_C \cos^2 \sigma_a}{(\varpi^2 - 1)1 \cos \omega t} \left[\frac{\varpi \tan(\varpi \sigma_a) - \tan(\sigma_a)}{\pi} \right]$$

Replacing the expression for $A = \dfrac{\omega_0^2}{\omega_0^2 - \omega^2}$ and $\sigma_a = \pi - \alpha$,

$$= \frac{4X_C \omega_0^2 \cos^2(\pi - \alpha)}{(\omega_0^2 - \omega^2)(\varpi^2 - 1)} \left[\frac{\varpi \tan(\varpi(\pi - \alpha)) - \tan(\pi - \alpha)}{\pi} \right]$$

$$\left[\because \cos \omega t|_{max} = 1, \because \varpi = \frac{\omega_0}{\omega}, \omega_0^2 = \frac{1}{LC} \right]$$

$$= \frac{4X_{LC}^2 \cos^2(\pi - \alpha)}{\pi X_L} [\varpi \tan(\varpi(\pi - \alpha)) - \tan(\pi - \alpha)]$$

$$= C_2 \cos^2(\pi - \alpha) [\varpi \tan(\varpi(\pi - \alpha)) - \tan(\pi - \alpha)] \tag{3.112}$$

where $X_{LC} = \dfrac{X_L X_C}{(X_C - X_L)}$ and $C_2 = \dfrac{4X_{LC}^2}{\pi X_L}$.

Combining Equations (3.111) and (3.112) with Equation (3.110), the TCSC fundamental impedance can be obtained as

$$Z_{TCSC} = j(-X_C + C_1(2(\pi - \alpha) + \sin 2(\pi - \alpha))$$
$$-C_2 \cos^2(\pi - \alpha)(\varpi \tan(\varpi(\pi - \alpha)) - \tan(\pi - \alpha)))   \quad (3.113)$$

Therefore, the TCSC equivalent reactance, as a function of the TCSC firing angle (α), which can be expressed from Equation (3.113), is

$$X_{TCSC} = -X_C + C_1(2(\pi - \alpha) + \sin(2(\pi - \alpha)))$$
$$- C_2 \cos^2(\pi - \alpha)(\varpi \tan(\varpi(\pi - \alpha)) - \tan(\pi - \alpha))   \quad (3.114)$$

The TCSC linearized equivalent reactance, which can then be obtained from Equation (3.114), is

$$\Delta X_{TCSC} = \left\{ -2C_1(1 + \cos(2\alpha)) + C_2 \sin(2\alpha)(\varpi \tan(\varpi(\pi - \alpha)) - \tan\alpha) \right.$$
$$\left. + C_2 \left(\varpi^2 \frac{\cos^2(\pi - \alpha)}{\cos^2(\varpi(\pi - \alpha))} - 1 \right) \right\} \Delta\alpha   \quad (3.115)$$

For a TCSC designed with $X_C = 5.7X_L\Omega$ at a base frequency of 50 Hz, its equivalent reactance (X_{TCSC}) as a function of the firing angle (α) has been plotted in Figure 3.20.

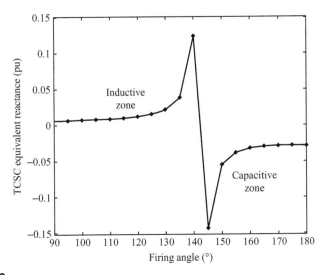

FIGURE 3.20

Variation of TCSC reactance (X_{TCSC}) with firing angle (α).

- **TCSC damping controller**

The transfer function model of a TCSC controller has been shown in Figure 3.21. The auxiliary input signal, $\Delta v\ (=\Delta\omega/\omega_s)$, is the normalized speed deviation, and output signal is the deviation in thyristor conduction angle ($\Delta\sigma$). It is composed of a gain block, a signal washout block, and a phase compensator block [9].

The signal washout block is a high-pass filter that prevents the steady changes in the speed with changes of frequency and has less importance in small-signal stability analysis. During steady-state conditions, $\Delta\sigma=0$ and the effective line impedance $Z_{st}=R_{st}+j(X_{st}-X_{TCSC}(\alpha_0))$. In dynamic conditions, $Z_{st}=R_{st}+j(X_{st}-X_{TCSC}(\alpha))$, where $\sigma=\sigma_0+\Delta\sigma$ and α, α_0, and σ_0 being initial values of firing angle and conduction angle, respectively. R_{st} and X_{st} are the resistance and reactance of the transmission line that connect a TCSC between the nodes 's' and 't' (Figure 3.17). Neglecting washout stage, the proposed TCSC controller model can be represented by the following state equations:

$$\Delta\dot{\alpha}=-\frac{1}{T_2}\Delta\alpha-\frac{K_{TCSC}}{\omega_s}\left(\frac{1}{T_2}\right)\Delta\omega-\frac{K_{TCSC}}{\omega_s}\left(\frac{T_1}{T_2}\right)\Delta\dot{\omega} \qquad (3.116)$$

$$\Delta\dot{X}_{TCSC}=-\frac{1}{T_{TCSC}}\Delta\alpha-\frac{1}{T_{TCSC}}\Delta X_{TCSC} \qquad (3.117)$$

When a TCSC controller is installed in a power system, the state variables corresponding to the TCSC controller $\Delta x_{tcsc}=[\Delta\alpha\ \ \Delta X_{TCSC}]^T$ are added with the generator state equations, and also, the TCSC power flow equations are included in the network equation. The TCSC linearized power flow equations, when it is controlling power flow in a branch between the nodes 's' and 't,' can be derived as follows:

The TCSC power flow from node 's' to node 't' is given by

$$S_{st}=V_sI_{st}^*=V_sY_{st}^*\left(V_s^*-V_t^*\right) \qquad (3.118)$$

with network admittance

$$Y_{st}^*=\frac{1}{R_{st}+j(X_{st}-X_{TCSC})}=\frac{R_{st}-j(X_{st}-X_{TCSC})}{R_{st}^2+(X_{st}-X_{TCSC})^2}$$

$$=g_{st}-jb_{st} \qquad (3.119)$$

and $V_s(V_s^*-V_t^*)=V_s^2-V_sV_te^{j\theta_{st}}$; $\theta_{st}=\theta_s-\theta_t$

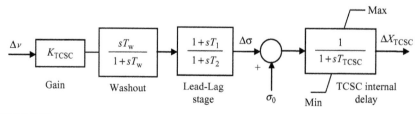

FIGURE 3.21

Transfer function model of a TCSC controller.

Therefore,

$$S_{st} = V_s^2(g_{st} - jb_{st}) - V_sV_t(g_{st} - jb_{st})(\cos\theta_{st} + j\sin\theta_{st}) \qquad (3.120)$$

Separating real and imaginary parts of Equation (3.120) gives

$$P_{st} = V_s^2 g_{st} - V_sV_t(g_{st}\cos\theta_{st} + b_{st}\sin\theta_{st}) \qquad (3.121)$$

$$Q_{st} = -V_s^2 b_{st} - V_sV_t(g_{st}\sin\theta_{st} - b_{st}\cos\theta_{st}) \qquad (3.122)$$

Here, P_{st} and Q_{st} are the active power flow and reactive power flow between nodes 's' and 't' where TCSC has been installed. The active power and reactive power injected at sending end bus (s) and receiving end bus (t) of the TCSC are obtained as

$$P_s = V_s^2 g_{ss} - V_sV_t(g_{st}\cos\theta_{st} + b_{st}\sin\theta_{st}) \qquad (3.123)$$

$$Q_s = -V_s^2 b_{ss} - V_sV_t(g_{st}\sin\theta_{st} - b_{st}\cos\theta_{st}) \qquad (3.124)$$

Also,

$$P_t = V_t^2 g_{tt} - V_tV_s(g_{ts}\cos\theta_{ts} + b_{ts}\sin\theta_{ts}) \qquad (3.125)$$

$$Q_t = -V_t^2 b_{tt} - V_tV_s(g_{ts}\sin\theta_{ts} - b_{ts}\cos\theta_{ts}) \qquad (3.126)$$

Therefore, the TCSC linearized power flow equations at the node 's' and at the node 't' are computed following Equations (3.121)–(3.126) and are given by

$$0 = \begin{bmatrix} \dfrac{\partial P_s}{\partial\theta_s} & \dfrac{\partial P_s}{\partial V_s} & \dfrac{\partial P_s}{\partial\alpha} \\[2mm] \dfrac{\partial Q_s}{\partial\theta_s} & \dfrac{\partial Q_s}{\partial V_s} & \dfrac{\partial Q_s}{\partial\alpha} \\[2mm] \dfrac{\partial P_{st}}{\partial\theta_s} & \dfrac{\partial P_{st}}{\partial V_s} & \dfrac{\partial P_{st}}{\partial\alpha} \end{bmatrix} \begin{bmatrix} \Delta\theta_s \\[2mm] \Delta V_s \\[2mm] \Delta\alpha \end{bmatrix} \qquad (3.127)$$

$$0 = \begin{bmatrix} \dfrac{\partial P_t}{\partial\theta_t} & \dfrac{\partial P_t}{\partial V_t} & \dfrac{\partial P_t}{\partial\alpha} \\[2mm] \dfrac{\partial Q_t}{\partial\theta_t} & \dfrac{\partial Q_t}{\partial V_t} & \dfrac{\partial Q_s}{\partial\alpha} \\[2mm] \dfrac{\partial P_{st}}{\partial\theta_t} & \dfrac{\partial P_{st}}{\partial V_t} & \dfrac{\partial P_{st}}{\partial\alpha} \end{bmatrix} \begin{bmatrix} \Delta\theta_t \\[2mm] \Delta V_t \\[2mm] \Delta\alpha \end{bmatrix} \qquad (3.128)$$

The TCSC linearized *real power* flow equations with respect to the firing angle (α) can be computed as

$$\frac{\partial P_s}{\partial\alpha} = -\frac{\partial P_{st}}{\partial\alpha}$$

$$= -V_s^2\frac{\partial g_{st}}{\partial\alpha} + V_sV_t\left(\cos\theta_{st}\frac{\partial g_{st}}{\partial\alpha} + \sin\theta_{st}\frac{\partial b_{st}}{\partial\alpha}\right) \qquad (3.129)$$

Also, $\dfrac{\partial P_{st}}{\partial \theta_s} = -\dfrac{\partial P_s}{\partial \theta_s}$ and $\dfrac{\partial P_{st}}{\partial V_s} = -\dfrac{\partial P_s}{\partial V_s}$ are true for the branch between nodes 's' and 't' where TCSC is installed. The linearized *reactive power* flow equations with respect to the firing angle 'α' can be derived in a similar way from Equations (3.122), (3.124), and (3.126).

The expression for $\dfrac{\partial g_{st}}{\partial \alpha}$ and $\dfrac{\partial b_{st}}{\partial \alpha}$ is obtained from (3.119) as $\dfrac{\partial g_{st}}{\partial \alpha} = G\dfrac{\partial X_{TCSC}}{\partial \alpha}$,

where $G = \dfrac{-2R_{st}(X_{st}-X_{TCSC})}{\left(R_{st}^2+(X_{st}-X_{TCSC})^2\right)^2}.$

Similarly, $\dfrac{\partial b_{st}}{\partial \alpha} = B\dfrac{\partial X_{TCSC}}{\partial \alpha}$, where $B = \dfrac{-R_{st}^2+(X_{st}-X_{TCSC})^2}{\left(R_{st}^2+(X_{st}-X_{TCSC})^2\right)^2}.$

The values of $\dfrac{\partial g_{st}}{\partial \alpha}$ and $\dfrac{\partial b_{st}}{\partial \alpha}$ are used in Equation (3.129) to obtain elements $\dfrac{\partial P_s}{\partial \alpha}$ and $\dfrac{\partial P_{st}}{\partial \alpha}$ of Equations (3.127) and (3.128). This model of TCSC controller has been used in Chapters 7 and 8 for small-signal stability analysis of SMIB as well as in multimachine power systems.

3.9.4 Static synchronous series compensator

The static synchronous series compensator (SSSC) is a series-connected FACTS controller based on VSC and can be viewed as an advanced type of controlled series compensation, just as a STATCOM is an advanced SVC. An SSSC has several advantages over a TCSC, such as (a) elimination of bulky passive components, capacitors and reactors; (b) improved technical characteristics; (c) symmetric capability in both inductive and capacitive operating modes; and (d) possibility of connecting an energy source on the dc side to exchange real power with the ac network. The major objective of SSSC is to control or regulate the power flow in the line in which it is connected, while a STATCOM is used to regulate the voltage at the bus where it is connected.

The schematic of an SSSC is shown in Figure 3.22a. The equivalent circuit of the SSSC is shown in Figure 3.22b. The magnitude of V_C can be controlled to regulate power flow. The winding resistance and leakage reactance of the connecting transformer appear is series with the voltage source V_C. If there is no energy source on the dc side, neglecting losses in the converter and dc capacitor, the power balance in steady state leads to

$$\text{Re}[V_C I^*] = 0 \tag{3.130}$$

The equation mentioned earlier shows that V_C is in quadrature with I. If V_C lags I by 90°, the operating mode is capacitive and the current (magnitude) in the line is increased with resultant increase in power flow. On the other hand, if V_C leads I by 90°, the operating mode is inductive, and the line current is decreased. Note that we are assuming the injected voltage is sinusoidal (neglecting harmonics).

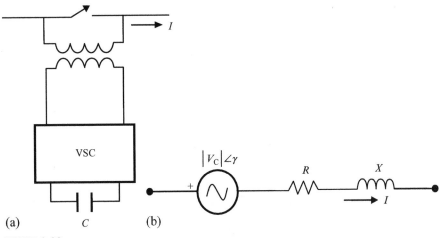

(a) C (b)

FIGURE 3.22

Static synchronous series compensator (SSSC). (a) Schematic of SSSC. (b) Equivalent circuit.

Since the losses are always present, the phase shift between I and V_C is less than $90°$ (in steady state). In general, we can write

$$\hat{V}_C = V_C(\cos\gamma - j\sin\gamma)e^{j\varphi} \qquad (3.131)$$

$$= (V_{Cp} - jV_{Cr})e^{j\varphi} \qquad (3.132)$$

where φ is the phase angle of the line current and γ is the angle by which $\hat{V}_C$ lags the current. V_{Cp} and V_{Cr} are the inphase and quadrature components of the injected voltage (with reference to the line current). We can also term them as active (or real) and reactive components. The real component is required to meet the losses in the converter and the dc capacitor. Since the losses are expected to be small (typically below 1%), the magnitude of V_{Cp} is very small and may be neglected to simplify the analysis. V_{Cp} will vary during a transient to increase or decrease the voltage across the dc capacitor.

The influence of injecting a reactive voltage on the power flow can be analyzed by considering a simple system with a transmission line (with the compensator) connected to constant voltage sources at the two ends (Figure 3.23). The transmission line is represented only by a series reactance.

If the magnitudes of V_S and V_R are equal to V, the voltage drop across the line (see phasor diagram) is given by [10]

$$IX - V_C = 2V\sin\frac{\delta}{2} \qquad (3.133)$$

The magnitude of the current is obtained as

$$I = \frac{\frac{V_C}{X} + 2V\sin\frac{\delta}{2}}{X} \qquad (3.134)$$

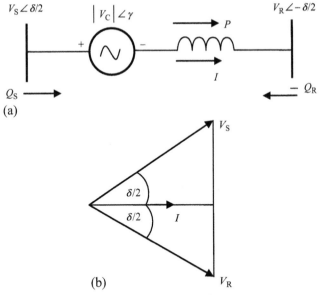

FIGURE 3.23

Representation of SSSC in a transmission line. (a) Single line diagram. (b) Phasor diagram.

The power flow (P) in the line is given by

$$P = VI\cos\frac{\delta}{2} = \frac{VV_C}{X}\cos\frac{\delta}{2} + V^2\frac{\sin\delta}{X} \tag{3.135}$$

The reactive power (Q) supplied at the two ends of the line is equal. The expression for Q is given by

$$Q = VI\sin\frac{\delta}{2} = \frac{V_C V}{X}\sin\frac{\delta}{2} + \frac{V^2}{X}(1 - \cos\delta) \tag{3.136}$$

From Equation (3.134), it is observed that I varies linearly with V_C for a specified value of δ. If V_C is negative (inductive), it is possible to reverse the line current phasor that leads to power reversal in the line. The property of reversal of power is not feasible with variable series compensation as shown later.

The expression (3.135) can be compared with that in the case of series compensation, given by

$$P = V^2\frac{\sin\delta}{X_L(1 - K_{se})} \tag{3.137}$$

where K_{se} is the degree of series compensation.

It is observed that for small values of δ, the increase in the power flow introduced by SSSC is constant and nearly independent δ. On the other hand, the power flow increased by series compensation is a percentage of the power flow in the uncompensated line. Thus, at $\delta = 0$, the series compensation has no effect on the power flow.

The linearized power flow equations of an SSSC can be obtained from Equations (3.135) and (3.136) as

$$
\begin{bmatrix} \Delta P \\ \Delta Q \\ \Delta P_{SR} \end{bmatrix} = \begin{bmatrix} \dfrac{\partial P}{\partial \delta} & \dfrac{\partial P}{\partial V_C} & \dfrac{\partial P}{\partial V} \\ \dfrac{\partial Q}{\partial \delta} & \dfrac{\partial Q}{\partial V_C} & \dfrac{\partial Q}{\partial V} \\ \dfrac{\partial P_{SR}}{\partial \delta} & \dfrac{\partial P_{SR}}{\partial V_C} & \dfrac{\partial P_{SR}}{\partial V} \end{bmatrix} \begin{bmatrix} \Delta \delta \\ \Delta V_C \\ \Delta V \end{bmatrix} \tag{3.138}
$$

Therefore, installations of an SSSC controller in a multimachine system combine the linearized power flow equations (3.138) with the network equations between the nodes denoted by 'S' and 'R.'

- **SSSC damping controller**

The structure of SSSC controller, to modulate the SSSC-injected voltage V_C, is shown in Figure 3.24. The input signal of the proposed controller is the speed deviation ($\Delta \omega$), and the output signal is the injected voltage V_C. The structure consists of a gain block with gain K_S, a signal washout block, and two stage phase compensation blocks [11]. The signal washout block serves as a high-pass filter, with the time constant T_w, high enough to allow signals associated with oscillations in input signal to pass unchanged. From the viewpoint of the washout function, the value of T_w is not critical and may be in the range of 1 to 20 seconds. The phase compensation block with time constants T_1, T_2, T_3, and T_4 provides the appropriate phase-lead characteristics to compensate for the phase lag between the input and the output. V_{Cref} represents the reference-injected voltage as desired by the steady-state power flow control loop. The steady-state power flow loop acts quite slowly in practice, and hence, in the present study, V_{Cref} is assumed to be constant during large disturbance transient period. The desired value of compensation is obtained according to the change in the SSSC-injected voltage ΔV_C, which is added to V_{Cref}. During steady-state conditions, ΔV_C and V_{Cref} are constant. During

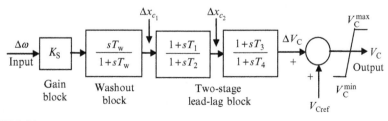

FIGURE 3.24

Structure of SSSC-based damping controller.

dynamic conditions, the series-injected voltage V_C is modulated to damp system oscillations. The effective V_C in dynamic conditions is

$$V_C = V_{Cref} + \Delta V_C \tag{3.139}$$

The block diagram model of the SSSC-based damping controller depicted in Figure 3.24 can be represented by the following equations:

$$\Delta \dot{x}_{c_1} = -\frac{1}{T_w}\Delta x_{c_1} + K_S \Delta \dot{\omega} \tag{3.140}$$

$$\Delta \dot{x}_{c_2} = -\frac{1}{T_2}\Delta x_{c_2} + \frac{1}{T_2}\Delta x_{c_1} + \frac{T_1}{T_2}\Delta \dot{x}_{c_1} \tag{3.141}$$

$$\Delta \dot{V}_C = -\frac{1}{T_4}\Delta V_C + \frac{1}{T_4}\Delta x_{c_2} + \frac{T_3}{T_4}\Delta \dot{x}_{c_2} \tag{3.142}$$

Further simplification of Equations (3.140)–(3.142) gives

$$\Delta \dot{x}_{c_1} = -\frac{1}{T_w}\Delta x_{c_1} + K_S \Delta \dot{\omega} \tag{3.143}$$

$$\Delta \dot{x}_{c_2} = \left(\frac{T_w - T_1}{T_w T_2}\right)\Delta x_{c_1} - \frac{1}{T_2}\Delta x_{c_2} + \frac{K_S T_1}{T_2}\Delta \dot{\omega} \tag{3.144}$$

$$\Delta \dot{V}_C = -\frac{1}{T_4}\Delta V_C + \frac{T_3}{T_2 T_4}\left(\frac{T_w - T_1}{T_w}\right)\Delta x_{c_1}$$
$$+ \left(\frac{T_2 - T_3}{T_2 T_4}\right)\Delta \dot{x}_{c_2} + \frac{K_S T_1 T_3}{T_2 T_4}\Delta \dot{\omega} \tag{3.145}$$

where the term $\Delta \dot{\omega}$ in the right-hand side of the equations mentioned earlier can be replaced using Equation (5.21) (Chapter 5, Section 5.2.1).

The small-signal model of a multimachine system with SSSC controller can be obtained by including additional Equations (3.143)–(3.145) with the DAEs of the multimachine system and the SSSC linearized power flow equations given by Equation (3.138).

3.9.5 Unified power flow controller

Unified power flow controller (UPFC) is one of the most advanced FACTS devices and is a combination of STATCOM and a SSSC. UPFC may be seen to consist of two VSCs sharing a common capacitor on their dc side and a unified control system. The two devices are coupled through the dc link and the combination allows bidirectional flow of real power between the series output of SSSC and the shunt output of STAT-COM. This controller (UPFC) has the facility to provide concurrent real and reactive series line compensation without any external electric energy source. UPFC can have angularly controlled series voltage injection for transmission voltage control in addition to line impedance and power angle control. Thus, UPFC is able to control real power flow, reactive power flow in a line, and the voltage magnitude at the UPFC terminals and may also be used as independently for shunt reactive compensation. The controller may be set to control one or more of these parameters in any combination.

Figure 3.25 represents the schematic of a UPFC that contains a STATCOM with an SSSC. The active power flow for the series unit (SSSC) is obtained from the line itself through the shunt unit (STATCOM). STATCOM is employed for voltage (or reactive power) control, while SSSC is utilized for real power control. UPFC is a complete FACTS controller for both real and reactive power flow controls in a line. The active power required for the series converter is drawn by the shunt converter from the ac bus (i) and supplied to bus j by the dc link. The inverted ac voltage (V_{ser}) at the output of series converter is added to the sending end node voltage V_i at line side to boost the nodal voltage at the jth bus. It may be noted here that the voltage magnitude of the output voltage $|V_{ser}|$ provides voltage regulation, while the phase angle δ_{ser} determines the mode of power flow control. Additional storage device (viz., a superconducting magnet connected to the dc link) through an electronic interface would provide the enhancement in capability of UPFC in real power flow control.

In addition to providing a supporting role in the active power exchange that takes place between the series converter and the ac system, the shunt converter may also generate or absorb reactive power in order to provide independent voltage regulation at its point of connection with the ac system.

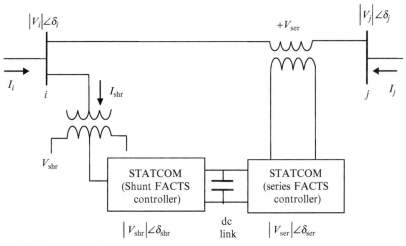

FIGURE 3.25

Schematic of a UPFC.

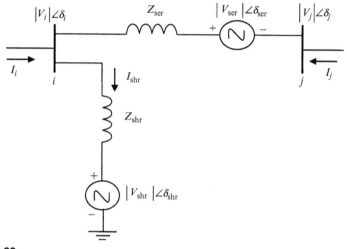

FIGURE 3.26

Equivalent circuit of a UPFC between two buses i and j.

The UPFC equivalent circuit shown in Figure 3.26 consists of a shunt-connected voltage source and a series-connected voltage source. The active power constraint equation links the two voltage sources. The two voltage sources are connected to the ac system through inductive reactance representing the VSC transformers. The expressions for the two voltage sources and the constraint equation would be

$$V_{shr} = |V_{shr}|(\cos\delta_{shr} + j\sin\delta_{shr})$$

$$V_{ser} = |V_{ser}|(\cos\delta_{ser} + j\sin\delta_{ser})$$

$$\text{Re}\left\{-V_{shr}I_{shr}^{*} + V_{ser}I_{j}^{*}\right\} = 0$$

Here, V_{shr} and δ_{shr} are the controllable magnitude and phase of the voltage source representing the shunt converter. The magnitude V_{ser} and phase angle δ_{ser} of the voltage source represent the series converter. Similar to the shunt and series voltage sources used to represent the STATCOM and the SSSC, respectively, the voltage sources used in the UPFC application would also have control limits, i.e., $V_{shrmin} \le V_{shr} \le V_{shrmax}$, $0 \le \delta_{shr} \le 2\pi$ and $V_{sermin} \le V_{ser} \le V_{sermax}$, $0 \le \delta_{ser} \le 2\pi$, respectively.

The phase angle of the series-injected voltage determines the mode of power flow control. The following conditions are important in understanding UPFC operation with reference to its equivalent circuit (Figure 3.26):

- If δ_{ser} is in phase with node voltage angle δ_i, the UPFC regulates the terminal voltage and no active power flow takes place between ith and jth buses. Reactive power flow can be controlled by varying $|V_{ser}|$.

- If δ_{ser} is in quadrature with δ_i, active power flow can be controlled between ith and jth buses by controlling δ_{ser} and acting as a phase shifter. No reactive power flow will occur between ith and jth buses.
- If δ_{ser} is in quadrature with line current angle, then it can also control active power flow, acting as a variable series compensator.
- If δ_{ser} is in between $0°$ and $90°$, it can control both real power flow and reactive power flow in the line. The magnitude of the series-injected voltage determines the amount of power flow to be controlled.

- **UPFC power flow modeling [2]**

  Based on equivalent circuit as shown in Figure 3.26, we have

  $$I_i = (V_i - V_j - V_{\text{ser}})Y_{\text{ser}} + (V_i - V_{\text{shr}})Y_{\text{shr}}$$
  $$= V_i(Y_{\text{ser}} + Y_{\text{shr}}) - V_jY_{\text{ser}} - V_{\text{ser}}Y_{\text{ser}} - V_{\text{shr}}Y_{\text{shr}}$$

  and $I_j = (-V_i + V_j + V_{\text{ser}})Y_{\text{ser}}$

  i.e.,
  $$\begin{bmatrix} I_i \\ I_j \end{bmatrix} = \begin{bmatrix} (Y_{\text{ser}} + Y_{\text{shr}}) & -Y_{\text{ser}} & -Y_{\text{ser}} & -Y_{\text{shr}} \\ -Y_{\text{ser}} & Y_{\text{ser}} & Y_{\text{ser}} & 0 \end{bmatrix} \begin{bmatrix} V_i \\ V_j \\ V_{\text{ser}} \\ V_{\text{shr}} \end{bmatrix}$$

  Also,
  $$\begin{bmatrix} S_i \\ S_j \end{bmatrix} = \begin{bmatrix} V_i & 0 \\ 0 & V_j \end{bmatrix} \begin{bmatrix} I_i^* \\ I_j^* \end{bmatrix}$$

  $$= \begin{bmatrix} V_i & 0 \\ 0 & V_j \end{bmatrix} \begin{bmatrix} (Y_{\text{ser}} + Y_{\text{shr}})^* & -Y_{\text{ser}}^* & -Y_{\text{ser}}^* & -Y_{\text{shr}}^* \\ -Y_{\text{ser}}^* & Y_{\text{ser}}^* & Y_{\text{ser}}^* & 0 \end{bmatrix} \begin{bmatrix} V_i^* \\ V_j^* \\ V_{\text{ser}}^* \\ V_{\text{shr}}^* \end{bmatrix}$$

  or
  $$\begin{bmatrix} P_i + jQ_i \\ P_j + jQ_j \end{bmatrix} = \begin{bmatrix} V_i & 0 \\ 0 & V_j \end{bmatrix} \begin{bmatrix} (G_{ii} - jB_{ii}) & (G_{ij} - jB_{ij}) & (G_{ij} - jB_{ij}) & (G_{i0} - jB_{i0}) \\ (G_{ji} - jB_{ji}) & (G_{jj} - jB_{jj}) & (G_{jj} - jB_{jj}) & 0 \end{bmatrix} \begin{bmatrix} V_i^* \\ V_j^* \\ V_{\text{ser}}^* \\ V_{\text{shr}}^* \end{bmatrix}$$

  $$\tag{3.146}$$

  $$\therefore P_i = |V_i|^2 G_{ii} + |V_i||V_j|\{G_{ij}\cos(\delta_i - \delta_j) + B_{ij}\sin(\delta_i - \delta_j)\}$$
  $$+ |V_i||V_{\text{ser}}|\{G_{ij}\cos(\delta_i - \delta_{\text{ser}}) + B_{ij}\sin(\delta_i - \delta_{\text{ser}})\}$$
  $$+ |V||V_{\text{shr}}|\{G_{i0}\cos(\delta_i - \delta_{\text{shr}}) + B_{i0}\sin(\delta_i - \delta_{\text{shr}})\} \tag{3.147a}$$

$$Q_i = -|V_i|^2 B_{ii} + |V_i||V_j|\{G_{ij}\sin(\delta_i - \delta_j) - B_{ij}\cos(\delta_i - \delta_j)\}$$
$$+ |V||V_{\text{ser}}|\{G_{ij}\sin(\delta_i - \delta_{\text{ser}}) - B_{ij}\cos(\delta_i - \delta_{\text{ser}})\}$$
$$+ |V_i||V_{\text{shr}}|\{G_{i0}\sin(\delta_i - \delta_{\text{shr}}) - B_{i0}\cos(\delta_i - \delta_{\text{shr}})\} \qquad (3.147\text{b})$$

$$P_j = |V_j|^2 G_{jj} + |V_j||V_i|\{G_{ji}\cos(\delta_j - \delta_i) + B_{ji}\sin(\delta_j - \delta_i)\}$$
$$+ |V_j||V_{\text{ser}}|\{G_{jj}\cos(\delta_j - \delta_{\text{ser}}) + B_{jj}\sin(\delta_j - \delta_{\text{ser}})\} \qquad (3.148\text{a})$$

$$Q_j = -|V_j|^2 B_{jj} + |V_j||V_i|\{G_{ji}\sin(\delta_j - \delta_i) - B_{ji}\cos(\delta_j - \delta_i)\}$$
$$+ |V_j||V_{\text{ser}}|\{G_{jj}\sin(\delta_j - \delta_{\text{ser}}) - B_{jj}\cos(\delta_j - \delta_{\text{ser}})\} \qquad (3.148\text{b})$$

The active power and reactive power of the series converter (SSSC) are as follows:

$$S_{\text{ser}} = P_{\text{ser}} + jQ_{\text{ser}} = V_{\text{ser}}I_j^* = V_{\text{ser}}\left[Y_{ji}^* V_i^* + Y_{jj}^* V_j^* + Y_{jj}^* V_{\text{ser}}^*\right]$$

$$\therefore P_{\text{ser}} = |V_{\text{ser}}|^2 G_{jj} + |V_{\text{ser}}||V_i|\{G_{ji}\cos(\delta_{\text{ser}} - \delta_i) + B_{ji}\sin(\delta_{\text{ser}} - \delta_i)\}$$
$$+ |V_{\text{ser}}||V_j|\{G_{jj}\cos(\delta_{\text{ser}} - \delta_j) + B_{jj}\sin(\delta_{\text{ser}} - \delta_j)\} \qquad (3.149\text{a})$$

$$Q_{\text{ser}} = -|V_{\text{ser}}|^2 B_{jj} + |V_{\text{ser}}||V_i|\{G_{ji}\sin(\delta_{\text{ser}} - \delta_i) - B_{ji}\cos(\delta_{\text{ser}} - \delta_i)\}$$
$$+ |V_{\text{ser}}||V_j|\{G_{jj}\sin(\delta_{\text{ser}} - \delta_j) - B_{jj}\cos(\delta_{\text{ser}} - \delta_j)\} \qquad (3.149\text{b})$$

The active power and reactive power for the shunt controller (STATCOM) are obtained as

$$S_{\text{shr}} = P_{\text{shr}} + jQ_{\text{shr}} = V_{\text{shr}}I_{\text{shr}}^* = -V_{\text{shr}}Y_{\text{shr}}^*\left[V_{\text{shr}}^* - V_i^*\right]$$
$$\therefore P_{\text{shr}} = -|V_{\text{shr}}|^2 G_{i0} + |V_{\text{shr}}||V_i|\{G_{i0}\cos(\delta_{\text{shr}} - \delta_i) + B_{i0}\sin(\delta_{\text{shr}} - \delta_i)\} \qquad (3.150\text{a})$$

$$Q_{\text{shr}} = |V_{\text{shr}}|^2 B_{i0} + |V_{\text{shr}}||V_i|\{G_{i0}\sin(\delta_{\text{shr}} - \delta_i) - B_{i0}\cos(\delta_{\text{shr}} - \delta_i)\} \qquad (3.150\text{b})$$

Since we assume lossless converters, the UPFC neither absorbs nor injects active power with respect to the ac system, that is, active power supplied to the shunt converter, P_{shr}, equals the active power demand by the series converter, P_{ser}. Hence, the constraint equation is

$$P_{\text{shr}} + P_{\text{ser}} = 0 \qquad (3.151)$$

Furthermore, if the coupling transformers are assumed to contain no resistance, then the active power at bus i matches the active power at bus j. Accordingly,

$$P_{\text{shr}} + P_{\text{ser}} = P_i + P_j$$

The UPFC power equations, in linearized form, are combined with those of the ac network. In order to get the linearized model of the system using power mismatch form, let us assume UPFC is connected to node i and the power system is connected to node j. UPFC is required to control voltage at the shunt converter terminal, node i, and active power flows from node j to node i. Assuming reactive power is injected at node j, the linearized system equations are as follows:

$$
\begin{bmatrix} \Delta P_i \\ \Delta P_j \\ \Delta Q_i \\ \Delta Q_j \\ \Delta P_{ji} \\ \Delta Q_{ji} \\ \Delta P \end{bmatrix} =
\begin{bmatrix}
\dfrac{\partial P_i}{\partial \delta_i} & \dfrac{\partial P_i}{\partial \delta_j} & \dfrac{\partial P_i}{\partial |V_{\text{shr}}|} & \dfrac{\partial P_i}{\partial |V_j|} & \dfrac{\partial P_i}{\partial \delta_{\text{ser}}} & \dfrac{\partial P_i}{\partial |V_{\text{ser}}|} & \dfrac{\partial P_i}{\partial \delta_{\text{shr}}} \\[2ex]
\dfrac{\partial P_j}{\partial \delta_i} & \dfrac{\partial P_j}{\partial \delta_j} & 0 & \dfrac{\partial P_j}{\partial |V_j|} & \dfrac{\partial P_j}{\partial \delta_{\text{ser}}} & \dfrac{\partial P_j}{\partial |V_{\text{ser}}|} & 0 \\[2ex]
\dfrac{\partial Q_i}{\partial \delta_i} & \dfrac{\partial Q_i}{\partial \delta_j} & \dfrac{\partial Q_i}{\partial |V_{\text{shr}}|} & \dfrac{\partial Q_i}{\partial |V_j|} & \dfrac{\partial Q_i}{\partial \delta_{\text{ser}}} & \dfrac{\partial Q_i}{\partial |V_{\text{ser}}|} & \dfrac{\partial Q_i}{\partial \delta_{\text{shr}}} \\[2ex]
\dfrac{\partial Q_j}{\partial \delta_i} & \dfrac{\partial Q_j}{\partial \delta_j} & 0 & \dfrac{\partial Q_j}{\partial |V_j|} & \dfrac{\partial Q_j}{\partial \delta_{\text{ser}}} & \dfrac{\partial Q_j}{\partial |V_{\text{ser}}|} & 0 \\[2ex]
\dfrac{\partial P_{ji}}{\partial \delta_i} & \dfrac{\partial P_{ji}}{\partial \delta_j} & 0 & \dfrac{\partial P_{ji}}{\partial |V_j|} & \dfrac{\partial P_{ji}}{\partial \delta_{\text{ser}}} & \dfrac{\partial P_{ji}}{\partial |V_{\text{ser}}|} & 0 \\[2ex]
\dfrac{\partial Q_{ji}}{\partial \delta_i} & \dfrac{\partial Q_{ji}}{\partial \delta_j} & 0 & \dfrac{\partial Q_{ji}}{\partial |V_j|} & \dfrac{\partial Q_{ji}}{\partial \delta_{\text{ser}}} & \dfrac{\partial Q_{ji}}{\partial |V_{\text{ser}}|} & 0 \\[2ex]
\dfrac{\partial P}{\partial \delta_i} & \dfrac{\partial P}{\partial \delta_j} & \dfrac{\partial P}{\partial |V_{\text{shr}}|} & \dfrac{\partial P}{\partial |V_j|} & \dfrac{\partial P}{\partial \delta_{\text{ser}}} & \dfrac{\partial P}{\partial |V_{\text{ser}}|} & \dfrac{\partial P}{\partial \delta_{\text{shr}}}
\end{bmatrix}
\begin{bmatrix} \Delta \delta_i \\ \Delta \delta_j \\ \Delta |V_{\text{shr}}| \\ \Delta |V_j| \\ \Delta \delta_{\text{ser}} \\ \Delta |V_{\text{ser}}| \\ \Delta \delta_{\text{shr}} \end{bmatrix}
\qquad (3.152)
$$

It has been assumed that node j is the PQ node, while ΔP is the power mismatch given by the constraint equation (3.151). If voltage control at bus i is deactivated, the third column of Equation (3.152) is replaced by partial derivatives of the bus and UPFC mismatch powers with respect to the bus voltage magnitude V_i. Moreover, the voltage magnitude increment of the shunt source, ΔV_{shr}, is replaced by the voltage magnitude increment at bus i, ΔV_i. Extensive algorithm is needed for the solution of these power flow equations of UPFC. Good starting conditions for all the UPFC state variables are also an important requirement to ensure convergence.

- **UPFC damping controllers**

The block diagram structure of UPFC damping controller is shown in Figure 3.27, where u can be V_{shr} and δ_{shr}, which are the controllable magnitude and phase of the voltage source representing the shunt converter. In order to maintain the power balance between the series and the shunt converters, a dc voltage regulator must be incorporated. The dc voltage is controlled through modulating the phase angle of the shunt transformer voltage, δ_{shr}. The dc voltage regulator is a PI controller. The other blocks of the controllers are gain block, washout block, and the lead-lag controller block. T_{upfc} represents the UPFC internal delay [12].

The functions of these blocks are already familiarized during discussions on other FACTS controllers. The linearized state-space model of the UPFC damping controller can be derived algebraically from the presented block diagram, which can be

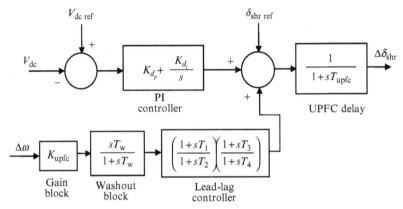

FIGURE 3.27

UPFC damping controller.

combined with the differential algebraic model of the multimachine system in order to study the small-signal stability problem.

EXERCISES

3.1. Draw the synchronous machine steady-state equivalent circuit, and hence, obtain its steady-state model in dqo coordinate system.

3.2. Consider a synchronous machine serving a load without saturation and with $\bar{V} = 1 \angle 10°$pu and $\bar{I} = 0.5 \angle -20°$pu. The parameters of the machines are given as $X_d = 1.2$, $X_q = 1.0$, $X_{md} = 1.1$, $X'_d = 0.232$, and $R_s = 0.0$ (all in pu). Find the following steady-state variables of the machine:

(i) δ and δ_T
(ii) I_d, I_q, V_d, and V_q
(iii) ψ_d, ψ_q, and E_q,
(iv) E_{fd} and I_{fd}
(all are in pu except angle in degree)

3.3. Derive the expression of fundamental component of SVC equivalent susceptance as

$$B_{svc} = -\frac{X_L - \frac{X_C}{\pi}(2(\pi - \alpha) + \sin(2\alpha))}{X_C X_L}$$

where X_L is the equivalent reactance of the TCR, X_C is the equivalent reactance of the fixed capacitor connected across with the TCR circuit, and α is the firing angle. Obtain its linearized version for application in small-signal stability analysis of a power system.

3.4. A TCSC is connected between the nodes 's' and 't.' The power flow between the nodes 's' and 't' is given by the equation

$$S_{st} = V_s^2(g_{st} - jb_{st}) - V_s V_t(g_{st} - jb_{st})(\cos\theta_{st} + j\sin\theta_{st})$$

Obtain the linearized power flow equations of the TCSC. V_s and V_t are the voltages at the nodes 's' and 't.' $Y_{st}^* = g_{st} - jb_{st}$ is the network admittance between the nodes where TCSC is being connected.

3.5. Show that the TCSC equivalent reactance, as a function of firing angle (α), can be expressed by the following equation:

$$X_{TCSC} = -X_C + C_1(2(\pi - \alpha) + \sin(2(\pi - \alpha)))$$
$$- C_2 \cos^2(\pi - \alpha)(\overline{\omega}\tan(\overline{\omega}(\pi - \alpha)) - \tan(\pi - \alpha))$$

where $C_1 = \dfrac{X_C + X_{LC}}{\pi}$, $C_2 = \dfrac{4X_{LC}^2}{\pi X_L}$, and $X_{LC} = \dfrac{X_C X_L}{X_C - X_L}$.

3.6. The block diagram of a UPFC damping controller is given in Figure 3.27. Derive a linearized state-space model of the controller, while the controller is modulating the phase angle δ_{shr} of the shunt VSC.

References

[1] P.W. Sauer, M.A. Pai, Power System Dynamics and Stability, Pearson Education Pte. Ltd., Singapore, 1998.

[2] A. Chakrabarti, S. Halder, Power System Analysis Operation and Control, PHI learning Pvt Ltd., India, New Delhi, 2010.

[3] P. Kundur, Power System Stability and Control, McGraw-Hill, New York, 1994.

[4] E.V. Larsen, D.A. Swann, Applying power system stabilizer, Part I: General concept, Part II: Performance objective and tuning concept, Part III: Practical considerations, IEEE Trans. Power Ap. Syst. PAS-100 (12) (1981) 3017–3046.

[5] R.A. Lawson, D.A. Swann, G.F. Wright, Minimization of power system stabilizer torsional interaction on large steam turbine-generators, IEEE Trans. Power Ap. Syst. PAS-97 (1) (1978) 183–190.

[6] E. Acha, V.G. Agelidis, O.A. Lara, T.J.E. Miller, Power Electronics Control in Electrical Systems, Newnes Power Engg. Series, Reed Educational and Professional Publishing Ltd., 2002.

[7] R.M. Mathur, R.K. Varma, Thyristor Based-FACTS Controllers for Electrical Transmission Systems, IEEE Press, Wiley & Sons, Inc. Publication, USA, 2002.

[8] C.R. Fuerte-Esquivel, E. Acha, H. Ambriz-Pe'rez, A thyristor controlled series compensator model for the power flow solution of practical power networks, IEEE Trans. Power Syst. 15 (1) (2000) 58–64.

[9] S. Panda, N.P. Padhy, R.N. Patel, Modelling, simulation and optimal tuning of TCSC controller, Int. J. Simulat. Model. 6 (1) (2007) 37–48.

[10] K.R. Padiyar, Analysis of Subsynchronous Resonance in Power Systems, Kluwer Academic Publishers, Boston/London, 1999.

[11] S. Panda, N.P. Padhy, A PSO-based SSSC Controller for Improvement of Transient Stability Performance, Int. J. Intell. Syst. Tech. 2 (1) (2007) 28–35.

[12] M.A. Abido, A.T. Al-Awami, Y.L. Abdel-Magid, Analysis and design of UPFC damping stabilizers for power system stability enhancement, IEEE Inter. Symp. Ind. Electron. 3 (2006) 2040–2045.

Small-Signal Stability Analysis in SMIB Power System

4.1 INTRODUCTION

Small-signal oscillations in a synchronous generator, particularly when it is connected to the power system through a long transmission line, are a matter of concern since before. As long transmission lines interconnect geographically vast areas, it is becoming difficult to maintain synchronism between different parts of the power system. Moreover, long lines reduce load ability of the power system and make the system weak, which is associated with interarea oscillations during heavy loading. The phenomenon of small signal or small disturbance stability of a synchronous machine connected to an infinite bus through external reactance has been studied in [1,2] by means of *block diagrams* and *frequency response analysis*. The objective of this analysis is to develop insights into the effects of excitation systems, voltage regulator gain, and stabilizing functions derived from generator speed and working through the voltage reference of the voltage regulator. The analysis based on linearization technique is ideally suitable for investigating problems associated with the small-signal oscillations. In this technique, the characteristics of a power system can be determined through a specific operating point and the stability of the system is clearly examined by the system eigenvalues. This chapter describes the linearized model of a single-machine infinite bus (SMIB) system given by *Heffron and Philips* that investigates the local mode of oscillations in the range of frequency 1-3 Hz. Voltage stability or *dynamic voltage stability* can be analyzed by monitoring the eigenvalues of the linearized power system with progressive loading. Instability occurs when a pair of complex conjugate eigenvalues crosses the right half of *s*-plane. This is referred to as dynamic voltage instability, and mathematically, this phenomenon is called Hopf bifurcation.

The following steps have been adopted sequentially to analyze the small-signal stability performance of an SMIB system:

1. The differential equations of the flux-decay model of the synchronous machine are linearized and a state-space model is constructed considering exciter output E_{fd} as input.
2. From the resulting linearized model, certain constants known as the K constants $(K_1–K_6)$ are derived. They are evaluated by small-perturbation analysis on the fundamental synchronous machine equations and hence are functions of machine and system impedances and operating point.

3. The model so obtained is put in a block diagram form and a fast-acting exciter between terminal voltage ΔV_t and exciter output ΔE_{fd} is introduced in the block diagram.
4. The state-space model is then used to examine the eigenvalues and to design supplementary controllers to ensure adequate damping of the dominant modes. The real parts of the electromechanical modes are associated with the *damping torque* and the imaginary parts contribute to the *synchronizing torque*.

4.2 HEFFRON–PHILIPS MODEL OF SMIB POWER SYSTEM

The Heffron–Phillips model for small-signal oscillations in synchronous machines connected to an infinite bus was first presented in 1952. For small-signal stability studies of an SMIB power system, the linear model of Heffron–Phillips has been used for many years, providing reliable results [3,4]. This section presents the small-signal model for a single machine connected to a large system through a transmission line (infinite bus) to analyze the local mode of oscillations in the range of frequency 1-3 Hz. A schematic representation of this system is shown in Figure 4.1.

The flux-decay model (Figure 4.2) of the equivalent circuit of the synchronous machine has been considered for the analysis.

The said model is known as the classical model of the synchronous machine. The following assumptions are generally made to analyze the small-signal stability problem in an SMIB power system:

(i) The mechanical power input remains constant during the period of transient.
(ii) Damping or asynchronous power is negligible.
(iii) Stator resistance is equal to zero.
(iv) The synchronous machine can be represented by a constant voltage source (electrically) behind the transient reactance.
(v) The mechanical angle of the synchronous machine rotor coincides with the electric phase angle of the voltage behind transient reactance.

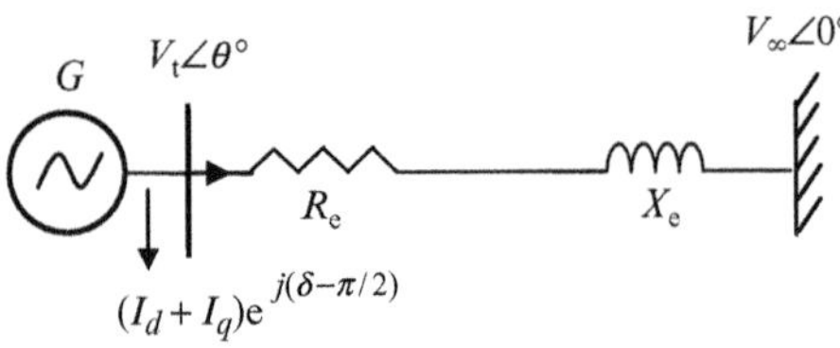

FIGURE 4.1

Single-machine infinite bus (SMIB) system. Here, $V_t \angle \theta°$, the terminal voltage of the synchronous machine; $V_\infty \angle 0°$, the voltage of the infinite bus, which is used as reference; R_e, X_e, the external equivalent resistance and reactance; and $\theta = \delta - \frac{\pi}{2}$, the angle by which V_t leads the infinite bus voltage V_∞.

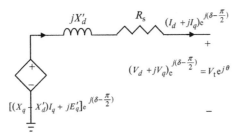

FIGURE 4.2

Dynamic circuit for the flux-decay model of the machine.

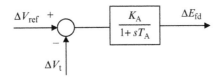

FIGURE 4.3

Fast (static) exciter model.

(vi) No local load is assumed at the generator bus; if a local load is fed at the terminal of the machine, it is to be represented by constant impedance (or admittance).

4.2.1 Fundamental equations

The differential algebraic equations of the synchronous machine of the flux-decay model with fast exciter (Figure 4.3) can be represented as follows:

- **Differential equations**

$$\frac{dE'_q}{dt} = -\frac{1}{T'_{do}}\left(E'_q + (X_d - X'_d)I_d - E_{fd}\right) \tag{4.1}$$

$$\frac{d\delta}{dt} = \omega - \omega_s \tag{4.2}$$

$$\frac{d\omega}{dt} = \frac{\omega_s}{2H}\left[T_M - \left(E'_q I_q + (X_q - X'_d)I_d I_q + D(\omega - \omega_s)\right)\right] \tag{4.3}$$

$$\frac{dE_{fd}}{dt} = -\frac{E_{fd}}{T_A} + \frac{K_A}{T_A}(V_{ref} - V_t) \tag{4.4}$$

- **Stator algebraic equations**

$$V_t \sin(\delta - \theta) + R_s I_d - X_q I_q = 0 \tag{4.5}$$

$$E'_q - V_t \cos(\delta - \theta) - R_s I_q - X'_d I_d = 0 \tag{4.6}$$

As it is assumed stator resistance $R_s = 0$ and V_t denote the magnitude of the generator terminal voltage, the earlier-mentioned equations are reduced to

$$X_q I_q - V_t \sin(\delta - \theta) = 0 \tag{4.7}$$

$$E'_q - V_t \cos(\delta - \theta) - X'_d I_d = 0 \tag{4.8}$$

Now, $(V_d + jV_q)e^{(\delta - \frac{\pi}{2})} = V_t e^{j\theta}$
Hence,

$$V_d + jV_q = V_t e^{j\theta} e^{-j(\delta - \frac{\pi}{2})} \tag{4.9}$$

Expansion of the right-hand side results in

$$V_d + jV_q = V_t \sin(\delta - \theta) + jV_t \cos(\delta - \theta)$$

Therefore, $V_d = V_t \sin(\delta - \theta)$ and $V_q = V_t \cos(\delta - \theta)$.
Substitution of V_d and V_q in Equations (4.5) and (4.6) gives

$$X_q I_q - V_d = 0 \tag{4.10}$$

$$E'_q - V_q - X'_d I_d = 0 \tag{4.11}$$

- **Network equations**

The network equation assuming zero phase angle at the infinite bus:

$$(I_d + jI_q)e^{j(\delta - \frac{\pi}{2})} = \frac{V_t \angle \theta° - V_\infty \angle 0°}{R_e + jX_e} \tag{4.12}$$

$$(I_d + jI_q)e^{j(\delta - \frac{\pi}{2})} = \frac{(V_d + jV_q)e^{j(\delta - \frac{\pi}{2})} - V_\infty \angle 0°}{R_e + jX_e} \tag{4.13}$$

After cross multiplication when real and imaginary parts are separated, Equation (4.13) becomes

$$I_d R_e + jI_q R_e + jI_d X_e - I_q X_e = (V_d + jV_q) - V_\infty e^{-j(\delta - \frac{\pi}{2})} \tag{4.14}$$

or

$$(R_e I_d - X_e I_q) + j(R_e I_q + X_e I_d) = (V_d + jV_q) - \left(V_\infty \cos\left(\delta - \frac{\pi}{2}\right) - jV_\infty \sin\left(\delta - \frac{\pi}{2}\right)\right) \tag{4.15}$$

$$\left(R_eI_d - X_eI_q\right) + j\left(R_eI_q + X_eI_d\right) = \left(V_d - V_\infty \sin \delta\right) + j\left(V_q - V_\infty \cos \delta\right) \qquad (4.16)$$

$$\therefore R_eI_d - X_eI_q = V_d - V_\infty \sin \delta \qquad (4.17)$$

$$R_eI_q + X_eI_d = V_q - V_\infty \cos \delta \qquad (4.18)$$

4.2.2 Linearization process and state-space model

Step I: The linearization of the stator algebraic equations (4.10) and (4.11) gives

$$X_q \Delta I_q - \Delta V_d = 0 \qquad (4.19)$$

$$\Delta E'_q - \Delta V_q - X'_d \Delta I_d = 0 \qquad (4.20)$$

Rearranging Equations (4.19) and (4.20) gives

$$\Delta V_d = X_q \Delta I_q \qquad (4.21)$$

$$\Delta V_q = -X'_d \Delta I_d + \Delta E'_q \qquad (4.22)$$

Writing Equations (4.21) and (4.22) in matrix form gives

$$\begin{bmatrix} \Delta V_d \\ \Delta V_q \end{bmatrix} = \begin{bmatrix} 0 & X_q \\ -X'_d & 0 \end{bmatrix} \begin{bmatrix} \Delta I_d \\ \Delta I_q \end{bmatrix} + \begin{bmatrix} 0 \\ \Delta E'_q \end{bmatrix} \qquad (4.23)$$

Step II: The linearization of the load-flow equations (4.17) and (4.18) results in

$$R_e \Delta I_d - X_e \Delta I_q = \Delta V_d - V_\infty \cos \delta \Delta \delta \qquad (4.24)$$

$$R_e \Delta I_q + X_e \Delta I_d = \Delta V_q + V_\infty \sin \delta \Delta \delta \qquad (4.25)$$

Rearranging Equations (4.24) and (4.25) gives

$$\Delta V_d = R_e \Delta I_d - X_e \Delta I_q + V_\infty \cos \delta \Delta \delta \qquad (4.26)$$

$$\Delta V_q = X_e \Delta I_d + R_e \Delta I_q - V_\infty \sin \delta \Delta \delta \qquad (4.27)$$

Writing Equations (4.26) and (4.27) in matrix form gives

$$\begin{bmatrix} \Delta V_d \\ \Delta V_q \end{bmatrix} = \begin{bmatrix} R_e & -X_e \\ X_e & R_e \end{bmatrix} \begin{bmatrix} \Delta I_d \\ \Delta I_q \end{bmatrix} + \begin{bmatrix} V_\infty \cos \delta \\ -V_\infty \sin \delta \end{bmatrix} \Delta \delta \qquad (4.28)$$

Step III: Equating the right-hand side of Equations (4.23) and (4.28) gives

$$\begin{bmatrix} R_e & -X_e \\ X_e & R_e \end{bmatrix} \begin{bmatrix} \Delta I_d \\ \Delta I_q \end{bmatrix} + \begin{bmatrix} V_\infty \cos \delta \\ -V_\infty \sin \delta \end{bmatrix} \Delta \delta = \begin{bmatrix} 0 & X_q \\ -X'_d & 0 \end{bmatrix} \begin{bmatrix} \Delta I_d \\ \Delta I_q \end{bmatrix} + \begin{bmatrix} 0 \\ \Delta E'_q \end{bmatrix} \qquad (4.29)$$

or

$$\left(\begin{bmatrix} R_e & -X_e \\ X_e & R_e \end{bmatrix} - \begin{bmatrix} 0 & X_q \\ -X'_d & 0 \end{bmatrix}\right)\begin{bmatrix} \Delta I_d \\ \Delta I_q \end{bmatrix} = \begin{bmatrix} 0 \\ \Delta E'_q \end{bmatrix} + \begin{bmatrix} -V_\infty \cos\delta \\ V_\infty \sin\delta \end{bmatrix}\Delta\delta \tag{4.30}$$

or

$$\begin{bmatrix} R_e & -(X_e+X_q) \\ (X_e+X'_d) & R_e \end{bmatrix}\begin{bmatrix} \Delta I_d \\ \Delta I_q \end{bmatrix} = \begin{bmatrix} 0 \\ \Delta E'_q \end{bmatrix} + \begin{bmatrix} -V_\infty \cos\delta \\ V_\infty \sin\delta \end{bmatrix}\Delta\delta \tag{4.31}$$

Now,

$$\begin{bmatrix} R_e & -(X_e+X_q) \\ (X_e+X'_d) & R_e \end{bmatrix}^{-1} = \frac{1}{\Delta_e}\begin{bmatrix} R_e & (X_e+X_q) \\ -(X_e+X'_d) & R_e \end{bmatrix}$$

where $\Delta_e = R_e^2 + (X_e+X_q)(X_e+X'_d)$.

Solving for ΔI_d and ΔI_q from Equation (4.31) results in

$$\begin{aligned}\begin{bmatrix} \Delta I_d \\ \Delta I_q \end{bmatrix} &= \begin{bmatrix} 0 \\ \Delta E'_q \end{bmatrix}\begin{bmatrix} R_e & -(X_e+X_q) \\ (X_e+X'_d) & R_e \end{bmatrix}^{-1} \\ &+ \begin{bmatrix} -V_\infty \cos\delta \\ V_\infty \sin\delta \end{bmatrix}\Delta\delta \cdot \begin{bmatrix} R_e & -(X_e+X_q) \\ (X_e+X'_d) & R_e \end{bmatrix}^{-1}\end{aligned} \tag{4.32}$$

$$\begin{aligned}\begin{bmatrix} \Delta I_d \\ \Delta I_q \end{bmatrix} &= \frac{1}{\Delta_e}\begin{bmatrix} 0 \\ \Delta E'_q \end{bmatrix}\begin{bmatrix} R_e & (X_e+X_q) \\ -(X_e+X'_d) & R_e \end{bmatrix} \\ &+ \frac{1}{\Delta_e}\begin{bmatrix} -V_\infty \cos\delta \\ V_\infty \sin\delta \end{bmatrix}\Delta\delta\begin{bmatrix} R_e & (X_e+X_q) \\ -(X_e+X'_d) & R_e \end{bmatrix}\end{aligned} \tag{4.33}$$

i.e.,

$$\begin{aligned}\begin{bmatrix} \Delta I_d \\ \Delta I_q \end{bmatrix} &= \frac{1}{\Delta_e}\begin{bmatrix} (X_e+X_q)\Delta E'_q \\ R_e\Delta E'_q \end{bmatrix} \\ &+ \frac{1}{\Delta_e}\begin{bmatrix} -R_e V_\infty \cos\delta + V_\infty \sin\delta(X_e+X_q) \\ R_e V_\infty \sin\delta + V_\infty \cos\delta(X_e+X'_d) \end{bmatrix}\Delta\delta\end{aligned} \tag{4.34}$$

Therefore,

$$\begin{bmatrix} \Delta I_d \\ \Delta I_q \end{bmatrix} = \frac{1}{\Delta_e}\begin{bmatrix} (X_e+X_q) & -R_e V_\infty \cos\delta + V_\infty \sin\delta(X_e+X_q) \\ R_e & R_e V_\infty \sin\delta + V_\infty \cos\delta(X_e+X'_d) \end{bmatrix}\begin{bmatrix} \Delta E'_q \\ \Delta\delta \end{bmatrix} \tag{4.35}$$

Step IV: The linearizations of the differential equations (4.1)–(4.4) are as follows. Here, the frequency is normalized as $v = \frac{\omega}{\omega_s}$ throughout our study:

$$\Delta\dot{E}'_q = -\frac{1}{T'_{do}}\Delta E'_q - \frac{1}{T'_{do}}(X_d - X'_d)\Delta I_d + \frac{1}{T'_{do}}\Delta E_{fd} \tag{4.36}$$

$$\Delta \dot{\delta} = \omega_s \Delta v \tag{4.37}$$

$$\Delta \dot{v} = \frac{1}{2H}\Delta T_M - \frac{1}{2H}\Delta E_q' I_q - \frac{1}{2H}E_q'\Delta I_q - \frac{(X_q - X_d')}{2H}\Delta I_d I_q$$

$$- \frac{(X_q - X_d')}{2H}I_d\Delta I_q - \frac{D\omega_s}{2H}\Delta v \tag{4.38}$$

$$T_A \Delta \dot{E}_{fd} = -\Delta E_{fd} + K_A(\Delta V_{ref} - \Delta V_t) \tag{4.39}$$

Writing Equations (4.36)–(4.39) in matrix form, the state-space model of the SMIB system without exciter is

$$
\begin{bmatrix} \Delta \dot{E}_q' \\ \Delta \dot{\delta} \\ \Delta \dot{v} \end{bmatrix} = \begin{bmatrix} -\dfrac{1}{T_{do}'} & 0 & 0 \\ 0 & 0 & \omega_s \\ -\dfrac{I_q}{2H} & 0 & -\dfrac{D\omega_s}{2H} \end{bmatrix} \begin{bmatrix} \Delta E_q' \\ \Delta \delta \\ \Delta v \end{bmatrix} + \begin{bmatrix} -\dfrac{(X_d - X_d')}{T_{do}'} & 0 \\ 0 & 0 \\ \dfrac{I_q(X_d' - X_q)}{2H} & \dfrac{(X_d' - X_q)}{2H} - \dfrac{E_q'}{2H} \end{bmatrix} \begin{bmatrix} \Delta I_d \\ \Delta I_q \end{bmatrix}
$$

$$
+ \begin{bmatrix} \dfrac{1}{T_{do}'} & 0 \\ 0 & 0 \\ 0 & \dfrac{1}{2H} \end{bmatrix} \begin{bmatrix} \Delta E_{fd} \\ \Delta T_M \end{bmatrix}
\tag{4.40}
$$

Step V: Obtain the linearized equations in terms of the K constants.
   The expressions for ΔI_d and ΔI_q obtained from Equation (4.35) are

$$\Delta I_d = \frac{1}{\Delta_e}\left[(X_e + X_q)\Delta E_q' + \{-R_e V_\infty \cos\delta + (X_e + X_q)V_\infty \sin\delta\}\Delta\delta\right] \tag{4.41}$$

and

$$\Delta I_q = \frac{1}{\Delta_e}\left[R_e\Delta E_q' + \{R_e V_\infty \sin\delta + (X_e + X_d')V_\infty \cos\delta\}\Delta\delta\right] \tag{4.42}$$

On substitution of ΔI_d and ΔI_q in Equation (4.40), the resultant equations relating the constants K_1, K_2, K_3, and K_4 can be expressed as

$$\Delta \dot{E}_q' = -\frac{1}{K_3 T_{do}'}\Delta E_q' - \frac{K_4}{T_{do}'}\Delta\delta + \frac{1}{T_{do}'}\Delta E_{fd} \tag{4.43}$$

$$\Delta\dot{\delta} = \omega_s \Delta v \tag{4.44}$$

$$\Delta\dot{v} = -\frac{K_2}{2H}\Delta E'_q - \frac{K_1}{2H}\Delta\delta - \frac{D\omega_s}{2H}\Delta v + \frac{1}{2H}\Delta T_M \tag{4.45}$$

Step VI: The linearization of generator terminal voltage is as follows:

The magnitude of the generator terminal voltage is

$$V_t = \sqrt{V_d^2 + V_q^2}$$

$$\therefore V_t^2 = V_d^2 + V_q^2 \tag{4.46}$$

The linearization of Equation (4.46) gives

$$2V_t \Delta V_t = 2V_d \Delta V_d + 2V_q \Delta V_q \tag{4.47}$$

Therefore,

$$\Delta V_t = \frac{V_d}{V_t}\Delta V_d + \frac{V_q}{V_t}\Delta V_q \tag{4.48}$$

Now, substituting Equation (4.35) into Equation (4.23),

$$\begin{bmatrix} \Delta V_d \\ \Delta V_q \end{bmatrix} = \frac{1}{\Delta_e} \begin{bmatrix} 0 & X_q \\ -X'_d & 0 \end{bmatrix}$$

$$\begin{bmatrix} (X_e + X_q) & -R_e V_\infty \cos\delta + V_\infty \sin\delta (X_e + X_q) \\ R_e & R_e V_\infty \sin\delta + V_\infty \cos\delta (X_e + X'_d) \end{bmatrix} \begin{bmatrix} \Delta E'_q \\ \Delta\delta \end{bmatrix} + \begin{bmatrix} 0 \\ \Delta E'_q \end{bmatrix} \tag{4.49}$$

or

$$\begin{bmatrix} \Delta V_d \\ \Delta V_q \end{bmatrix} = \frac{1}{\Delta_e} \begin{bmatrix} R_e X_q & X_q(R_e V_\infty \sin\delta + V_\infty(X'_d + X_e)\cos\delta) \\ -X'_d(X_e + X_q) & -X'_d(-R_e V_\infty \cos\delta + V_\infty(X_e + X_q)\sin\delta) \end{bmatrix} \begin{bmatrix} \Delta E'_q \\ \Delta\delta \end{bmatrix}$$

$$+ \begin{bmatrix} 0 \\ \Delta E'_q \end{bmatrix}$$

Therefore,

$$\Delta V_d = \frac{1}{\Delta_e} \left[R_e X_q \Delta E'_q + X_q R_e V_\infty \sin\delta + V_\infty X_q (X'_d + X_e)\cos\delta\Delta\delta \right] \tag{4.50}$$

and

$$\Delta V_q = \frac{1}{\Delta_e} \left[-X'_d(X_e + X_q)\Delta E'_q + (X'_d R_e V_\infty \cos\delta - V_\infty X'_d(X_e + X_q)\sin\delta\Delta\delta) \right] + \Delta E'_q \tag{4.51}$$

Replacing ΔV_d and ΔV_q from Equations (4.50) and (4.51) in Equation (4.48) results in

$$\Delta V_t = K_5 \Delta \delta + K_6 \Delta E'_q \qquad (4.52)$$

4.2.3 Derivation of K constants: K_1, K_2, K_3, K_4, K_5, and K_6

From Equation (4.36), the expression of $\Delta \dot{E}'_q$ on substitution of ΔI_d is

$$\Delta \dot{E}'_q = -\frac{1}{T'_{do}} \Delta E'_q - \frac{1}{T'_{do}}(X_d - X'_d)\left(\frac{1}{\Delta_e}\left[(X_e + X_q)\Delta E'_q + \{-R_e V_\infty \cos\delta\right.\right.$$

$$\left.\left. + (X_e + X_q)V_\infty \sin\delta\}\Delta\delta\right]\right) + \frac{1}{T'_{do}}\Delta E_{fd} \qquad (4.53)$$

$$\Delta \dot{E}'_q = -\frac{1}{T'_{do}}\left[1 + \frac{(X_d - X'_d)(X_e + X_q)}{\Delta_e}\right]\Delta E'_q$$

$$- \frac{1}{T'_{do}}\frac{V_\infty(X_d - X'_d)}{\Delta_e}\{(X_e + X_q)\sin\delta - R_e \cos\delta\}\Delta\delta + \frac{1}{T'_{do}}\Delta E_{fd} \qquad (4.54)$$

$$\therefore \Delta \dot{E}'_q = -\frac{1}{K_3 T'_{do}}\Delta E'_q - \frac{K_4}{T'_{do}}\Delta\delta + \frac{1}{T'_{do}}\Delta E_{fd} \qquad (4.55)$$

where

$$\frac{1}{K_3} = 1 + \frac{(X_d - X'_d)(X_q + X_e)}{\Delta_e} \qquad (4.56)$$

$$K_4 = \frac{V_\infty(X_d - X'_d)}{\Delta_e}\left[(X_q + X_e)\sin\delta - R_e \cos\delta\right] \qquad (4.57)$$

Again from Equation (4.38), the expression of $\Delta \dot{v}$ on substitution of ΔI_d and ΔI_q is

$$\Delta \dot{v} = -\frac{1}{2H}\Delta E'_q I_q - \frac{(X_q - X'_d)I_q}{2H}\frac{1}{\Delta_e}\left[(X_e + X_q)\Delta E'_q + \{-R_e V_\infty \cos\delta\right.$$

$$\left. + (X_e + X_q)V_\infty \sin\delta\}\Delta\delta\right]$$

$$+ \left(\frac{(X'_d - X_q)}{2H}I_d - \frac{1}{2H}E'_q\right)\frac{1}{\Delta_e}\left[R_e \Delta E'_q + \{R_e V_\infty \sin\delta\right.$$

$$\left. + (X_e + X'_d)V_\infty \cos\delta\}\Delta\delta\right] - \frac{D\omega_s}{2H}\Delta v + \frac{1}{2H}\Delta T_M \qquad (4.58)$$

$$\Delta \dot{v} = -\frac{1}{2H}\frac{1}{\Delta_e}\left[I_q\Delta_e - I_q(X_d' - X_q)(X_e + X_q) - R_e I_d(X_d' - X_q) + R_e E_q'\right]\Delta E_q'$$

$$+\frac{V_\infty I_q}{2H\Delta_e}(X_d' - X_q)\left[(X_e + X_q)\sin\delta - R_e\cos\delta\right]\Delta\delta$$

$$+\frac{V_\infty}{\Delta_e}\left[\left\{I_d(X_d' - X_q) - E_q'\right\}\left\{(X_e + X_d')\cos\delta + R_e\sin\delta\right\}\right]$$

$$-\frac{D\omega_s}{2H}\Delta v + \frac{1}{2H}\Delta T_M$$

(4.59)

This can be written in terms of K constants as

$$\Delta \dot{v} = -\frac{K_2}{2H}\Delta E_q' - \frac{K_1}{2H}\Delta\delta - \frac{D\omega_s}{2H}\Delta v + \frac{1}{2H}\Delta T_M$$ (4.60)

where

$$K_2 = \frac{1}{\Delta_e}\left[I_q\Delta_e - I_q(X_d' - X_q)(X_q + X_e) - R_e(X_d' - X_q)I_d + R_e E_q'\right]$$ (4.61)

and

$$K_1 = -\frac{1}{\Delta_e}\left[I_q V_\infty(X_d' - X_q)\left\{(X_q + X_e)\sin\delta - R_e\cos\delta\right\}\right.$$

$$\left. + V_\infty\left\{(X_d' - X_q)I_d - E_q'\right\}\left\{(X_d' + X_e)\cos\delta + R_e\sin\delta\right\}\right]$$ (4.62)

On substitution of ΔV_d and ΔV_q in Equation (4.46), it reduces to

$$\Delta V_t = \frac{V_d}{V_t}\left[\frac{1}{\Delta_e}\left\{R_e X_q\Delta E_q' + X_q R_e V_\infty\sin\delta + V_\infty X_q(X_d' + X_e)\cos\delta\Delta\delta\right\}\right]$$

$$+\frac{V_q}{V_t}\left[\frac{1}{\Delta_e}\left\{-X_d'(X_e + X_q)\Delta E_q'\right.\right.$$

$$\left.\left. +(X_d' R_e V_\infty\cos\delta - V_\infty X_d'(X_e + X_q)\sin\delta\Delta\delta\right\} + \Delta E_q'\right]$$

or

$$\Delta V_t = \left[\frac{1}{\Delta_e}\left\{\frac{V_d}{V_t}R_e X_q - \frac{V_d}{V_t}X_d'(X_q + X_e)\right\} + \frac{V_d}{V_t}\right]\Delta E_q'$$

$$+\left[\frac{1}{\Delta_e}\left\{\frac{V_d}{V_t}X_q(R_e V_\infty\sin\delta + V_\infty\cos\delta(X_d' + X_e))\right.\right.$$

$$\left.\left. +\frac{V_q}{V_t}X_d'(R_e V_\infty\cos\delta - V_\infty(X_e + X_q)\sin\delta)\right\}\right]\Delta\delta$$ (4.63)

Therefore, Equation (4.63) can be written in terms of K constants as

$$\Delta V_t = K_5 \Delta \delta + K_6 \Delta E'_q \tag{4.64}$$

where

$$K_5 = \frac{1}{\Delta_e}\left[\frac{V_d}{V_t}X_q\{R_eV_\infty \sin\delta + V_\infty \cos\delta(X'_d + X_e)\} \right.$$
$$\left. + \frac{V_q}{V_t}X'_d\{R_eV_\infty \cos\delta - V_\infty(X_e + X_q)\sin\delta\} \right] \tag{4.65}$$

$$K_6 = \frac{1}{\Delta_e}\left[\frac{V_d}{V_t}R_eX_q - \frac{V_d}{V_t}X'_d(X_q + X_e) \right] + \frac{V_d}{V_t} \tag{4.66}$$

Now, the overall linearized machine differential equations (4.43)–(4.45) and the linearized exciter equation (4.39) together can be put in a block diagram shown in Figure 4.4. In this representation, the dynamic characteristics of the system can be expressed in terms of the K constants. These constants (K_1–K_6) and the block diagram representation were developed first by Heffron–Phillips in Ref. [5] and later by de Mello in Ref. [2] to study the synchronous machine stability as affected by local low-frequency oscillations and its control through excitation system.

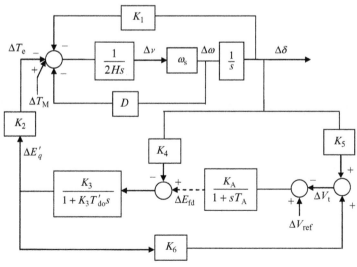

FIGURE 4.4

Block diagram representation of the synchronous machine flux-decay model.

The K constants presented in the block diagram (Figure 4.4) are defined as follows:

$K_1 = \dfrac{\Delta T_e}{\Delta \delta}\Big|_{E_q'}$ change in electric torque for a change in rotor angle with constant flux linkages in the d-axis

$K_2 = \dfrac{\Delta T_e}{\Delta E_q'}\Big|_{\delta}$ change in electric torque for a change in d-axis flux linkages withconstant rotor angle

$K_3 = \dfrac{X_d' + X_e}{X_d + X_e}$ for the case where the external impedance is a pure reactance X_e

$K_4 = \dfrac{1}{K_3}\dfrac{\Delta E_q'}{\Delta \delta}$ demagnetizing effect of a change in rotor angle

$K_5 = \dfrac{\Delta V_t}{\Delta \delta}\Big|_{E_q'}$ change in terminal voltage with change in rotor angle for constant E_q'

$K_6 = \dfrac{\Delta V_t}{\Delta E_q'}\Big|_{\delta}$ change in terminal voltage with change in E_q' for constant rotor angle

It is evident that the K constants are dependent on various system parameters such as system loading and the external network resistance (R_e) and reactance (X_e). Generally, the value of the K constants is greater than zero (>0), but under heavy loading condition (high generator output) and for high value of external system reactance, K_5 might be negative, contributing to negative damping and causing system instability. This phenomenon has been discussed in the following sections based on state-space model.

4.3 SMALL-SIGNAL STABILITY ANALYSIS USING STATE-SPACE MODEL AND BLOCK DIAGRAM

The state-space representation of the synchronous machine can be obtained when Equations (4.43)–(4.45) and (4.52) are written together in matrix form. Assuming $\Delta T_M = 0$, the state-space model of the SMIB system without exciter is therefore

$$\begin{bmatrix} \Delta \dot{E}_q' \\ \Delta \dot{\delta} \\ \Delta \dot{v} \end{bmatrix} = \begin{bmatrix} -\dfrac{1}{K_3 T_{do}'} & -\dfrac{K_4}{T_{do}'} & 0 \\ 0 & 0 & \omega_s \\ -\dfrac{K_2}{2H} & -\dfrac{K_1}{2H} & \dfrac{D\omega_s}{2H} \end{bmatrix} \begin{bmatrix} \Delta E_q' \\ \Delta \delta \\ \Delta v \end{bmatrix} + \begin{bmatrix} \dfrac{1}{T_{do}'} \\ 0 \\ 0 \end{bmatrix} \Delta E_{fd} \qquad (4.67)$$

$$\Delta V_t = [K_6 \quad K_5 \quad 0] \begin{bmatrix} \Delta E'_q \\ \Delta \delta \\ \Delta v \end{bmatrix} \tag{4.68}$$

Here, the input ΔE_{fd} is the perturbed field voltage without excitation, i.e., the machine is said to be operating under manual control action. It has been found that the previously mentioned system matrix has a pair of complex conjugate eigenvalues and a stable real eigenvalue. The complex eigenvalues are associated with the electromechanical mode (1-3 Hz range) and the real eigenvalue corresponds to the flux-decay mode.

From the block diagram (Figure 4.4), it has been observed that, without the exciter (i.e., $K_A=0$), there are three loops in the system. The above two loops are termed as torque-angle loops corresponding to the complex pair of eigenvalues and the bottom loop due to $\Delta E'_q$ via K_4 is for the real eigenvalue. Without damping (i.e., $D=0$), the torque-angle loop is purely oscillatory. However, the positive feedback introduced by the bottom loop tends to push the torque-angle loop eigenvalues to the left half of s-plane and the negative real eigenvalue to the right. Thus, with constant field (i.e., $\Delta E_{\text{fd}}=0$) without excitation, there is an inherent damping in the system.

But with high gain, the real pole may move to the right half of the s-plane, pushing the system toward monotonically unstable situation. For such an operating point, stable operation can only be achieved by superimposing the effects of excitation control without deteriorating damping that can cancel out the unstable monotonic component. Operation in this region under automatic voltage regulator has been referred to as operation with dynamic stability or conditional stability.

To study the *effect of excitation system*, the exciter is now added with the state-space model (4.67) and (4.68). In this case, the state-space equations will be modified by making ΔE_{fd} as a state variable. Equation (4.39) for ΔE_{fd} is given by

$$T_A \Delta \dot{E}_{\text{fd}} = -\Delta E_{\text{fd}} + K_A(\Delta V_{\text{ref}} - \Delta V_t) \tag{4.69}$$

Replacing ΔV_t from Equation (4.52) results in

$$\Delta \dot{E}_{\text{fd}} = -\frac{1}{T_A}\Delta E_{\text{fd}} - \frac{K_A K_5}{T_A}\Delta \delta - \frac{K_A K_6}{T_A}\Delta E'_q + \frac{K_A}{T_A}\Delta V_{\text{ref}} \tag{4.70}$$

Therefore, the overall state-space model for Figure 4.4 becomes

$$\begin{bmatrix} \Delta \dot{E}'_q \\ \Delta \dot{\delta} \\ \Delta \dot{v} \\ \Delta \dot{E}_{\text{fd}} \end{bmatrix} = \begin{bmatrix} -\dfrac{1}{K_3 T'_{\text{do}}} & -\dfrac{K_4}{T'_{\text{do}}} & 0 & \dfrac{1}{T'_{\text{do}}} \\ 0 & 0 & \omega_s & 0 \\ -\dfrac{K_2}{2H} & -\dfrac{K_1}{2H} & -\dfrac{D\omega_s}{2H} & 0 \\ -\dfrac{K_A K_6}{T_A} & -\dfrac{K_A K_5}{T_A} & 0 & -\dfrac{1}{T_A} \end{bmatrix} \begin{bmatrix} \Delta E'_q \\ \Delta \delta \\ \Delta v \\ \Delta E_{\text{fd}} \end{bmatrix} + \begin{bmatrix} 0 \\ 0 \\ 0 \\ \dfrac{K_A}{T_A} \end{bmatrix} \Delta V_{\text{ref}} \tag{4.71}$$

The exciter introduces an additional negative real eigenvalue to the system. Ignoring the dynamics of the exciter for the moment (i.e., assuming $T_A=0$), if $K_5<0$ and K_A are sufficiently large, then the gain through T'_{do} is $-(K_4+K_AK_5)K_3$ (approximately) and may become positive, which introduces negative feedback to the torque-angle loop. As a result, the complex pair of eigenvalues moves to the right half of s-plane and the real eigenvalues to the left, which pushes the system toward instability. This critical situation is now being analyzed through an illustration for the following two test systems.

4.4 AN ILLUSTRATION

For the following two test systems whose K_1–K_6 constants and other parameters are given, find the eigenvalues for $K_A=50$. Plot the root locus to study the system stability for varying K_A. Notice that in system 1, the value of $K_5>0$ and, in the system 2, $K_5<0$.

Test system 1
$K_1=3.7585$, $K_2=3.6816$, $K_3=0.2162$, $K_4=2.6582$, $K_5=0.0544$, $K_6=0.3616$, $T'_{do}=5$ s, $H=6$ s, $T_A=0.2$ s, and $D=0$.

Test system 2
$K_1=0.9831$, $K_2=1.0923$, $K_3=0.3864$, $K_4=1.4746$, $K_5=-0.1103$, $K_6=0.4477$, $T'_{do}=5$ s, $H=6$ s, $T_A=0.2$ s, and $D=0$.

Solution
The system matrix with exciter dynamics is

$$A_{sys} = \begin{bmatrix} -\dfrac{1}{K_3T'_{do}} & -\dfrac{K_4}{T'_{do}} & 0 & \dfrac{1}{T'_{do}} \\[2mm] 0 & 0 & \omega_s & 0 \\[2mm] -\dfrac{K_2}{2H} & -\dfrac{K_1}{2H} & -\dfrac{D\omega_s}{2H} & 0 \\[2mm] -\dfrac{K_AK_6}{T_A} & -\dfrac{K_AK_5}{T_A} & 0 & -\dfrac{1}{T_A} \end{bmatrix}$$

For the system 1,

$$A_{sys} = \begin{bmatrix} -0.9250 & -0.5316 & 0 & 0.2 \\ 0 & 0 & 377 & 0 \\ -0.3068 & -0.3132 & 0 & 0 \\ -90.4 & -13.6 & 0 & -5.0 \end{bmatrix}$$

The eigenvalues of the systems are calculated in MATLAB.
The eigenvalues of test system 1 are

$$\lambda_1 = -0.3527 + j10.9457; \quad \lambda_2 = -0.3527 - j10.9457$$

$$\lambda_3 = -2.6098 + j3.2180; \quad \lambda_4 = -2.6098 - j3.2180$$

and of test system 2 are

$$\lambda_1 = -3.1194 + j4.6197; \quad \lambda_2 = -3.1194 - j4.6197$$

$$\lambda_3 = 0.3603 + j5.3995; \quad \lambda_4 = 0.3603 - j5.3995$$

It is clear from these results that system 2 is unstable at the exciter gain $K_A = 50$. This is due to the fact that under heavy loading or a high value of transmission line reactance when K_5 becomes negative pushes one pair of eigenvalues to the right half of the s-plane.

- **Stability analysis by root-locus method**

    Consider the exciter transfer function as $G_A(s) = \dfrac{K_A}{1+0.2s}$.
    The general block diagram of SMIB system (Figure 4.4) can be reduced to Figure 4.5. Here, all machine dynamics except that of the exciter are included in the feedback transfer function *$B(s)$. The closed-loop characteristic equation of the system is given by $1 + G_A(s)B(s) = 0$.
    Where $B(s)$ can be obtained as

$$B(s) = \frac{K_3 K_6 \left(\dfrac{2H}{\omega_s} s^2 + Ds + K_1 \right) - K_2 K_3 K_5}{\left(1 + K_3 T_{do}' s\right)\left(\dfrac{2H}{\omega_s} s^2 + Ds + K_1 \right) - K_2 K_3 K_4}$$

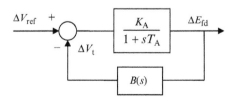

FIGURE 4.5

Small-signal model represented as a feedback system.

Putting different values of system parameters for system 1,

$$B(s) = \frac{(0.2162 \times 0.3616)\left(\frac{2 \times 6}{377}s^2 + 0 + 3.7585\right) - 0.2162 \times 3.6816 \times 0.0544}{(1 + 0.2162 \times 5 \times s)\left(\frac{2 \times 6}{377}s^2 + 0 + 3.7585\right) - 0.2162 \times 3.6816 \times 2.6582}$$

$$= \frac{0.002488s^2 + 0.25050}{0.03437s^3 + 0.0318s^2 + 4.0629s + 1.64268}$$

$$\therefore B(s) = \frac{0.0699s^2 + 7.2883}{s^3 + 0.9271s^2 + 118.48s + 47.79}$$

and for system 2,

$$B(s) = \frac{(0.3864 \times 0.4477)\left(\frac{2 \times 6}{377}s^2 + 0 + 0.9831\right) - 1.0923 \times 0.3864 \times (-0.1103)}{(1 + 0.3864 \times 5 \times s)\left(\frac{2 \times 6}{377}s^2 + 0 + 0.9831\right) - 1.0923 \times 0.3864 \times 1.4746}$$

$$= \frac{0.005506s^2 + 0.216616}{0.06143s^3 + 0.03183s^2 + 1.8993s + 0.36073}$$

$$\therefore B(s) = \frac{0.0896s^2 + 3.5262}{s^3 + 0.5181s^2 + 30.918s + 5.872}$$

The loop transfer function $G_A(s)B(s)$ for system 1 is

$$G_A(s)B(s) = \frac{50}{1 + 0.2s} \cdot \frac{0.0699s^2 + 7.2883}{s^3 + 0.9271s^2 + 118.48s + 47.79}$$

$$= \frac{3.495s^2 + 364.415}{0.2s^4 + 1.1854s^3 + 24.6681s^2 + 128.058s + 47.89}$$

and for system 2,

$$G_A(s)B(s) = \frac{50}{1 + 0.2s} \cdot \frac{0.0896s^2 + 3.5262}{s^3 + 0.5181s^2 + 30.918s + 5.872}$$

$$= \frac{4.475s^2 + 176.385}{0.2s^4 + 1.10358s^3 + 6.7039s^2 + 32.1052s + 5.876}$$

The root-locus plot of $G_A(s)B(s)$ is shown in Figure 4.6a and b for systems 1 and 2, respectively. The plots are interpreted as that system 1 is stable at exciter gain $K_A = 50$ and that system 2 becomes unstable with marginal enhancement of gain at $K_A = 22.108$.

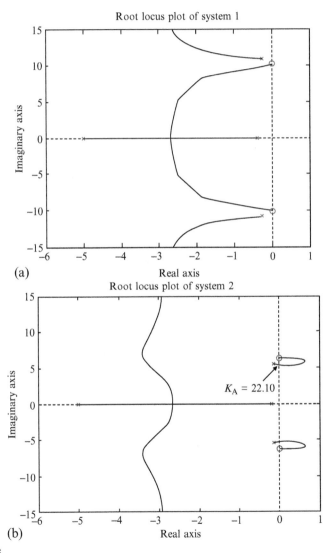

(a)

(b)

FIGURE 4.6

Root-locus analysis (a) test system 1 and (b) test system 2. (For color version of this figure, the reader is referred to the online version of this chapter.)

*Computation of $B(s)$ through block diagram reduction approach

Step 1: Removing exciter block from Figure 4.4, it becomes

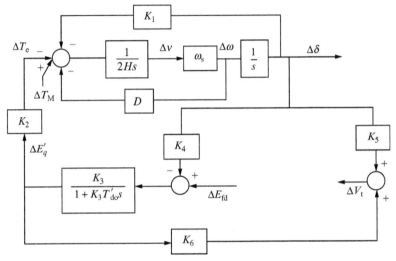

FIGURE (I)

Step 2:
Reduction of the torque-angle loop

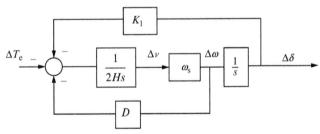

FIGURE (II)

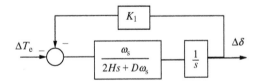

FIGURE (III)

$$\Delta T_e \longrightarrow \boxed{\dfrac{-1}{\dfrac{2H}{\omega_s}s^2 + Ds + K_1}} \longrightarrow \Delta\delta$$

FIGURE (IV)

Step 3:
  Overall reduced block diagram

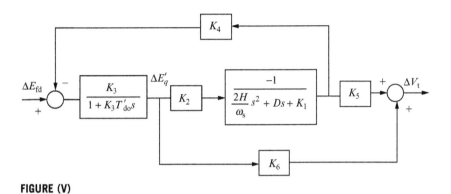

FIGURE (V)

Let $K_G = \dfrac{K_3}{1 + K_3 T'_{do} s}$ and $K_T = \dfrac{K_2}{\frac{2H}{\omega_s} s^2 + Ds + K_1}$

Step 4:

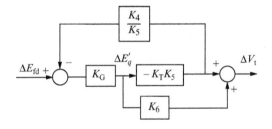

FIGURE (VI)

Step 5:
  Signal-flow graph of Figure (vi)

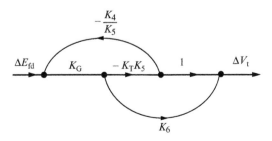

FIGURE (VII)

(i) There are two nos. of forward paths in Figure (vii) with path gain;
$P_1 = -K_G K_T K_5$ and $P_2 = K_G K_6$
 (ii) one no. individual loop of loop gain $= K_G K_T K_4$
(iii) $\Delta = 1 - K_G K_T K_4$.

Applying *Masson's gain rule*

$$B(s) = \frac{\Delta V_t}{\Delta E_{fd}} = \frac{1}{\Delta} \sum_{k=1}^{2} P_k \Delta_k$$

where k is the no. of forward paths. As there is no non-touching loops with the kth forward path, $\Delta_1 = \Delta_2 = 1$.

$$\therefore \quad B(s) = \frac{P_1 \Delta_1 + P_2 \Delta_2}{\Delta} = \frac{K_G K_6 - K_G K_T K_5}{1 - K_G K_T K_4}.$$

Replacing the expression for K_G and K_T

$$B(s) = \frac{K_3 K_6 \left(\dfrac{2H}{\omega_s} s^2 + Ds + K_1 \right) - K_2 K_3 K_5}{\left(1 + K_3 T'_{do} s \right) \left(\dfrac{2H}{\omega_s} s^2 + Ds + K_1 \right) - K_2 K_3 K_4}$$

4.5 EFFECT OF GENERATOR FIELD

From the block diagram shown in Figure 4.4, it is clear that the change of air gap torque (ΔT_e) can be expressed as a function of $\Delta \delta$ and $\Delta E'_q$. The variation of field flux $\Delta E'_q$ is determined from the field circuit dynamic equation

$$\Delta \dot{E}'_q = -\frac{1}{K_3 T'_{do}} \Delta E'_q - \frac{K_4}{T'_{do}} \Delta \delta + \frac{1}{T'_{do}} \Delta E_{fd} \tag{4.72}$$

Therefore, with constant field voltage ($\Delta E_{fd} = 0$), the field flux variation is caused only by feedback of $\Delta \delta$ through the coefficient K_4. This represents the demagnetizing effect of the armature reaction. The change of air-gap torque due to field flux linkage variation by rotor angle change is given by

$$\frac{\Delta T_e}{\Delta \delta} \bigg| \text{ due to } \Delta E'_q = -K_2 \, \Delta E'_q$$

$$\therefore \Delta T_e = -\frac{K_2 K_3 K_4}{1 + K_3 T'_{do} s} \Delta \delta \tag{4.73}$$

The constants K_2, K_3, and K_4 are usually positive. The contribution of $\Delta E'_q$ to synchronizing and damping torque components depends on the oscillating frequency as discussed below.

In the steady state or at very low frequency ($s = j\omega = 0$). Therefore

$$\Delta T_e | \text{due to } \Delta E'_q = -K_2 K_3 K_4 \Delta \delta \tag{4.74}$$

Thus the field flux variation due to feedback $\Delta\delta$ (i.e., due to armature reaction) introduces a negative synchronizing torque component.

At oscillating frequency much higher than $\dfrac{1}{K_3 T'_{do}}$

$$\Delta T_e \approx -\frac{K_2 K_3 K_4}{j\omega K_3 T'_{do}} \Delta\delta$$

$$= \frac{K_2 K_4}{\omega T'_{do}} j \Delta\delta \tag{4.75}$$

Thus the component of air gap torque due to $\Delta E'_q$ is 90° ahead of $\Delta\delta$ or in phase with $\Delta\omega$. Hence $\Delta E'_q$ results in a positive damping torque component.

At a typical machine oscillating frequency of about 1 Hz (2π rad/s), $\Delta E'_q$ results in a positive damping torque component and a negative synchronizing torque component (Figure 4.7). The net effect is to reduce slightly the synchronizing torque component and increase the damping torque component.

- **Special situation with K_4 negative**

The coefficient K_4 is normally positive. As long as it is positive, the effect of field flux variation due to armature reaction is to introduce a positive damping torque component. However, there can be situations where K_4 is negative (when $\{(X_q + X_e)\sin\delta - R_e\cos\delta\} < 0$) due to the fact that a hydraulic generator without

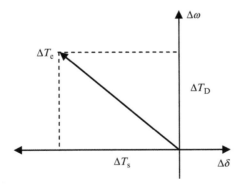

FIGURE 4.7

Positive damping torque and negative synchronizing torque due to field flux variation.

damper winding is operating at light load and is connected by a transmission line of relatively high resistance to reactance ratio to a large system.

Also, K_4 can be negative when a machine is connected to a large local load, supplied by the generator and partly by the remote large system [6]. Under such conditions, the torque produced by the induced current in the field due to armature reaction has components out of phase with $\Delta\omega$ and produces negative damping.

EXAMPLE 4.1

This example illustrates the effects of the generator field circuit dynamics on the small-signal stability performance at low frequency and at rotor oscillation frequency of a system. The machine parameters and K constants are given as follows:

$K_1 = 0.7643$, $K_2 = 0.8644$, $K_3 = 0.3230$, $K_4 = 1.4187$, $H = 3.5s$, $T'_{do} = 7.32s$, and $D = 0$

Determine the following:

(i) The elements of the state matrix "A" representing the small-signal performance of the system
(ii) The eigenvalues of "A" and the corresponding eigenvectors and participation matrix
(iii) Steady-state synchronizing torque coefficient
(iv) The damping and synchronizing torque coefficient at the rotor oscillation frequency

---

Solution

(i) The state matrix is

$$A = \begin{bmatrix} -\dfrac{1}{K_3 T'_{do}} & -\dfrac{K_4}{T'_{do}} & 0 \\ 0 & 0 & \omega_s \\ -\dfrac{K_2}{2H} & -\dfrac{K_1}{2H} & -\dfrac{D\omega_s}{2H} \end{bmatrix}$$

$$= \begin{bmatrix} -0.4228 & -0.1938 & 0 \\ 0 & 0 & 377 \\ -0.1234 & -0.1092 & 0 \end{bmatrix}$$

(ii) $\lambda_1 = -0.204$; $\lambda_2, \lambda_3 = -0.1094 \pm j6.41$.
The right eigenvectors are given by

$$(A - \lambda_p I)\phi = 0$$

Therefore,

$$\begin{bmatrix} -0.4228 - \lambda_p & 0.1938 & 0 \\ 0 & -\lambda_p & 377 \\ -0.1234 & -0.1092 & -\lambda_p \end{bmatrix} \begin{bmatrix} \phi_{1p} \\ \phi_{2p} \\ \phi_{3p} \end{bmatrix} = 0$$

$$(-0.4228 - \lambda_p)\phi_{1p} + 0.1938\phi_{2p} = 0$$

$$-\lambda_p\phi_{2p} + 377\phi_{3p} = 0$$

$$-0.1234\phi_{1p} - 0.1092\phi_{2p} - \lambda_p\phi_{3p} = 0$$

Solving this homogeneous equations for $p = 1, 2, 3$ for the eigenvalues $\lambda_1, \lambda_2, \lambda_3$, respectively, the right eigenvectors,

$$\phi = \begin{bmatrix} 0.6631 & -0.0015 + j0.0302 & -0.0015 - j0.0302 \\ -0.7485 & 0.9994 & 0.9994 \\ 0.0004 & -0.0003 + j0.0170 & -0.0003 - j0.0170 \end{bmatrix}$$

The left eigenvectors normalized so that $\phi\psi = I$ and are given by

$$\psi = \phi^{-1} = \frac{\text{adj}(\phi)}{|\phi|}$$

$$\psi = \begin{bmatrix} 1.5114 & 0.0015 & -2.679 \\ 0.5660 + j0.0084 & 0.5008 - j0.0085 & -1.0035 - j29.3922 \\ 0.5660 - j0.0084 & 0.5008 + j0.0085 & -1.0035 + j29.3922 \end{bmatrix}$$

Therefore, the *participation matrix* is

$$P = \begin{bmatrix} \phi_{11}\psi_{11} & \phi_{12}\psi_{21} & \phi_{13}\psi_{31} \\ \phi_{21}\psi_{12} & \phi_{22}\psi_{22} & \phi_{23}\psi_{32} \\ \phi_{31}\psi_{13} & \phi_{32}\psi_{23} & \phi_{33}\psi_{33} \end{bmatrix}$$

i.e.,

$$P = \begin{bmatrix} 1.0022 & -0.0011 + j0.0171 & -0.0011 - j0.0171 \\ -0.0011 & 0.5005 - j0.0085 & 0.5005 + j0.0085 \\ -0.0011 & 0.5005 - j0.0085 & 0.5005 + j0.0085 \end{bmatrix}$$

Taking only the magnitudes,

$$P = \begin{bmatrix} 1.0022 & 0.0171 & 0.0171 \\ 0.0011 & 0.5006 & 0.5006 \\ 0.0011 & 0.5006 & 0.5006 \end{bmatrix} \begin{matrix} \Delta E'_q \\ \Delta\delta \\ \Delta v \end{matrix}$$
$$\quad\quad\quad \lambda_1 \quad\; \lambda_2 \quad\;\; \lambda_3$$

From the participation matrix, it is found that $\Delta\delta$ and Δv have high participation in the oscillatory modes corresponding to the eigenvalues λ_2 and λ_3. The field flux linkage $\Delta E'_q$ has a high participation in the nonoscillatory mode λ_1.

(iii) The steady-state synchronizing torque coefficient due to field flux linkage $\Delta E'_q$ at very low frequency (refer Equation 4.74) is

$$-K_2K_3K_4 = -0.8644 \times 0.3230 \times 1.4187$$
$$= -0.3961$$

The total steady-state synchronizing torque coefficient is

$$K_s = K_1 + (-K_2K_3K_4)$$
$$= 0.7643 + (-0.3961) \;.$$
$$= 0.3679 \text{ pu torque/rad}$$

(iv) The computation of damping torque and synchronizing torque components at rotor oscillation frequency is as follows. Referring to Equation (4.75)

$$\frac{\Delta T_e}{\Delta\delta}\bigg| \text{ due to } \Delta E'_q = -\frac{K_2K_3K_4}{1+K_3T'_{do}s}. \text{ Let } K_3T'_{do}=T_3$$

$$\therefore \frac{\Delta T_e}{\Delta\delta}\bigg|_{\Delta E'_q} = -\frac{K_2K_3K_4(1-sT_3)}{1-s^2T_3^2}$$

Therefore,

$$\Delta T_e|_{\Delta E'_q} = \frac{-K_2K_3K_4}{1-s^2T_3^2}\Delta\delta + \frac{K_2K_3K_4T_3}{1-s^2T_3^2}s\Delta\delta$$

As $s\Delta\delta = \Delta\dot{\delta} \Rightarrow \Delta\omega = \omega_s\Delta v$

$$\Delta T_e|_{\Delta E'_q} = \frac{-K_2K_3K_4}{1-s^2T_3^2}\Delta\delta + \frac{K_2K_3K_4T_3}{1-s^2T_3^2}\omega_s\Delta v$$

this can be written as

$$\Delta T_e|_{\Delta E'_q} = K_S\left(\Delta E'_q\right)\Delta\delta + K_D\left(\Delta E'_q\right)\Delta v$$

where $K_S\left(\Delta E'_q\right) = \dfrac{-K_2K_3K_4}{1-s^2T_3^2}$, the synchronizing torque coefficient and $K_D\left(\Delta E'_q\right) = \dfrac{K_2K_3K_4T_3\omega_s}{1-s^2T_3^2}$, the damping torque coefficient.

The eigenvalues $\lambda_2, \lambda_3 = -0.1094 \pm j6.41$ denote the complex frequency of rotor oscillations. Since the real part is much smaller than the imaginary part, K_S and K_D can be evaluated by setting $s = j6.41$ without loss of much accuracy. Therefore,

$$K_S\left(\Delta E'_q\right) = \frac{-K_2 K_3 K_4}{1 - s^2 T_3^2} = -\frac{0.8644 \times 0.3230 \times 1.4187}{1 - (j6.41)^2 (2.365)^2}$$

$$= -0.00172 \text{ pu torque/rad}$$
$$[\because T_3 = 0.3230 \times 7.32 = 2.365]$$

and

$$K_D\left(\Delta E'_q\right) = \frac{K_2 K_3 K_4 T_3 \omega_s}{1 - s^2 T_3^2} = \frac{0.3963 \times 0.3230 \times 1.4187 \times 2.365 \times 377}{1 - (j6.41)^2 (2.365)^2}$$

$$= 1.53 \text{ pu torque/pu speed change}$$

Thus, the effect of field flux variation reduces the synchronizing torque component slightly and adds a damping torque component.

The net synchronizing torque component is

$$K_S = K_1 + K_S\left(\Delta E'_q\right) = 0.7643 + (-0.00172)$$
$$= 0.7626 \text{ pu torque/rad}$$

Here, the only source of damping is the damping produced by the field flux variation. Hence, the net damping torque coefficient is

$$K_D = K_D\left(\Delta E'_q\right) = 1.53 \text{ pu torque/pu speed change}$$

4.6 EFFECT OF EXCITATION SYSTEM

In this section, the effect of excitation system on the small-signal stability performance of the SMIB system is examined in frequency domain [3]. For simplicity, let us assume that the exciter is simply a high constant gain K_A, i.e., $T_A = 0$.

Now from Equation (4.70), $T_A \Delta E_{fd} = -\Delta E_{fd} - K_A K_5 \Delta\delta - K_A K_6 \Delta E'_q + K_A \Delta V_{ref}$ with $T_A = 0$ and $\Delta V_{ref} = 0$ gives

$$0 = -\Delta E_{fd} - K_A K_5 \Delta\delta - K_A K_6 \Delta E'_q \tag{4.76}$$

$$\therefore \Delta E_{fd} = -K_A K_5 \Delta\delta - K_A K_6 \Delta E'_q \tag{4.77}$$

Again, from the block diagram (Figure 4.4),

$$\Delta E'_q = \frac{K_3}{1 + K_3 T'_{do} S}[\Delta E_{fd} - K_4 \Delta\delta] \tag{4.78}$$

or

$$\Delta E_q' = \frac{K_3}{1 + K_3 T_{do}' s} \left[-K_A K_5 \Delta\delta - K_A K_6 \Delta E_q' - K_4 \Delta\delta \right] \tag{4.79}$$

$$\Delta E_q' = \frac{K_3}{1 + K_3 T_{do}' s} \left[-(K_A K_5 + K_4)\Delta\delta - K_A K_6 \Delta E_q' \right] \tag{4.80}$$

or

$$\Delta E_q' \left[1 + \frac{K_3 K_A K_6}{1 + K_3 T_{do}' s} \right] = \frac{-K_3(K_4 + K_A K_5)}{1 + K_3 T_{do}' s} \Delta\delta \tag{4.81}$$

$$\therefore \frac{\Delta E_q'}{\Delta\delta} = \frac{-K_3(K_4 + K_A K_5)}{1 + K_3 K_A K_6 + s K_3 T_{do}'} \tag{4.82}$$

The effect of the feedback around T_{do}' is to reduce the time constant. If $K_5 > 0$, the overall situation does not differ qualitatively from the case without the exciter, i.e., the system has three open-loop poles, with one of them being complex and having a positive feedback. Thus, the real pole tends to move into the right-half plane. If $K_5 < 0$ and, consequently, $K_4 + K_A K_5 < 0$, the feedback from $\Delta\delta$ to ΔT_e changes from positive to negative, and with a large enough gain K_A, the electromechanical modes may move to the right-half plane and the real eigenvalue to the left on the real axis. Thus, a fast-acting exciter is bad for damping, but it has beneficial effects also. It minimizes voltage fluctuations, increases the synchronizing torque, and improves transient stability.

Taking the exciter dynamics into account, i.e., with $T_A \neq 0$ and $\Delta V_{ref} = 0$, Equation (4.70) becomes

$$T_A \Delta E_{fd} = -\Delta E_{fd} - K_A K_5 \Delta\delta - K_A K_6 \Delta E_q' \tag{4.83}$$

or

$$\Delta E_{fd}(1 + sT_A) = -K_A K_5 \Delta\delta - K_A K_6 \Delta E_q' \tag{4.84}$$

or

$$\Delta E_{fd} = \frac{-K_A K_5}{1 + sT_A} \Delta\delta - \frac{K_A K_6}{1 + sT_A} \Delta E_q' \tag{4.85}$$

Using the block diagram (Figure 4.4), the expression for $\Delta E_q'$ is

$$\Delta E_q' = \frac{K_3}{1 + K_3 T_{do}' s} [\Delta E_{fd} - K_4 \Delta\delta] \tag{4.86}$$

$$\Delta E' = \frac{K_3}{1 + K_3 T_{do}' s} \left[\frac{-K_A K_5}{1 + sT_A} \Delta\delta - \frac{K_A K_6}{1 + sT_A} \Delta E_q' - K_4 \Delta\delta \right] \tag{4.87}$$

Or

$$\Delta E_q' \left[1 + \frac{K_3 K_A K_6}{(1 + K_3 T_{do}' s)(1 + sT_A)} \right] = \frac{-K_3 \{K_4(1 + sT_A) + K_A K_5\}}{(1 + K_3 T_{do}' s)(1 + sT_A)} \cdot \Delta \delta$$

$$\therefore \frac{\Delta E_q'}{\Delta \delta} = \frac{-K_3 \{K_4(1 + sT_A) + K_A K_5\}}{K_3 K_A K_6 + (1 + K_3 T_{do}' s)(1 + sT_A)} \tag{4.88}$$

The contribution of this expression to the torque-angle loop is given by

$$\frac{\Delta T_e(s)}{\Delta \delta(s)} = K_2 \frac{\Delta E_q'}{\Delta \delta} = \frac{-K_2 K_3 \{K_4(1 + sT_A) + K_A K_5\}}{K_3 K_A K_6 + (1 + K_3 T_{do}' s)(1 + sT_A)} \tag{4.89}$$

Therefore, considering exciter system dynamics, the change of air-gap torque due to field flux variation by rotor angle change is given by

$$H(s) = \frac{\Delta T_e(s)}{\Delta \delta(s)} = \frac{-K_2 K_3 \{K_4(1 + sT_A) + K_A K_5\}}{K_3 K_A K_6 + (1 + K_3 T_{do}' s)(1 + sT_A)} \tag{4.90}$$

4.6.1 Effect of excitation system in torque-angle loop

Assuming $\Delta T_M = 0$, the torque-angle loop is given by Figure (ii). The undamped natural frequency of the torque-angle loop ($D \equiv 0$) is given by the roots of the characteristic equation:

$$\frac{2H(s)}{\omega_s} s^2 + K_1 = 0 \tag{4.91}$$

from which we get

$$s_1 = +j \sqrt{\frac{K_1 \omega_s}{2H(s)}} \text{ rad/s}; \quad s_2 = -j \sqrt{\frac{K_1 \omega_s}{2H(s)}} \text{ rad/s}$$

With a higher synchronizing torque coefficient K_1 and lower $H(s)$, s_1 and s_2 will be high. The complicated expression for K_1 involving loading conditions and external reactance is obtained from Equation (4.62). The synchronizing and damping torque component due to field flux linkage ($\Delta E_q'$) with exciter dynamics will now be computed in the following section.

By separating the torque-angle loop, in the block diagram of Figure 4.4, the reduced block diagram can be shown as in Figure 4.8. The characteristic equation of the closed-loop transfer function is obtained as

$$\frac{2H(s)}{\omega_s} s^2 \Delta \delta + K_1 \Delta \delta + H(s) \Delta \delta = 0 \tag{4.92}$$

as $\dfrac{\Delta T_e(s)}{\Delta \delta(s)} = H(s)$.

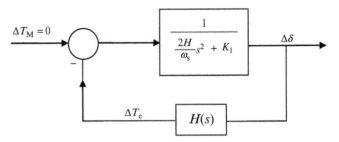

FIGURE 4.8

Block diagram representation of torque-angle loop.

Therefore, $\Delta T_e(s)|$ due to $\Delta E'_q = H(s)\Delta\delta$ gives the contribution to both the synchronizing torque and the damping torque. At oscillation frequency of 1-3 Hz. ranges, it can be shown that the constant K_4 has negligible effect on system performance.

Therefore, Equation (4.90) becomes

$$H(s) = \frac{-K_2 K_A K_5}{\frac{1}{K_3} + K_A K_6 + s\left(\frac{T_A}{K_3} + T'_{do}\right) + s^2 T'_{do} T_A} \tag{4.93}$$

Putting $s = j\omega$,

$$H(j\omega) = \frac{-K_2 K_A K_5}{\frac{1}{K_3} + K_A K_6 + j\omega\left(\frac{T_A}{K_3} + T'_{do}\right) + (j\omega)^2 T'_{do} T_A} \tag{4.94}$$

$$H(j\omega) = \frac{-K_2 K_A K_5}{\left(\frac{1}{K_3} + K_A K_6 - \omega^2 T'_{do} T_A\right) + j\omega\left(\frac{T_A}{K_3} + T'_{do}\right)} \tag{4.95}$$

Let $x = \frac{1}{K_3} + K_A K_6 - \omega^2 T'_{do} T_A$ and $y = \omega\left(\frac{T_A}{K_3} + T'_{do}\right)$.

$$\therefore H(j\omega) = \frac{-K_2 K_A K_5(x - jy)}{x^2 + y^2} \tag{4.96}$$

Therefore,

$$\Delta T_e = \frac{-K_2 K_A K_5 x}{x^2 + y^2} \Delta\delta + \frac{K_2 K_A K_5 y}{x^2 + y^2}(j\Delta\delta) \tag{4.97}$$

From Equation (4.97), it is clear that, at oscillation frequency, $\text{Re}[H(j\omega)]$ represents the synchronizing torque component and $\text{Im}[H(j\omega)]$ is the damping torque component.

If $Im[H(j\omega)] > 0$, positive damping is implied, i.e., the roots move to the left half of s-plane. If $Im[H(j\omega)] < 0$, there is a negative damping torque component, which tends to make the system unstable.

Thus,

$$\text{the synchronizing torque component} = \frac{-K_2 K_A K_5 x}{x^2 + y^2} \qquad (4.98)$$

and

$$\text{the damping torque component} = \frac{+K_2 K_A K_5 y}{x^2 + y^2} \qquad (4.99)$$

4.6.2 Calculation of steady-state synchronizing and damping torque

The steady-state synchronizing torque component is calculated at a very low oscillation frequency (setting $\omega \approx 0$).

Thus,

$$Re[H(j\omega)] = \frac{-K_2 K_A K_5}{\frac{1}{K_3} + K_A K_6} \qquad (4.100)$$

The total synchronizing torque coefficient is

$$K_S = K_1 + \left(\frac{-K_2 K_A K_5}{\frac{1}{K_3} + K_A K_6} \right) > 0 \qquad (4.101)$$

As K_1 is usually high, so even with $K_5 > 0$ for the case of low to medium external impedance and low to medium loading, the term $K_1 - \frac{K_2 K_A K_5}{\frac{1}{K_3} + K_A K_6}$ is positive. With $K_5 < 0$ for moderate to high external impedance and heavy loading, the synchronizing torque component K_S is enhanced positively.

4.6.3 Synchronizing and damping torque at rotor oscillation frequency

The synchronizing torque coefficient at rotor oscillation frequency (1-3 Hz) is

$$Re[H(j\omega)] = \frac{-K_2 K_A K_5 x}{x^2 + y^2} \qquad (4.102)$$

The net synchronizing torque coefficient is

$$K_S = K_1 + \left(\frac{-K_2 K_A K_5 x}{x^2 + y^2} \right) \text{ pu torque/rad} \qquad (4.103)$$

The damping torque coefficient is

$$K_D = \text{Im}[H(j\omega)] = \frac{K_2 K_A K_5 \left(\frac{T_A}{K_3} + T'_{do}\right)\omega}{x^2 + y^2} \tag{4.104}$$

The expression (4.104) contributes to positive damping for $K_5 > 0$ but negative damping for $K_5 < 0$, which is a cause of concern. Further, with $K_5 < 0$, a higher value of exciter gain K_A makes the system unstable as shown in Figure 4.6b.

4.7 AN ILLUSTRATION

In this illustration, the synchronizing torque coefficient and damping torque coefficient for test systems 1 and 2 are to be computed by considering the specific case illustrated in this section. The exciter gain $K_A = 50$ and the system parameters are taken as given in Section 4.4.

Solution
It is assumed that the rotor oscillation frequency is 10 rad/s (1.6 Hz) with $s = j\omega = j10$.

At $s = j\omega$, considering Equation (4.96),

$$H(j\omega) = \frac{-K_2 K_A K_5 (x - jy)}{x^2 + y^2}$$

where $x = \frac{1}{K_3} + K_A K_6 - \omega^2 T'_{do} T_A$ and $y = \omega\left(\frac{T_A}{K_3} + T'_{do}\right)$.

For system 1,

$$x = \frac{1}{0.2162} + 50 \times 0.3616 - (10)^2 \times 5 \times 0.2$$

$$= 4.6253 + 18.08 - 100 = -77.294$$

and $y = 10\left(\frac{0.2}{0.2162} + 5\right) = 59.250$.

Therefore,

$$x^2 + y^2 = (-77.294)^2 + (59.250)^2$$
$$= 5974.362 + 3510.562 = 9484.924$$

Now, the synchronizing torque coefficient is

$$\mathrm{Re}[H(j\omega)] = \frac{-K_2 K_A K_5 x}{x^2 + y^2}$$

$$= \frac{-3.6816 \times 50 \times 0.0544 \times (-77.294)}{9484.924}$$

$$= 0.0816 \ \text{pu torque/rad}$$

∴ Net the synchronizing torque component as

$$K_S = K_1 + \mathrm{Re}[H(j\omega)] = 3.7585 + 0.0816$$
$$= 3.8401 \ \text{pu torque/rad}$$

The damping torque coefficient is

$$K_D = \frac{K_2 K_A K_5 \left(\frac{T_A}{K_3} + T'_{do}\right)\omega}{x^2 + y^2}$$

$$= \frac{3.6816 \times 50 \times 0.0544 \times 59.250}{9484.924}$$

$$= 0.0625 \ \text{pu torque/pu speed change}$$

For system 2,

$$x = \frac{1}{0.3864} + 50 \times 0.4477 - (10)^2 \times 5 \times 0.2 = 2.5879 + 22.385 - 100$$

$$= -75.0271$$

and $y = 10\left(\dfrac{0.2}{0.3864}\right) + 5 = 55.1759.$

Therefore,

$$x^2 + y^2 = (-75.0271)^2 + (55.1759)^2$$
$$= 5629.065 + 3044.3891 = 8673.454$$

Now, the synchronizing torque coefficient is equivalent to

$$\frac{-1.0923 \times 50 \times -0.1103 \times (-75.0271)}{8673.454} = -0.0521 \ \text{pu torque/rad}$$

∴ Net the synchronizing torque component as

$$K_S = 0.9831 + (-0.521) = 0.931 \ \text{pu torque/rad}$$

and the damping torque component as

$$K_D = \frac{1.0923 \times 50 \times (-0.1103) \times 55.1759}{8673.454}$$

$$= -0.0383 \text{ pu torque/pu speed change}$$

Therefore, it is readily clear that when K_5 is negative, the damping torque component due to $\Delta E'_q$ would be opposite in sign to the value when K_5 is positive.

For the systems under consideration, Table 4.1 summarizes the effect of excitation on K_S and K_D at $\omega = 10$ rad/s for different values of gain K_A. It is observed from the table that except in very high value of exciter gain as the exciter gain increases, the stability of system 1 ($K_5 > 0$) increases, and, for system 2 ($K_5 < 0$), stability decreases. It has been further observed that for test system 1, with an increase of exciter gain up to a certain value ($K_A = 400$), the stability of the system increases because of improvement of damping torque (T_D). With a further increase

Table 4.1 Effect of Exciter Gain on Synchronizing Torque and Damping Torque on Systems 1 and 2

Exciter Gain (K_A)	Synchronizing Torque Due to $\Delta E'_q$	Net Synchronizing Torque (K_S)	Damping Torque (K_D)
System 1			
0	0	3.7585	0
10	0.0154	3.7739	0.0099
15	0.0233	3.7818	0.0153
25	0.0394	3.7979	0.0271
50	0.0816	3.8401	0.0625
100	0.1690	3.9275	0.1691
200	0.2285	3.9870	0.5871
400	−0.6647	3.0938	0.7994
1000	−0.7168	3.0417	0.1595
System 2			
0	0	0.9831	0
10	−0.0096	0.9735	−0.0057
15	−0.0145	0.9686	−0.0088
25	−0.0248	0.9583	−0.0159
50	−0.0521	0.9310	−0.0383
100	−0.1091	0.8740	−0.1143
200	−0.0611	0.9220	−0.4280
400	0.4052	1.3883	−0.2737
1000	0.3356	1.3187	−0.0529

of exciter gain, both the torques start decreasing and become negative with a very high gain, resulting in instability in the system, whereas for test system 2, the results are the opposite, with an increase of exciter gain, and it has been found that for high values of external system reactance and high generator output, when K_5 is negative, a high gain exciter introduces positive damping but decreasing synchronizing torque.

Thus, the situation is conflicting with regard to exciter response, and there is a requirement of optimum setting of the exciter gain so that it results in sufficient synchronizing and damping torque components in the expected range of operating conditions. This may not be always possible. Again, it may be necessary to use a high-response exciter to provide the required synchronizing torque and transient stability performance. In this situation, the introduction of additional damping to the system has been achieved by installing *power system stabilizer (PSS)* with the machine. The input stabilizing signal of the PSS may be derived from the system as speed change (Δv) or change in electric accelerating power (ΔP_{acc}) or a combination of both.

The small-signal stability performance of an SMIB power system and of a multi-machine power system employing speed input PSS has been described in Chapter 6.

EXERCISES

4.1 A synchronous machine is connected to an infinite bus. Draw the equivalent circuit diagram of the machine considering flux-decay model. Derive the state-space model of the system with and without exciter dynamics in terms of K constants.

4.2 Draw block diagram representation of the synchronous machine flux-decay model. Define different K constants presented in the block diagram. Explain the effects of K_4 and K_5 constants on synchronous machine stability.

4.3 For an SMIB system, considering exciter system dynamics with exciter gain K_A and time constant T_A, show that the change of air-gap torque ΔT_e due to field flux variation by rotor angle change $\Delta\delta$ is given by

$$H(s) = \frac{\Delta T_e(s)}{\Delta\delta(s)} = \frac{-K_2 K_3 \{K_4(1+sT_A) + K_A K_5\}}{K_3 K_A K_6 + (1+K_3 T'_{do}s)(1+sT_A)}$$

4.4 In Figure 4.1, assume that $R_e=0$, $X_e=0.5$ pu, $V_t \angle \theta° = 1 \angle 15°$ pu, and $V_\infty \angle \theta° = 1.05 \angle 0°$ pu. The machine data are $H=3.2$ s, $T'_{do}=9.6$ s, $K_A=400$, $T_A=0.2$ s, $R_s=0.0$ pu, $X_q=2.1$ pu, $X_d=2.5$ pu, $X'_d=0.39$ pu, and $D=0$. Using flux-decay model, find the following:
(a) The initial values of state and algebraic variables
(b) K_1–K_6 constants

4.5   Consider a single machine connected to an infinite bus. Assume that $V_\infty \angle \theta° = 1 \angle 0°$. The parameters are as follows:

Transmission line: $R_e = 0$ pu and $X_e = 0.5$ pu

Generator: $R_s = 0.0$ pu, $X_d = 1.6$ pu, and $X_q = 1.55$ pu, $X'_d = 0.32$ pu, $D = 0$, $H = 3.0$ s, and $T'_{do} = 6.0$ s

Exciter: $K_A = 50$ and $T_A = 0.05$ s

(a) Compute the eigenvalues of the system.

(b) Obtain the K_1–K_6 constants of the system.

References

[1] F.R. Schleif, H.D. Hunkins, G.E. Martin, E.E. Hattan, Excitation control to improve powerline stability, IEEE Trans. Power Apparatus Syst. PAS-87 (6) (1968) 1426–1434.

[2] F.P. demello, C. Concordia, Concepts of synchronous machine stability as effected by excitation control, IEEE Trans. Power Apparatus Syst. PAS-88 (4) (1969) 316–329.

[3] P.W. Sauer, M.A. Pai, Power System Dynamics and Stability, Pearson Education Pte. Ltd., Singapore, 1998.

[4] M.A. Pai, D.P. Sengupta, K.R. Padiyar, Small Signal Analysis of Power Systems, Narosa Publishing House, India, 2004.

[5] W.G. Heffron, R.A. Phillips, Effect of modern amplidyne voltage regulators on under excited operation of large turbine generators, AIEE Trans. Power Apparatus Systems PAS-71 (1952) 692–697.

[6] P. Kundur, Power System Stability and Control, McGraw-Hill, New York, 1994.

Small-Signal Stability Analysis in Multimachine System

5.1 INTRODUCTION

Because of the rapid growth of power systems and the continual reduction in inherent stability margins, increased reliance is being placed on analysis of system dynamic performance. The utilization of excitation control for improved steady-state or dynamic stability of multimachine power systems has received much attention since decades ago and will become much more significant in view of the increasing size and complexity of today's modern power systems. In practical power systems, there exist a number of poorly damped electromechanical modes of oscillations, and these oscillations between interconnected synchronous generators are inherent. The stability of these oscillations is of vital concern for secure system operation. The eigenvalue analysis techniques are widely applied in the literatures [1–4] to the analysis of dynamic or small-signal performance of power systems.

The formulation of state equations for small-signal stability analysis in a multi-generator system involves the development of linearized equations about any operating point and eliminating all variables other than the desired state variables. The general procedure is similar to that used for a single-machine infinite-bus system in Chapter 4. However, the need to allow for the representation of the extensive transmission networks, loads, excitation systems, prime movers, etc., makes the system very complex.

In this chapter, the *two-axis multimachine model with IEEE Type I exciter* considering all network bus dynamics is taken for small-signal stability analysis [1]. A detailed description of the method of multimachine simulations and case studies are illustrated as follows.

5.2 MULTIMACHINE SMALL-SIGNAL MODEL

To formulate multimachine small-signal model, the following assumptions are made without loss of generality:

(i) The stator and the network transient are neglected.
(ii) The turbine governor dynamics are neglected resulting in constant mechanical torque T_{M_i} ($i =$ no. of machines).

(iii) The limit constraints on AVR output (V_{R_i}) are deleted as the focus of interest is on modeling and simulation.

(iv) The damping torque $T_{F_i} = D_i(\omega_i - \omega_s)$ is assumed linear.

5.2.1 Two-axis model of multimachine system

- **The differential-algebraic equations**

$$\frac{d\delta_i}{dt} = \omega_i - \omega_s \tag{5.1}$$

$$\frac{d\omega_i}{dt} = \frac{T_{M_i}}{M_i} - \frac{\left[E'_{q_i} - X'_{d_i}I_{d_i}\right]I_{q_i}}{M_i} - \frac{\left[E'_{d_i} + X'_{q_i}I_{q_i}\right]I_{d_i}}{M_i}$$

$$- \frac{D_i(\omega_i - \omega_s)}{M_i} \tag{5.2}$$

$$\frac{dE'_{q_i}}{dt} = -\frac{E'_{q_i}}{T'_{do_i}} - \frac{\left(X_{d_i} - X'_{d_i}\right)I_{d_i}}{T'_{do_i}} + \frac{E_{fd_i}}{T'_{do_i}} \tag{5.3}$$

$$\frac{dE'_{d_i}}{dt} = -\frac{E'_{d_i}}{T'_{qo_i}} + \frac{I_{q_i}}{T'_{qo_i}}\left(X_{q_i} - X'_{q_i}\right) \tag{5.4}$$

$$\frac{dE_{fd_i}}{dt} = -\frac{K_{E_i} + S_{E_i}(E_{fd_i})}{T_{E_i}}E_{fd_i} + \frac{V_{R_i}}{T_{E_i}} \tag{5.5}$$

$$\frac{dV_{R_i}}{dt} = -\frac{V_{R_i}}{T_{A_i}} + \frac{K_{A_i}}{T_{A_i}}R_{F_i} - \frac{K_{A_i}K_{F_i}}{T_{A_i}T_{F_i}}E_{fd_i} + \frac{K_{A_i}}{T_{A_i}}(V_{ref_i} - V_i) \tag{5.6}$$

$$\frac{dR_{F_i}}{dt} = -\frac{R_{F_i}}{T_{F_i}} + \frac{K_{F_i}}{(T_{F_i})^2}E_{fd_i} \tag{5.7}$$

for $i = 1, 2, 3, \ldots, m$.

Equation (5.2) has dimensions of torque in pu. When the stator transients are neglected, the electric torque becomes equal to the per unit power associated with the internal voltage source.

- **Algebraic equations**

The algebraic equations consist of the stator algebraic equations and the network equations. The stator algebraic equations directly follow from the synchronous

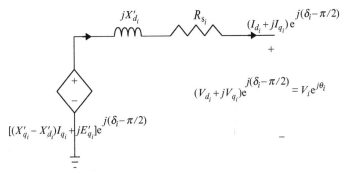

FIGURE 5.1

Dynamic circuit synchronous machine two-axis model ($i = 1, 2, 3, \ldots, m$).

machine dynamic equivalent circuit (Figure 5.1). The application of KVL yields the stator algebraic equations:

(i) Stator algebraic equations

$$0 = V_i e^{j\theta_i} + \left(R_{s_i} + jX'_{d_i}\right)\left(I_{d_i} + jI_{q_i}\right)e^{j(\delta_i - \pi/2)}$$
$$- \left[E'_{d_i} + \left(X'_{q_i} - X'_{d_i}\right)I_{q_i} + jE'_{q_i}\right]e^{j(\delta_i - \pi/2)} \tag{5.8}$$

for $i = 1, 2, 3, \ldots, m$.

The stator algebraic equations in polar form then are

$$E'_{d_i} - V_i \sin\left(\delta_i - \theta_i\right) - R_{s_i}I_{q_i} + X'_{q_i}I_{q_i} = 0 \tag{5.9}$$

$$E'_{q_i} - V_i \cos\left(\delta_i - \theta_i\right) - R_{s_i}I_{d_i} - X'_{d_i}I_{d_i} = 0 \tag{5.10}$$

(ii) Network equations

The dynamic circuit together with the stator networks and the loads is given in Figure 3.8, Chapter 3. The network equations for an n-bus system can be written in complex form.

The network equations for the *generator buses* are

$$V_i e^{j\theta_i} + \left(I_{d_i} - I_{q_i}\right)e^{-j(\delta_i - \pi/2)} + P_{L_i}(V_i) + jQ_{L_i}(V_i) = \sum_{k=1}^{n} V_i V_k Y_{ik} e^{j(\theta_i - \theta_k - \alpha_{ik})}$$

for $i = 1, 2, 3, \ldots, m$.   (5.11)

The network equations for the *load buses* are

$$P_{L_i}(V_i) + jQ_{L_i}(V_i) = \sum_{k=1}^{n} V_i V_k Y_{ik} e^{j(\theta_i - \theta_k - \alpha_{ik})} \quad \text{for } i = m+1, m+2, m+3, \ldots, n.$$

$$\tag{5.12}$$

Again, $V_i e^{j\theta_i} + (I_{d_i} - I_{q_i})e^{-j(\delta_i - \pi/2)} = P_{G_i} + jQ_{G_i}$ is the complex power "injected" into bus i due to the generator. Thus, Equations (5.11) and (5.12) are only the real and reactive power-balance equation at all the n buses.

Equations (5.11) and (5.12) can be written further in the power-balance form as

$$I_{d_i} V_i \sin(\delta_i - \theta_i) + I_{q_i} V_i \cos(\delta_i - \theta_i) + P_{L_i}(V_i) - \sum_{k=1}^{n} V_i V_k Y_{ik} \cos(\theta_i - \theta_k - \alpha_{ik}) = 0$$

$$(5.13)$$

$$I_{d_i} V_i \cos(\delta_i - \theta_i) - I_{q_i} V_i \sin(\delta_i - \theta_i) + Q_{L_i}(V_i) - \sum_{k=1}^{n} V_i V_k Y_{ik} \sin(\theta_i - \theta_k - \alpha_{ik}) = 0$$

$$(5.14)$$

where $i = 1, 2, 3, \ldots, m$ for generator buses and

$$P_{L_i}(V_i) - \sum_{k=1}^{n} V_i V_k Y_{ik} \cos(\theta_i - \theta_k - \alpha_{ik}) = 0 \qquad (5.15)$$

$$Q_{L_i}(V_i) - \sum_{k=1}^{n} V_i V_k Y_{ik} \sin(\theta_i - \theta_k - \alpha_{ik}) = 0 \qquad (5.16)$$

where $i = m+1, m+2, m+3, \ldots, n$ for load buses.
Thus, there are

(i) seven differential equations for each machine, i.e., $7m$ differential equations (5.1)–(5.7);
(ii) two real equations of stator for each machine, i.e., $2m$ real equations (5.9) and (5.10);
(iii) two real equations for each generator bus and two real equations for each load bus, i.e., $2n$ real equations from (5.13) to (5.16).

Therefore, Equations (5.1)–(5.16) constitute $7m + 2m + 2n$ equations with $X = \begin{bmatrix} X_1^T & X_2^T & \ldots & X_m^T \end{bmatrix}^T$ as the state vector where $X_i = \begin{bmatrix} \delta_i & \omega_i & E'_{q_i} & E'_{d_i} & E_{fd_i} & V_{R_i} & R_{F_i} \end{bmatrix}^T$ represent the state vectors for each machine.

$Y = \begin{bmatrix} I_{d-q}^T & V^T & \theta^T \end{bmatrix}^T$ is the set of algebraic variables, where $I_{d-q} = \begin{bmatrix} I_{d_1} & I_{q_1} & I_{d_2} & I_{q_2} & \ldots & I_{d_m} & I_{qm} \end{bmatrix}^T$, $V = \begin{bmatrix} V_1 & V_2 & \ldots & V_n \end{bmatrix}^T$, and $\theta = \begin{bmatrix} \theta_1 & \theta_2 & \ldots & \theta_n \end{bmatrix}^T$.

Therefore, the differential equations, together with the stator algebraic equations and the network equations, form a set of differential-algebraic equations (DAEs) as

$$\dot{X} = f(X, Y, U) \qquad (5.17)$$

$$0 = g(X, Y) \qquad (5.18)$$

where $U = \begin{bmatrix} U_1^\mathrm{T} & U_2^\mathrm{T} & \dots & U_m^\mathrm{T} \end{bmatrix}^\mathrm{T}$ with $U_i = \begin{bmatrix} \omega_\mathrm{s} & T_{\mathrm{M}_i} & V_{\mathrm{ref}_i} \end{bmatrix}^\mathrm{T}$ as the input vectors for each machine. Equation (5.17) is the dimension of $7m$ and Equation (5.18) is of dimension $2(m+n)$.

5.2.2 Linearization process and multimachine state-space model

Linearization of the *DAEs* (5.1)–(5.7) about any operating point yields

$$\Delta \dot{\delta}_i = \Delta \omega_i \tag{5.19}$$

$$\Delta \dot{\omega}_i = \frac{1}{M_i}\Delta T_{\mathrm{M}_i} - \frac{E'_{q_i}}{M_i}\Delta I_{q_i} + \frac{X'_{d_i} I_{d_i}}{M_i}\Delta I_{q_i} + \frac{X'_{d_i} I_{q_i}}{M_i}\Delta I_{d_i}$$

$$- \frac{I_{q_i}}{M_i}\Delta E'_{q_i} - \frac{E'_{d_i}}{M_i}\Delta I_{d_i} - \frac{I_{d_i}}{M_i}\Delta E'_{d_i} - \frac{X'_{q_i} I_{d_i}}{M_i}\Delta I_{q_i}$$

$$- \frac{X'_{q_i} I_{q_i}}{M_i}\Delta I_{d_i} - \frac{D_i}{M_i}\Delta \omega_i \tag{5.20}$$

$$\Delta \dot{E}'_{q_i} = -\frac{\Delta E'_{q_i}}{T'_{d_i}} - \frac{\left(X_{d_i} - X'_{d_i}\right)\Delta I_{d_i}}{T'_{d_i}} + \frac{\Delta E_{\mathrm{fd}_i}}{T'_{d_i}} \tag{5.21}$$

$$\Delta \dot{E}'_{d_i} = -\frac{\Delta E'_{d_i}}{T'_{q_i}} - \frac{\left(X_{q_i} - X'_{q_i}\right)\Delta I_{q_i}}{T'_{q_i}} \tag{5.22}$$

$$\Delta \dot{E}_{\mathrm{fd}_i} = f_{\mathrm{s}_i}(E_{\mathrm{fd}_i})\Delta E_{\mathrm{fd}_i} + \frac{\Delta V_{\mathrm{R}_i}}{T_{\mathrm{E}_i}} \tag{5.23}$$

$$\Delta \dot{V}_{\mathrm{R}_i} = -\frac{\Delta V_{\mathrm{R}_i}}{T_{\mathrm{A}_i}} + \frac{K_{\mathrm{A}_i}}{T_{\mathrm{A}_i}}\Delta R_{\mathrm{F}_i} - \frac{K_{\mathrm{A}_i} K_{\mathrm{F}_i}}{T_{\mathrm{A}_i} T_{\mathrm{F}_i}}\Delta E_{\mathrm{fd}_i}$$

$$- \frac{K_{\mathrm{A}_i}}{T_{\mathrm{A}_i}}\Delta V_i + \frac{K_{\mathrm{A}_i}}{T_{\mathrm{A}_i}}\Delta V_{\mathrm{ref}_i} \tag{5.24}$$

$$\Delta \dot{R}_{\mathrm{F}_i} = -\frac{\Delta R_{\mathrm{F}_i}}{T_{\mathrm{F}_i}} + \frac{K_{\mathrm{F}_i}}{\left(T_{\mathrm{F}_i}\right)^2}\Delta E_{\mathrm{fd}_i} \tag{5.25}$$

for $i = 1, 2, 3, \dots, m$ (no. of machines).

Here, $f_{\mathrm{s}_i}(E_{\mathrm{fd}_i}) = -\dfrac{K_{\mathrm{E}_i} + E_{\mathrm{fd}_i}\partial S_{\mathrm{E}}(E_{\mathrm{fd}_i}) + S_{\mathrm{E}}(E_{\mathrm{fd}_i})}{T_{\mathrm{E}_i}}$, where the symbol ∂ stands for partial derivative.

Writing Equations (5.19)–(5.25) together in matrix form,

$$
\begin{bmatrix} \Delta\dot{\delta}_i \\ \Delta\dot{\omega}_i \\ \Delta\dot{E}'_{q_i} \\ \Delta\dot{E}'_{d_i} \\ \Delta\dot{E}_{\mathrm{fd}_i} \\ \Delta\dot{V}_{\mathrm{R}_i} \\ \Delta\dot{R}_{\mathrm{F}_i} \end{bmatrix}
=
\begin{bmatrix}
0 & 1 & 0 & 0 & 0 & 0 & 0 \\
0 & -\dfrac{D_i}{M_i} & -\dfrac{I_{q_i}}{M_i} & -\dfrac{I_{d_i}}{M_i} & 0 & 0 & 0 \\
0 & 0 & -\dfrac{1}{T'_{\mathrm{do}_i}} & 0 & \dfrac{1}{T'_{\mathrm{do}}} & 0 & 0 \\
0 & 0 & 0 & -\dfrac{1}{T'_{\mathrm{qo}_i}} & 0 & 0 & 0 \\
0 & 0 & 0 & 0 & f_{\mathrm{s}_i}(E_{\mathrm{fd}_i}) & \dfrac{1}{T_{\mathrm{E}_i}} & 0 \\
0 & 0 & 0 & 0 & -\dfrac{K_{\mathrm{A}_i}K_{\mathrm{F}_i}}{T_{\mathrm{A}_i}T_{\mathrm{F}_i}} & -\dfrac{1}{T_{\mathrm{A}_i}} & \dfrac{K_{\mathrm{A}_i}}{T_{\mathrm{A}_i}} \\
0 & 0 & 0 & 0 & \dfrac{K_{\mathrm{F}_i}}{(T_{\mathrm{F}_i})^2} & 0 & -\dfrac{1}{T_{\mathrm{F}_i}}
\end{bmatrix}
\begin{bmatrix} \Delta\delta_i \\ \Delta\omega_i \\ \Delta E'_{q_i} \\ \Delta E'_{d_i} \\ \Delta E_{\mathrm{fd}_i} \\ \Delta V_{\mathrm{R}_i} \\ \Delta R_{\mathrm{F}_i} \end{bmatrix}
$$

$$
+
\begin{bmatrix}
0 & 0 \\
\dfrac{I_{q_i}\left(X'_{d_i}-X'_{q_i}\right)-E'_{d_i}}{M_i} & \dfrac{I_{d_i}\left(X'_{d_i}-X'_{q_i}\right)-E'_{q_i}}{M_i} \\
-\dfrac{\left(X_{d_i}-X'_{d_i}\right)}{T'_{\mathrm{do}_i}} & 0 \\
0 & \dfrac{\left(X_{q_i}-X'_{q_i}\right)}{T'_{\mathrm{qo}_i}} \\
0 & 0 \\
0 & 0 \\
0 & 0
\end{bmatrix}
\begin{bmatrix} \Delta I_{d_i} \\ \Delta I_{q_i} \end{bmatrix}
$$

$$
+
\begin{bmatrix}
0 & 0 \\
0 & 0 \\
0 & 0 \\
0 & 0 \\
0 & 0 \\
0 & -\dfrac{K_{\mathrm{A}_i}}{T_{\mathrm{A}_i}} \\
0 & 0
\end{bmatrix}
\begin{bmatrix} \Delta\theta_i \\ \Delta V_i \end{bmatrix}
+
\begin{bmatrix}
0 & 0 \\
\dfrac{1}{M_i} & 0 \\
0 & 0 \\
0 & 0 \\
0 & 0 \\
0 & -\dfrac{K_{\mathrm{A}_i}}{T_{\mathrm{A}_i}} \\
0 & 0
\end{bmatrix}
\begin{bmatrix} \Delta T_{\mathrm{M}_i} \\ \Delta V_{\mathrm{ref}_i} \end{bmatrix}
\qquad (5.26)
$$

for $i=1, 2, 3,\ldots, m$ (no. of machines). Here, $M_i = \dfrac{2H_i}{\omega_{\mathrm{s}_i}}$.

Denoting $\left[\Delta I_{d_i} \quad \Delta I_{q_i}\right]^{\mathrm{T}} = \Delta I_{g_i}$, and $\left[\Delta T_{M_i} \quad \Delta V_{\mathrm{ref}_i}\right]^{\mathrm{T}} = \Delta U_i$, Equation (5.26) can be written as

$$\Delta \dot{X}_i = A_{1i}\Delta X_i + B_{1i}\Delta I_{g_i} + B_{2i}\Delta V_{g_i} + E_{1i}\Delta U_i \tag{5.27}$$

i.e., for the m-machine system,

$$\Delta \dot{X} = A_1\Delta X + B_1\Delta I_g + B_2\Delta V_g + E_1\Delta U \tag{5.28}$$

where $[A_1]_{7m \times 7m}$, $[B_1]_{7m \times 2m}$, $[B_2]_{7m \times 2m}$, and $[E_1]_{7m \times 2m}$ are the block-diagonal matrices of the form as represented by

$$A_1 = \begin{bmatrix} A_{11} & 0 & 0 & \cdots & 0 \\ 0 & A_{12} & 0 & \cdots & 0 \\ 0 & 0 & A_{13} & \cdots & 0 \\ \cdot & \cdot & \cdot & \cdot & \cdot \\ \cdot & \cdot & \cdot & \cdot & \cdot \\ \cdot & \cdot & \cdot & \cdot & \cdot \\ 0 & 0 & 0 & \cdots & A_{1m} \end{bmatrix}_{7m \times 7m} \quad B_1 = \begin{bmatrix} B_{21} & 0 & 0 & \cdots & 0 \\ 0 & B_{22} & 0 & \cdots & 0 \\ 0 & 0 & B_{23} & \cdots & 0 \\ \cdot & \cdot & \cdot & \cdot & \cdot \\ \cdot & \cdot & \cdot & \cdot & \cdot \\ \cdot & \cdot & \cdot & \cdot & \cdot \\ 0 & 0 & 0 & \cdots & B_{2m} \end{bmatrix}_{7m \times 2m}$$

$$B_2 = \begin{bmatrix} B_{11} & 0 & 0 & \cdots & 0 \\ 0 & B_{12} & 0 & \cdots & 0 \\ 0 & 0 & B_{13} & \cdots & 0 \\ \cdot & \cdot & \cdot & \cdot & \cdot \\ \cdot & \cdot & \cdot & \cdot & \cdot \\ \cdot & \cdot & \cdot & \cdot & \cdot \\ 0 & 0 & 0 & \cdots & B_{1m} \end{bmatrix}_{7m \times 2m} \quad E_1 = \begin{bmatrix} E_{11} & 0 & 0 & \cdots & 0 \\ 0 & E_{12} & 0 & \cdots & 0 \\ 0 & 0 & E_{13} & \cdots & 0 \\ \cdot & \cdot & \cdot & \cdot & \cdot \\ \cdot & \cdot & \cdot & \cdot & \cdot \\ \cdot & \cdot & \cdot & \cdot & \cdot \\ 0 & 0 & 0 & \cdots & E_{1m} \end{bmatrix}_{7m \times 2m}$$

Linearization of the *stator algebraic equations* (5.9) and (5.10) gives

$$\Delta E'_{d_i} - \sin(\delta_i - \theta_i)\Delta V_i - V_i\cos(\delta_i - \theta_i)\Delta\delta_i + V_i\cos \\ (\delta_i - \theta_i)\Delta\theta_i - R_{s_i}\Delta I_{d_i} + X'_{q_i}\Delta I_{q_i} = 0 \tag{5.29}$$

$$\Delta E'_{q_i} - \cos(\delta_i - \theta_i)\Delta V_i + V_i\sin(\delta_i - \theta_i)\Delta\delta_i - V_i\sin(\delta_i - \theta_i)\Delta\theta_i \\ - R_{s_i}\Delta I_{q_i} - X'_{d_i}\Delta I_{d_i} = 0 \tag{5.30}$$

Writing Equations (5.29) and (5.30) in matrix notation,

$$
\begin{bmatrix} -V_i\cos(\delta_i-\theta_i) & 0 & 0 & 1 & 0 & 0 & 0 \\ V_i\sin(\delta_i-\theta_i) & 0 & 1 & 0 & 0 & 0 & 0 \end{bmatrix}
\begin{bmatrix} \Delta\delta_i \\ \Delta\omega_i \\ \Delta E'_{q_i} \\ \Delta E'_{d_i} \\ \Delta E_{fd_i} \\ \Delta V_{R_i} \\ \Delta R_{F_i} \end{bmatrix}
\tag{5.31}
$$

$$
+\begin{bmatrix} -R_{s_i} & X'_{q_i} \\ -X'_{d_i} & -R_{s_i} \end{bmatrix}
\begin{bmatrix} \Delta I_{d_i} \\ \Delta I_{q_i} \end{bmatrix}
$$

$$
+\begin{bmatrix} V_i\cos(\delta_i-\theta_i) & -\sin(\delta_i-\theta_i) \\ -V_i\sin(\delta_i-\theta_i) & -\cos(\delta_i-\theta_i) \end{bmatrix}
\begin{bmatrix} \Delta\theta_i \\ \Delta V_i \end{bmatrix} = 0
$$

for $i=1, 2, 3,\ldots,m$, which can be further represented as

$$
0 = C_{1i}\Delta X_i + D_{1i}\Delta I_{g_i} + D_{2i}\Delta V_{g_i}
\tag{5.32}
$$

i.e., for the m-machine system Equation (5.32) can be written as

$$
0 = C_1\Delta X + D_1\Delta I_g + D_2\Delta V_g
\tag{5.33}
$$

where $[C_1]_{(2m)\times(7m)}$, $[D_1]_{(2m)\times(2m)}$, and $[D_2]_{(2m)\times(2m)}$ are block-diagonal matrices of the form as shown:

$$
C_1 = \begin{bmatrix}
C_{11} & 0 & 0 & \cdots & 0 \\
0 & C_{12} & 0 & \cdots & 0 \\
0 & 0 & C_{13} & \cdots & 0 \\
\cdot & \cdot & \cdot & \cdot & \cdot \\
\cdot & \cdot & \cdot & \cdot & \cdot \\
\cdot & \cdot & \cdot & \cdot & \cdot \\
0 & 0 & 0 & \cdots & C_{1m}
\end{bmatrix}_{2m\times 7m}
\qquad
D_1 = \begin{bmatrix}
D_{11} & 0 & 0 & \cdots & 0 \\
0 & D_{12} & 0 & \cdots & 0 \\
0 & 0 & D_{13} & \cdots & 0 \\
\cdot & \cdot & \cdot & \cdot & \cdot \\
\cdot & \cdot & \cdot & \cdot & \cdot \\
\cdot & \cdot & \cdot & \cdot & \cdot \\
0 & 0 & 0 & \cdots & D_{1m}
\end{bmatrix}_{2m\times 2m}
$$

$$
\text{and } D_2 = \begin{bmatrix}
D_{21} & 0 & 0 & \cdots & 0 \\
0 & D_{22} & 0 & \cdots & 0 \\
0 & 0 & D_{23} & \cdots & 0 \\
\cdot & \cdot & \cdot & \cdot & \cdot \\
\cdot & \cdot & \cdot & \cdot & \cdot \\
\cdot & \cdot & \cdot & \cdot & \cdot \\
0 & 0 & 0 & \cdots & D_{2m}
\end{bmatrix}_{2m\times 2m}
$$

Linearization of the network *equations* that pertain to *generator buses* (PV buses) gives

$$0 = \left(I_{d_i}V_i\cos\left(\delta_i - \theta_i\right) - I_{q_i}V_i\sin\left(\delta_i - \theta_i\right)\right)\Delta\delta_i$$

$$+V_i\sin\left(\delta_i - \theta_i\right)\Delta I_{d_i} + V_i\cos\left(\delta_i - \theta_i\right)\Delta I_{q_i}$$

$$+\left(I_{d_i}\sin\left(\delta_i - \theta_i\right) + I_{q_i}\cos\left(\delta_i - \theta_i\right) - \left[\sum_{k=1}^{n}V_kY_{ik}\cos\left(\theta_i - \theta_k - \alpha_{ik}\right)\right] + \frac{\partial P_{L_i}(V_i)}{\partial V_i}\right)\Delta V_i$$

$$+\left(-I_{d_i}V_i\cos\left(\delta_i - \theta_i\right) + I_{q_i}V_i\sin\left(\delta_i - \theta_i\right) + \left[V_i\sum_{\substack{k=1\\ \neq i}}^{n}V_kY_{ik}\sin\left(\theta_i - \theta_k - \alpha_{ik}\right)\right]\right)\Delta\theta_i$$

$$-V_i\sum_{k=1}^{n}[Y_{ik}\cos\left(\theta_i - \theta_k - \alpha_{ik}\right)]\Delta V_k - V_i\sum_{\substack{k=1\\ \neq i}}^{n}[V_kY_{ik}\sin\left(\theta_i - \theta_k - \alpha_{ik}\right)]\Delta\theta_k \quad (5.34)$$

$$0 = \left(-I_{d_i}V_i\sin\left(\delta_i - \theta_i\right) - I_{q_i}V_i\cos\left(\delta_i - \theta_i\right)\right)\Delta\delta_i + V_i\cos\left(\delta_i - \theta_i\right)\Delta I_{d_i} - V_i\sin\left(\delta_i - \theta_i\right)\Delta I_{q_i}$$

$$+\left(I_{d_i}\cos\left(\delta_i - \theta_i\right) - I_{q_i}\sin\left(\delta_i - \theta_i\right) - \left[\sum_{k=1}^{n}V_kY_{ik}\sin\left(\theta_i - \theta_k - \alpha_{ik}\right)\right] + \frac{\partial Q_{L_i}(V_i)}{\partial V_i}\right)\Delta V_i$$

$$+\left(I_{d_i}V_i\sin\left(\delta_i - \theta_i\right) + I_{q_i}V_i\cos\left(\delta_i - \theta_i\right) - \left[V_i\sum_{\substack{k=1\\ \neq i}}^{n}V_kY_{ik}\cos\left(\theta_i - \theta_k - \alpha_{ik}\right)\right]\right)\Delta\theta_i$$

$$-V_i\sum_{k=1}^{n}[Y_{ik}\sin\left(\theta_i - \theta_k - \alpha_{ik}\right)]\Delta V_k + V_i\sum_{\substack{k=1\\ \neq i}}^{n}[V_kY_{ik}\cos\left(\theta_i - \theta_k - \alpha_{ik}\right)]\Delta\theta_k$$

$$(5.35)$$

for $i = 1, 2, 3, \ldots, m$ (for the generator buses). Here, the load is treated as constant power type. Combining Equations (5.34) and (5.35) in matrix form gives

$$\begin{bmatrix} \{I_{d_i}V_i\cos\left(\delta_i - \theta_i\right) - I_{q_i}V_i\sin\left(\delta_i - \theta_i\right)\} & 0 & 0 & 0 & 0 & 0 & 0 \\ \{-I_{d_i}V_i\sin\left(\delta_i - \theta_i\right) - I_{q_i}V_i\cos\left(\delta_i - \theta_i\right)\} & 0 & 0 & 0 & 0 & 0 & 0 \end{bmatrix} \begin{bmatrix} \Delta\delta_i \\ \Delta\omega_i \\ \Delta E'_{q_i} \\ \Delta E'_{d_i} \\ \Delta E_{fd_i} \\ \Delta V_{R_i} \\ \Delta R_{F_i} \end{bmatrix}$$

$$+\begin{bmatrix} V_i\sin\left(\delta_i - \theta_i\right) & -V_i\cos\left(\delta_i - \theta_i\right) \\ V_i\cos\left(\delta_i - \theta_i\right) & -V_i\sin\left(\delta_i - \theta_i\right) \end{bmatrix}\begin{bmatrix} \Delta I_{d_i} \\ \Delta I_{q_i} \end{bmatrix}$$

$$+D_4(i,k)\Delta V_{g_i} + D_5(i,k)\Delta V_{l_i} = 0 \quad (5.36)$$

where in $D_4(i,k)$, $k=1, 2,\ldots,m$ (for generator buses) and in $D_5(i,k)$, $k=m+1$, $m+2,\ldots,n$ (for non generator buses).

As $[\Delta\theta_i \ \Delta V_i]^{\mathrm{T}} = \Delta V_{g_i}$, where in ΔV_{g_i}, $i=1, 2,\ldots,m$ (for generator buses) and $[\Delta\theta_i \ \Delta V_i]^{\mathrm{T}} = \Delta V_{l_i}$, where in ΔV_{l_i}, $i=m+1$, $m+2,\ldots,n$ (for nongenerator buses), Equation (5.36) can further be represented as

$$0 = C_{2i}\Delta X_i + D_{3i}\Delta I_{g_i} + D_4(i,k)\Delta V_{g_i} + D_5(i,k)\Delta V_{l_i} \tag{5.37}$$

This for m-machine system now becomes

$$0 = C_2\Delta X + D_3\Delta I_g + D_4\Delta V_g + D_5\Delta V_l \tag{5.38}$$

Here, $[C_2]_{2m\times 7m}$ and $[D_3]_{2m\times 2m}$ are block-diagonal matrices and $[D_4]_{2m\times 2m}$ and $[D_5]_{2m\times 2n}$ are full matrices. Therefore, Equation (5.38) in matrix form can be written as

$$
0 =
\begin{bmatrix}
C_{21} & 0 & 0 & \cdots & 0 \\
0 & C_{22} & 0 & \cdots & 0 \\
0 & 0 & C_{23} & \cdots & 0 \\
\cdot & \cdot & \cdot & \cdot & \cdot \\
\cdot & \cdot & \cdot & \cdot & \cdot \\
\cdot & \cdot & \cdot & \cdot & \cdot \\
0 & 0 & 0 & \cdots & C_{2m}
\end{bmatrix}
\begin{bmatrix}
\Delta X_1 \\
\Delta X_2 \\
\cdot \\
\cdot \\
\cdot \\
\cdot \\
\Delta X_m
\end{bmatrix}
+
\begin{bmatrix}
D_{31} & 0 & 0 & \cdots & 0 \\
0 & D_{32} & 0 & \cdots & 0 \\
0 & 0 & D_{33} & \cdots & 0 \\
\cdot & \cdot & \cdot & \cdot & \cdot \\
\cdot & \cdot & \cdot & \cdot & \cdot \\
\cdot & \cdot & \cdot & \cdot & \cdot \\
0 & 0 & 0 & \cdots & D_{3m}
\end{bmatrix}
\begin{bmatrix}
\Delta I_{g_1} \\
\Delta I_{g_2} \\
\cdot \\
\cdot \\
\cdot \\
\cdot \\
\Delta I_{g_m}
\end{bmatrix}
$$

$$
+
\begin{bmatrix}
D_{41,1} & D_{41,2} & \cdots & D_{41,m} \\
D_{42,1} & D_{42,2} & \cdots & D_{42,m} \\
\cdot & \cdot & \cdot & \cdot \\
\cdot & \cdot & \cdot & \cdot \\
\cdot & \cdot & \cdot & \cdot \\
D_{4m,1} & D_{4m,2} & \cdots & D_{4m,m}
\end{bmatrix}
\begin{bmatrix}
\Delta V_{g_1} \\
\Delta V_{g_2} \\
\cdot \\
\cdot \\
\cdot \\
\Delta V_{g_m}
\end{bmatrix}
$$

$$
+
\begin{bmatrix}
D_{51,m+1} & D_{51,m+2} & \cdots & D_{51,n} \\
D_{52,m+1} & D_{52,m+2} & \cdots & D_{52,n} \\
\cdot & \cdot & \cdot & \cdot \\
\cdot & \cdot & \cdot & \cdot \\
\cdot & \cdot & \cdot & \cdot \\
D_{5m,m+1} & D_{5m,m+2} & \cdots & D_{5m,n}
\end{bmatrix}
\begin{bmatrix}
\Delta V_{l_{m+1}} \\
\Delta V_{l_{m+2}} \\
\cdot \\
\cdot \\
\cdot \\
\Delta V_{l_n}
\end{bmatrix}
$$

$$\tag{5.39}$$

- The submatrices of D_4 and D_5 can be derived as follows:

To obtain the submatrices of the full matrices D_4 and D_5 from Equations (5.34) and (5.35), the terms involving the variables $\Delta\theta_i$, ΔV_i, $\Delta\theta_k$, and ΔV_k are only considered. Here, $i = 1, 2, \ldots, m$ (for generator buses) and $k = m+1, m+2, \ldots n$ (for non-generator buses).

For $i = 1$, Equation (5.34) gives

$$0 = \cdots + \left(-I_{d_i}V_i\cos(\delta_i-\theta_i) + I_{q_i}V_i\sin(\delta_i-\theta_i) + \left[V_i\sum_{\substack{k=1\\\neq i}}^{n}V_kY_{ik}\sin(\theta_i-\theta_k-\alpha_{ik})\right]\right)\Delta\theta_1$$

$$+ \left(I_{d_i}\sin(\delta_i-\theta_i) + I_{q_i}\cos(\delta_i-\theta_i) - \left[\sum_{k=1}^{n}V_kY_{ik}\cos(\theta_i-\theta_k-\alpha_{ik})\right] + \frac{\partial P_{L_i}(V_i)}{\partial V_i}\right)\Delta V_1$$

$$-V_iY_{ik}\cos(\theta_i-\theta_k-\alpha_{ik})\Delta V_1 \cdots - V_iY_{ik}\cos(\theta_i-\theta_k-\alpha_{ik})\Delta V_m$$

$$-V_iY_{ik}\cos(\theta_i-\theta_k-\alpha_{ik})\Delta V_{m+1} \cdots - V_iY_{ik}\cos(\theta_i-\theta_k-\alpha_{ik})\Delta V_n$$

$$-V_iV_kY_{ik}\sin(\theta_i-\theta_k-\alpha_{ik})\Delta\theta_2 - V_iV_kY_{ik}\sin(\theta_i-\theta_k-\alpha_{ik})\Delta\theta_3 \cdots$$

$$-V_iV_kY_{ik}\sin(\theta_i-\theta_k-\alpha_{ik})\Delta\theta_m - V_iV_kY_{ik}\sin(\theta_i-\theta_k-\alpha_{ik})\Delta\theta_{m+1} \cdots$$

$$-V_iV_kY_{ik}\sin(\theta_i-\theta_k-\alpha_{ik})\Delta\theta_n$$

$$(5.40)$$

Rearranging the terms containing the variables $\Delta V_{g_1} \ldots \Delta V_{g_m}$ and $\Delta V_{g_{m+1}} \ldots \Delta V_{g_n}$, Equation (5.40) can be written as

$$0 = \cdots + \left[-I_{d_i}V_i\cos(\delta_i-\theta_i) + I_{q_i}V_i\sin(\delta_i-\theta_i) + \left[V_i\sum_{\substack{k=1\\\neq i}}^{n}V_kY_{ik}\sin(\theta_i-\theta_k-\alpha_{ik})\right]\right.$$

$$I_{d_i}\sin(\delta_i-\theta_i) + I_{q_i}\cos(\delta_i-\theta_i) - \left[\sum_{k=1}^{n}V_kY_{ik}\cos(\theta_i-\theta_k-\alpha_{ik})\right]$$

$$\left.-V_iY_{ik}\cos(\theta_i-\theta_k-\alpha_{ik}) + \frac{\partial P_{L_i}(V_i)}{\partial V_i}\right]\begin{bmatrix}\Delta\theta_1\\\Delta V_1\end{bmatrix}$$

$$+[-V_iV_kY_{ik}\sin(\theta_i-\theta_k-\alpha_{ik}) \quad -V_iY_{ik}\cos(\theta_i-\theta_k-\alpha_{ik})]\begin{bmatrix}\Delta\theta_2\\\Delta V_2\end{bmatrix} + \cdots$$

$$+[-V_iV_kY_{ik}\sin(\theta_i-\theta_k-\alpha_{ik}) \quad -V_iY_{ik}\cos(\theta_i-\theta_k-\alpha_{ik})]\begin{bmatrix}\Delta\theta_m\\\Delta V_m\end{bmatrix} +$$

$$+[-V_iV_kY_{ik}\sin(\theta_i-\theta_k-\alpha_{ik}) \quad -V_iY_{ik}\cos(\theta_i-\theta_k-\alpha_{ik})]\begin{bmatrix}\Delta\theta_{m+1}\\\Delta V_{m+1}\end{bmatrix} + \cdots$$

$$+[-V_iV_kY_{ik}\sin(\theta_i-\theta_k-\alpha_{ik}) \quad -V_iY_{ik}\cos(\theta_i-\theta_k-\alpha_{ik})]\begin{bmatrix}\Delta\theta_n\\\Delta V_n\end{bmatrix}$$

$$(5.41)$$

Similarly, for $i=1$, Equation (5.35) gives

$$0 = \cdots + \left(I_{d_i} V_i \sin(\delta_i - \theta_i) + I_{q_i} V_i \cos(\delta_i - \theta_i) - \left[V_i \sum_{\substack{k=1 \\ \neq i}}^{n} V_k Y_{ik} \cos(\theta_i - \theta_k - \alpha_{ik}) \right] \right) \Delta\theta_1$$

$$+ \left(I_{d_i} \cos(\delta_i - \theta_i) - I_{q_i} \sin(\delta_i - \theta_i) - \left[\sum_{k=1}^{n} V_k Y_{ik} \sin(\theta_i - \theta_k - \alpha_{ik}) \right] + \frac{\partial Q_{L_i}(V_i)}{\partial V_i} \right) \Delta V_1$$

$$- V_i Y_{ik} \sin(\theta_i - \theta_k - \alpha_{ik}) \Delta V_1 \cdots - V_i Y_{ik} \sin(\theta_i - \theta_k - \alpha_{ik}) \Delta V_m$$

$$- V_i Y_{ik} \sin(\theta_i - \theta_k - \alpha_{ik}) \Delta V_{m+1} \cdots - V_i Y_{ik} \sin(\theta_i - \theta_k - \alpha_{ik}) \Delta V_n$$

$$+ V_i V_k Y_{ik} \cos(\theta_i - \theta_k - \alpha_{ik}) \Delta\theta_2 + V_i V_k Y_{ik} \cos(\theta_i - \theta_k - \alpha_{ik}) \Delta\theta_3 + \cdots$$

$$+ V_i V_k Y_{ik} \cos(\theta_i - \theta_k - \alpha_{ik}) \Delta\theta_m + V_i V_k Y_{ik} \cos(\theta_i - \theta_k - \alpha_{ik}) \Delta\theta_{m+1} + \cdots$$

$$+ V_i V_k Y_{ik} \cos(\theta_i - \theta_k - \alpha_{ik}) \Delta\theta_n$$

$$\text{(5.42)}$$

Rearranging the terms containing the variables $\Delta V_{g_1} \ldots \Delta V_{g_m}$ and $\Delta V_{g_{m+1}} \ldots \Delta V_{g_n}$, Equation (5.42) can be written as

$$0 = \cdots + \left[I_{d_i} V_i \sin(\delta_i - \theta_i) + I_{q_i} V_i \cos(\delta_i - \theta_i) - \left[V_i \sum_{\substack{k=1 \\ \neq i}}^{n} V_k Y_{ik} \cos(\theta_i - \theta_k - \alpha_{ik}) \right] \right.$$

$$I_{d_i} \cos(\delta_i - \theta_i) - I_{q_i} \sin(\delta_i - \theta_i) - \left[\sum_{k=1}^{n} V_k Y_{ik} \sin(\theta_i - \theta_k - \alpha_{ik}) \right]$$

$$\left. - V_i Y_{ik} \sin(\theta_i - \theta_k - \alpha_{ik}) + \frac{\partial Q_{L_i}(V_i)}{\partial V_i} \right] \begin{bmatrix} \Delta\theta_1 \\ \Delta V_1 \end{bmatrix}$$

$$+ [V_i V_k Y_{ik} \cos(\theta_i - \theta_k - \alpha_{ik}) \quad - V_i Y_{ik} \sin(\theta_i - \theta_k - \alpha_{ik})] \begin{bmatrix} \Delta\theta_2 \\ \Delta V_2 \end{bmatrix} + \cdots$$

$$+ [V_i V_k Y_{ik} \cos(\theta_i - \theta_k - \alpha_{ik}) \quad - V_i Y_{ik} \sin(\theta_i - \theta_k - \alpha_{ik})] \begin{bmatrix} \Delta\theta_m \\ \Delta V_m \end{bmatrix}$$

$$+ [V_i V_k Y_{ik} \cos(\theta_i - \theta_k - \alpha_{ik}) \quad - V_i Y_{ik} \sin(\theta_i - \theta_k - \alpha_{ik})] \begin{bmatrix} \Delta\theta_{m+1} \\ \Delta V_{m+1} \end{bmatrix} + \cdots$$

$$+ [V_i V_k Y_{ik} \cos(\theta_i - \theta_k - \alpha_{ik}) \quad - V_i Y_{ik} \sin(\theta_i - \theta_k - \alpha_{ik})] \begin{bmatrix} \Delta\theta_n \\ \Delta V_n \end{bmatrix}$$

$$\text{(5.43)}$$

Combining Equations (5.41) and (5.43) together in matrix form for $i=1$ results in

$$
0 = \cdots + \begin{bmatrix} -I_{d_i}V_i\cos(\delta_i-\theta_i)+I_{q_i}V_i\sin(\delta_i-\theta_i)+\left[V_i\displaystyle\sum_{\substack{k=1\\\neq i}}^{n}V_kY_{ik}\sin(\theta_i-\theta_k-\alpha_{ik})\right] \\[4ex] I_{d_i}V_i\sin(\delta_i-\theta_i)+I_{q_i}V_i\cos(\delta_i-\theta_i)-\left[V_i\displaystyle\sum_{\substack{k=1\\\neq i}}^{n}V_kY_{ik}\cos(\theta_i-\theta_k-\alpha_{ik})\right] \end{bmatrix}
$$

$$
\begin{bmatrix} I_{d_i}\sin(\delta_i-\theta_i)+I_{q_i}\cos(\delta_i-\theta_i)-\left[\displaystyle\sum_{k=1}^{n}V_kY_{ik}\cos(\theta_i-\theta_k-\alpha_{ik})\right] \\ \qquad\qquad -V_iY_{ik}\cos(\theta_i-\theta_k-\alpha_{ik})+\dfrac{\partial P_{L_i}(V_i)}{\partial V_i} \\[3ex] I_{d_i}\cos(\delta_i-\theta_i)-I_{q_i}\sin(\delta_i-\theta_i)-\left[\displaystyle\sum_{k=1}^{n}V_kY_{ik}\sin(\theta_i-\theta_k-\alpha_{ik})\right] \\ \qquad\qquad -V_iY_{ik}\sin(\theta_i-\theta_k-\alpha_{ik})+\dfrac{\partial Q_{L_i}(V_i)}{\partial V_i} \end{bmatrix}\begin{bmatrix}\Delta\theta_1\\\Delta V_1\end{bmatrix}
$$

$$
+\begin{bmatrix} -V_iV_kY_{ik}\sin(\theta_i-\theta_k-\alpha_{ik}) & -V_iY_{ik}\cos(\theta_i-\theta_k-\alpha_{ik})\\ V_iV_kY_{ik}\cos(\theta_i-\theta_k-\alpha_{ik}) & -V_iY_{ik}\sin(\theta_i-\theta_k-\alpha_{ik})\end{bmatrix}\begin{bmatrix}\Delta\theta_2\\\Delta V_2\end{bmatrix}
$$

$$
+\begin{bmatrix} -V_iV_kY_{ik}\sin(\theta_i-\theta_k-\alpha_{ik}) & -V_iY_{ik}\cos(\theta_i-\theta_k-\alpha_{ik})\\ V_iV_kY_{ik}\cos(\theta_i-\theta_k-\alpha_{ik}) & -V_iY_{ik}\sin(\theta_i-\theta_k-\alpha_{ik})\end{bmatrix}\begin{bmatrix}\Delta\theta_m\\\Delta V_m\end{bmatrix}
$$

$$
+\begin{bmatrix} -V_iV_kY_{ik}\sin(\theta_i-\theta_k-\alpha_{ik}) & -V_iY_{ik}\cos(\theta_i-\theta_k-\alpha_{ik})\\ V_iV_kY_{ik}\cos(\theta_i-\theta_k-\alpha_{ik}) & -V_iY_{ik}\sin(\theta_i-\theta_k-\alpha_{ik})\end{bmatrix}\begin{bmatrix}\Delta\theta_{m+1}\\\Delta V_{m+1}\end{bmatrix}+\cdots
$$

$$
+\begin{bmatrix} -V_iV_kY_{ik}\sin(\theta_i-\theta_k-\alpha_{ik}) & -V_iY_{ik}\cos(\theta_i-\theta_k-\alpha_{ik})\\ V_iV_kY_{ik}\cos(\theta_i-\theta_k-\alpha_{ik}) & -V_iY_{ik}\sin(\theta_i-\theta_k-\alpha_{ik})\end{bmatrix}\begin{bmatrix}\Delta\theta_n\\\Delta V_n\end{bmatrix}
$$

$$
\tag{5.44}
$$

Therefore, for $i=1$ and $k=1, 2,\ldots,m,$ the submatrix of the *first row* of the D_4 matrix can be obtained from Equation (5.44). Similarly, for $i=1$ and $k=m+1, m+2,\ldots,n,$ another submatrix of the first row of the D_5 matrix can also be derived from Equation (5.44).

For $i = m$, Equation (5.34) gives

$$0 = \cdots + \left(-I_{d_i} V_i \cos(\delta_i - \theta_i) + I_{q_i} V_i \sin(\delta_i - \theta_i) + \left[V_i \sum_{\substack{k=1 \\ \neq i}}^{n} V_k Y_{ik} \sin(\theta_i - \theta_k - \alpha_{ik}) \right] \right) \Delta\theta_m$$

$$+ \left(I_{d_i} \sin(\delta_i - \theta_i) + I_{q_i} \cos(\delta_i - \theta_i) - \left[\sum_{k=1}^{n} V_k Y_{ik} \cos(\theta_i - \theta_k - \alpha_{ik}) \right] + \frac{\partial P_{L_i}(V_i)}{\partial V_i} \right) \Delta V_m$$

$$- V_i Y_{ik} \cos(\theta_i - \theta_k - \alpha_{ik}) \Delta V_1 - V_i Y_{ik} \cos(\theta_i - \theta_k - \alpha_{ik}) \Delta V_2 - \cdots$$

$$- V_i Y_{ik} \cos(\theta_i - \theta_k - \alpha_{ik}) \Delta V_m - V_i Y_{ik} \cos(\theta_i - \theta_k - \alpha_{ik}) \Delta V_{m+1} - \cdots$$

$$- V_i Y_{ik} \cos(\theta_i - \theta_k - \alpha_{ik}) \Delta V_n - V_i V_k Y_{ik} \sin(\theta_i - \theta_k - \alpha_{ik}) \Delta\theta_1$$

$$- V_i V_k Y_{ik} \sin(\theta_i - \theta_k - \alpha_{ik}) \Delta\theta_2 \cdots - V_i V_k Y_{ik} \sin(\theta_i - \theta_k - \alpha_{ik}) \Delta\theta_{m-1}$$

$$- V_i V_k Y_{ik} \sin(\theta_i - \theta_k - \alpha_{ik}) \Delta\theta_{m+1} \cdots - V_i V_k Y_{ik} \sin(\theta_i - \theta_k - \alpha_{ik}) \Delta\theta_n$$

$$(5.45)$$

Rearranging the terms according to the variables $\Delta V_{g_1} \ldots \Delta V_{g_m}$ and $\Delta V_{g_{m+1}} \ldots \Delta V_{g_n}$, Equation (5.45) can be written as

$$0 = \cdots + \left[-V_i V_k Y_{ik} \sin(\theta_i - \theta_k - \alpha_{ik}) \quad -V_i Y_{ik} \cos(\theta_i - \theta_k - \alpha_{ik}) \right] \begin{bmatrix} \Delta\theta_1 \\ \Delta V_1 \end{bmatrix}$$

$$+ \left[-V_i V_k Y_{ik} \sin(\theta_i - \theta_k - \alpha_{ik}) \quad -V_i Y_{ik} \cos(\theta_i - \theta_k - \alpha_{ik}) \right] \begin{bmatrix} \Delta\theta_2 \\ \Delta V_2 \end{bmatrix} + \cdots$$

$$+ \left[-I_{d_i} V_i \cos(\delta_i - \theta_i) + I_{q_i} V_i \sin(\delta_i - \theta_i) + \left[V_i \sum_{\substack{k=1 \\ \neq i}}^{n} V_k Y_{ik} \sin(\theta_i - \theta_k - \alpha_{ik}) \right] \right.$$

$$I_{d_i} \sin(\delta_i - \theta_i) + I_{q_i} \cos(\delta_i - \theta_i) - \left[\sum_{k=1}^{n} V_k Y_{ik} \cos(\theta_i - \theta_k - \alpha_{ik}) \right]$$

$$\left. -V_i Y_{ik} \cos(\theta_i - \theta_k - \alpha_{ik}) + \frac{\partial P_{L_i}(V_i)}{\partial V_i} \right] \begin{bmatrix} \Delta\theta_m \\ \Delta V_m \end{bmatrix} +$$

$$+ \left[-V_i V_k Y_{ik} \sin(\theta_i - \theta_k - \alpha_{ik}) \quad -V_i Y_{ik} \cos(\theta_i - \theta_k - \alpha_{ik}) \right] \begin{bmatrix} \Delta\theta_{m+1} \\ \Delta V_{m+1} \end{bmatrix} + \cdots$$

$$+ \left[-V_i V_k Y_{ik} \sin(\theta_i - \theta_k - \alpha_{ik}) \quad -V_i Y_{ik} \cos(\theta_i - \theta_k - \alpha_{ik}) \right] \begin{bmatrix} \Delta\theta_n \\ \Delta V_n \end{bmatrix}$$

$$(5.46)$$

Similarly, for $i = m$, Equation (5.35) gives

$$0 = \cdots + \left(I_{d_i} V_i \sin(\delta_i - \theta_i) + I_{q_i} V_i \cos(\delta_i - \theta_i) - \left[V_i \sum_{\substack{k=1 \\ \neq i}}^{n} V_k Y_{ik} \cos(\theta_i - \theta_k - \alpha_{ik}) \right] \right) \Delta\theta_m$$

$$+ \left(I_{d_i} \cos(\delta_i - \theta_i) - I_{q_i} \sin(\delta_i - \theta_i) - \left[\sum_{k=1}^{n} V_k Y_{ik} \sin(\theta_i - \theta_k - \alpha_{ik}) \right] + \frac{\partial Q_{L_i}(V_i)}{\partial V_i} \right) \Delta V_m$$

$$- V_i Y_{ik} \sin(\theta_i - \theta_k - \alpha_{ik}) \Delta V_1 - V_i Y_{ik} \sin(\theta_i - \theta_k - \alpha_{ik}) \Delta V_2 - \cdots$$

$$- V_i Y_{ik} \sin(\theta_i - \theta_k - \alpha_{ik}) \Delta V_m - V_i Y_{ik} \sin(\theta_i - \theta_k - \alpha_{ik}) \Delta V_{m+1} - \cdots$$

$$- V_i Y_{ik} \sin(\theta_i - \theta_k - \alpha_{ik}) \Delta V_n + V_i V_k Y_{ik} \cos(\theta_i - \theta_k - \alpha_{ik}) \Delta\theta_1$$

$$+ V_i V_k Y_{ik} \cos(\theta_i - \theta_k - \alpha_{ik}) \Delta\theta_2 + \cdots + V_i V_k Y_{ik} \cos(\theta_i - \theta_k - \alpha_{ik}) \Delta\theta_{m-1}$$

$$+ V_i V_k Y_{ik} \cos(\theta_i - \theta_k - \alpha_{ik}) \Delta\theta_{m+1} + \cdots + V_i V_k Y_{ik} \cos(\theta_i - \theta_k - \alpha_{ik}) \Delta\theta_n$$

$$(5.47)$$

Rearranging the terms associated with the variables $\Delta V_{g_1} \ldots \Delta V_{g_m}$ and $\Delta V_{g_{m+1}} \ldots \Delta V_{g_n}$, Equation (5.47) can be written as

$$0 = \cdots + \left[V_i V_k Y_{ik} \cos(\theta_i - \theta_k - \alpha_{ik}) \quad - V_i Y_{ik} \sin(\theta_i - \theta_k - \alpha_{ik}) \right] \begin{bmatrix} \Delta\theta_1 \\ \Delta V_1 \end{bmatrix} +$$

$$+ \left[V_i V_k Y_{ik} \cos(\theta_i - \theta_k - \alpha_{ik}) \quad - V_i Y_{ik} \sin(\theta_i - \theta_k - \alpha_{ik}) \right] \begin{bmatrix} \Delta\theta_2 \\ \Delta V_2 \end{bmatrix} + \cdots$$

$$+ \left[I_{d_i} V_i \sin(\delta_i - \theta_i) + I_{q_i} V_i \cos(\delta_i - \theta_i) - \left[V_i \sum_{\substack{k=1 \\ \neq i}}^{n} V_k Y_{ik} \cos(\theta_i - \theta_k - \alpha_{ik}) \right] \right.$$

$$I_{d_i} \cos(\delta_i - \theta_i) - I_{q_i} \sin(\delta_i - \theta_i) - \left[\sum_{k=1}^{n} V_k Y_{ik} \sin(\theta_i - \theta_k - \alpha_{ik}) \right]$$

$$\left. - V_i Y_{ik} \sin(\theta_i - \theta_k - \alpha_{ik}) + \frac{\partial Q_{L_i}(V_i)}{\partial V_i} \right] \begin{bmatrix} \Delta\theta_m \\ \Delta V_m \end{bmatrix}$$

$$+ \left[V_i V_k Y_{ik} \cos(\theta_i - \theta_k - \alpha_{ik}) \quad - V_i Y_{ik} \sin(\theta_i - \theta_k - \alpha_{ik}) \right] \begin{bmatrix} \Delta\theta_{m+1} \\ \Delta V_{m+1} \end{bmatrix} + \cdots$$

$$+ \left[V_i V_k Y_{ik} \cos(\theta_i - \theta_k - \alpha_{ik}) \quad - V_i Y_{ik} \sin(\theta_i - \theta_k - \alpha_{ik}) \right] \begin{bmatrix} \Delta\theta_n \\ \Delta V_n \end{bmatrix} \qquad (5.48)$$

Combining Equations (5.46) and (5.48) together in matrix form for $i=m$ results in

$$0 = \cdots + \begin{bmatrix} -V_iV_kY_{ik}\sin(\theta_i-\theta_k-\alpha_{ik}) & -V_iY_{ik}\cos(\theta_i-\theta_k-\alpha_{ik}) \\ V_iV_kY_{ik}\cos(\theta_i-\theta_k-\alpha_{ik}) & -V_iY_{ik}\sin(\theta_i-\theta_k-\alpha_{ik}) \end{bmatrix} \begin{bmatrix} \Delta\theta_1 \\ \Delta V_1 \end{bmatrix}$$

$$+ \begin{bmatrix} -V_iV_kY_{ik}\sin(\theta_i-\theta_k-\alpha_{ik}) & -V_iY_{ik}\cos(\theta_i-\theta_k-\alpha_{ik}) \\ V_iV_kY_{ik}\cos(\theta_i-\theta_k-\alpha_{ik}) & -V_iY_{ik}\sin(\theta_i-\theta_k-\alpha_{ik}) \end{bmatrix} \begin{bmatrix} \Delta\theta_2 \\ \Delta V_2 \end{bmatrix} + \cdots$$

$$+ \begin{bmatrix} -I_{d_i}V_i\cos(\delta_i-\theta_i)+I_{q_i}V_i\sin(\delta_i-\theta_i)+\left[V_i\displaystyle\sum_{\substack{k=1\\ \ne i}}^{n}V_kY_{ik}\sin(\theta_i-\theta_k-\alpha_{ik})\right] \\ I_{d_i}V_i\sin(\delta_i-\theta_i)+I_{q_i}V_i\cos(\delta_i-\theta_i)-\left[V_i\displaystyle\sum_{\substack{k=1\\ \ne i}}^{n}V_kY_{ik}\cos(\theta_i-\theta_k-\alpha_{ik})\right] \end{bmatrix}$$

$$\begin{bmatrix} I_{d_i}\sin(\delta_i-\theta_i)+I_{q_i}\cos(\delta_i-\theta_i)-\left[\displaystyle\sum_{k=1}^{n}V_kY_{ik}\cos(\theta_i-\theta_k-\alpha_{ik})\right] \\ -V_iY_{ik}\cos(\theta_i-\theta_k-\alpha_{ik})+\dfrac{\partial P_{L_i}(V_i)}{\partial V_i} \\ I_{d_i}\cos(\delta_i-\theta_i)-I_{q_i}\sin(\delta_i-\theta_i)-\left[\displaystyle\sum_{k=1}^{n}V_kY_{ik}\sin(\theta_i-\theta_k-\alpha_{ik})\right] \\ -V_iY_{ik}\sin(\theta_i-\theta_k-\alpha_{ik})+\dfrac{\partial Q_{L_i}(V_i)}{\partial V_i} \end{bmatrix} \begin{bmatrix} \Delta\theta_m \\ \Delta V_m \end{bmatrix}$$

$$+ \begin{bmatrix} -V_iV_kY_{ik}\sin(\theta_i-\theta_k-\alpha_{ik}) & -V_iY_{ik}\cos(\theta_i-\theta_k-\alpha_{ik}) \\ V_iV_kY_{ik}\cos(\theta_i-\theta_k-\alpha_{ik}) & -V_iY_{ik}\sin(\theta_i-\theta_k-\alpha_{ik}) \end{bmatrix} \begin{bmatrix} \Delta\theta_{m+1} \\ \Delta V_{m+1} \end{bmatrix} + \cdots$$

$$+ \begin{bmatrix} -V_iV_kY_{ik}\sin(\theta_i-\theta_k-\alpha_{ik}) & -V_iY_{ik}\cos(\theta_i-\theta_k-\alpha_{ik}) \\ V_iV_kY_{ik}\cos(\theta_i-\theta_k-\alpha_{ik}) & -V_iY_{ik}\sin(\theta_i-\theta_k-\alpha_{ik}) \end{bmatrix} \begin{bmatrix} \Delta\theta_n \\ \Delta V_n \end{bmatrix} \tag{5.49}$$

Thus, for $i=m$ and $k=1, 2,\ldots,m$, the submatrix of the *last row* of the D_4 matrix can be obtained from the equation. Again, for $i=m$ and $k=m+1, m+2,\ldots,n$, other submatrix of the *last row* of the D_5 matrix can also be computed from Equation (5.49).

The submatrices of D_4 and D_5 for the intermediate rows corresponding to $i=2,\ldots,(m-1)$ can be determined by the similar procedure.

Linearization of the network *equations* that pertain to *load buses* (*PQ* buses) results in

$$0 = \frac{\partial P_{1_i}(V_i)}{\partial V_i}\Delta V_i - \left[\sum_{k=1}^{n}V_kY_{ik}\cos(\theta_i-\theta_k-\alpha_{ik})\right]\Delta V_i$$

$$+ \left[\sum_{\substack{k=1\\ \ne i}}^{n}V_iV_kY_{ik}\sin(\theta_i-\theta_k-\alpha_{ik})\right]\Delta\theta_i - V_i\sum_{k=1}^{n}[Y_{ik}\cos(\theta_i-\theta_k-\alpha_{ik})]\Delta V_k$$

$$- V_i\sum_{\substack{k=1\\ \ne i}}^{n}[V_kY_{ik}\sin(\theta_i-\theta_k-\alpha_{ik})]\Delta\theta_k \tag{5.50}$$

$$0 = \frac{\partial Q_{1_i}(V_i)}{\partial V_i}\Delta V_i - \left[\sum_{k=1}^{n}V_kY_{ik}\sin\left(\theta_i - \theta_k - \alpha_{ik}\right)\right]\Delta V_i$$

$$- \left[\sum_{\substack{k=1\\ \neq i}}^{n}V_iV_kY_{ik}\cos\left(\theta_i - \theta_k - \alpha_{ik}\right)\right]\Delta\theta_i - V_i\sum_{k=1}^{n}[Y_{ik}\sin\left(\theta_i - \theta_k - \alpha_{ik}\right)]\Delta V_k$$

$$+ V_i\sum_{\substack{k=1\\ \neq i}}^{n}[V_kY_{ik}\cos\left(\theta_i - \theta_k - \alpha_{ik}\right)]\Delta\theta_k$$

$$(5.51)$$

for $i = m+1, m+2, \ldots, n$ (for the load buses). Equations (5.50) and (5.51) together can be represented as

$$0 = D_6(i,k)\Delta V_{g_i} + D_7(i,k)\Delta V_{1_i} \qquad (5.52)$$

where in $D_6(i,k)$, $k = 1, 2, \ldots, m$ (for generator buses) and in $D_7(i,k)$, $k = m+1$, $m+2, \ldots, n$ (for load buses), i.e., in matrix form,

$$0 = D_6\Delta V_g + D_7\Delta V_1 \qquad (5.53)$$

Here, $[D_6]_{2n\times 2m}$ and $[D_7]_{2n\times 2n}$ are full matrices of the form as given by

$$0 = \begin{bmatrix} D_{6m+1,1} & D_{m+1,2} & \cdots & D_{m+1,m} \\ D_{6m+2,1} & D_{m+2,2} & \cdots & D_{m+2,m} \\ \cdot & \cdot & \cdot & \cdot \\ \cdot & \cdot & \cdot & \cdot \\ \cdot & \cdot & \cdot & \cdot \\ \cdot & \cdot & \cdot & \cdot \\ D_{6n,1} & D_{6n,2} & \cdots & D_{6n,m} \end{bmatrix}\begin{bmatrix} \Delta V_{g_1} \\ \Delta V_{g_2} \\ \cdot \\ \cdot \\ \cdot \\ \cdot \\ \Delta V_{g_m} \end{bmatrix}$$

$$+ \begin{bmatrix} D_{7m+1,m+1} & D_{7m+1,m+2} & \cdots & D_{7m+1,n} \\ D_{7m+2,m+1} & D_{7m+2,m+2} & \cdots & D_{7m+2,n} \\ \cdot & \cdot & \cdot & \cdot \\ \cdot & \cdot & \cdot & \cdot \\ \cdot & \cdot & \cdot & \cdot \\ \cdot & \cdot & \cdot & \cdot \\ D_{7n,m+1} & D_{7n,m+2} & \cdots & D_{7n,n} \end{bmatrix}\begin{bmatrix} \Delta V_{1_{m+1}} \\ \Delta V_{1_{m+2}} \\ \cdot \\ \cdot \\ \cdot \\ \cdot \\ \Delta V_{1_n} \end{bmatrix} \qquad (5.54)$$

- The submatrices of D_6 and D_7 can be derived as follows:

To obtain the submatrices of D_6 and D_7, Equations (5.50) and (5.51) are explored for $i = m+1, m+2, \ldots, n$ (for the load buses).

For $i = m + 1$, Equation (5.50) gives

$$0 = \left[\sum_{\substack{k=1 \\ \neq i}}^{n} V_i V_k Y_{ik} \sin(\theta_i - \theta_k - \alpha_{ik}) \right] \Delta\theta_{m+1}$$

$$+ \left(\frac{\partial P_{L_i}(V_i)}{\partial V_i} - \left[\sum_{k=1}^{n} V_k Y_{ik} \cos(\theta_i - \theta_k - \alpha_{ik}) \right] \right) \Delta V_{m+1}$$

$$- V_i Y_{ik} \cos(\theta_i - \theta_k - \alpha_{ik}) \Delta V_1 - V_i Y_{ik} \cos(\theta_i - \theta_k - \alpha_{ik}) \Delta V_2 - \cdots$$

$$- V_i Y_{ik} \cos(\theta_i - \theta_k - \alpha_{ik}) \Delta V_m - V_i Y_{ik} \cos(\theta_i - \theta_k - \alpha_{ik}) \Delta V_{m+1} - \cdots$$

$$- V_i Y_{ik} \cos(\theta_i - \theta_k - \alpha_{ik}) \Delta V_n - V_i V_k Y_{ik} \sin(\theta_i - \theta_k - \alpha_{ik}) \Delta\theta_1$$

$$- V_i V_k Y_{ik} \sin(\theta_i - \theta_k - \alpha_{ik}) \Delta\theta_2 - \cdots - V_i V_k Y_{ik} \sin(\theta_i - \theta_k - \alpha_{ik}) \Delta\theta_m$$

$$- V_i V_k Y_{ik} \sin(\theta_i - \theta_k - \alpha_{ik}) \Delta\theta_{m+2} - \cdots - V_i V_k Y_{ik} \sin(\theta_i - \theta_k - \alpha_{ik}) \Delta\theta_n \quad (5.55)$$

Rearranging the terms containing the variables $\Delta V_{g_1} \ldots \Delta V_{g_m}$ and $\Delta V_{1_{m+1}} \ldots \Delta V_{1_n}$, Equation (5.55) can be written as

$$0 = \left[-V_i V_k Y_{ik} \sin(\theta_i - \theta_k - \alpha_{ik}) \quad -V_i Y_{ik} \cos(\theta_i - \theta_k - \alpha_{ik}) \right] \begin{bmatrix} \Delta\theta_1 \\ \Delta V_1 \end{bmatrix}$$

$$+ \left[-V_i V_k Y_{ik} \sin(\theta_i - \theta_k - \alpha_{ik}) \quad -V_i Y_{ik} \cos(\theta_i - \theta_k - \alpha_{ik}) \right] \begin{bmatrix} \Delta\theta_2 \\ \Delta V_2 \end{bmatrix}$$

$$+ \left[-V_i V_k Y_{ik} \sin(\theta_i - \theta_k - \alpha_{ik}) \quad -V_i Y_{ik} \cos(\theta_i - \theta_k - \alpha_{ik}) \right] \begin{bmatrix} \Delta\theta_m \\ \Delta V_m \end{bmatrix}$$

$$+ \left[\sum_{\substack{k=1 \\ \neq i}}^{n} V_i V_k Y_{ik} \sin(\theta_i - \theta_k - \alpha_{ik}) \quad \frac{\partial P_{L_i}(V_i)}{\partial V_i} - \left[\sum_{k=1}^{n} V_k Y_{ik} \cos(\theta_i - \theta_k - \alpha_{ik}) \right] \right.$$

$$\left. -V_1 Y_{ik} \cos(\theta_i - \theta_k - \alpha_{ik}) \right] \begin{bmatrix} \Delta\theta_{m+1} \\ \Delta V_{m+1} \end{bmatrix}$$

$$+ \left[-V_i V_k Y_{ik} \sin(\theta_i - \theta_k - \alpha_{ik}) \quad -V_i Y_{ik} \cos(\theta_i - \theta_k - \alpha_{ik}) \right] \begin{bmatrix} \Delta\theta_{m+2} \\ \Delta V_{m+2} \end{bmatrix} + \cdots$$

$$+ \left[-V_i V_k Y_{ik} \sin(\theta_i - \theta_k - \alpha_{ik}) \quad -V_i Y_{ik} \cos(\theta_i - \theta_k - \alpha_{ik}) \right] \begin{bmatrix} \Delta\theta_n \\ \Delta V_n \end{bmatrix}$$

$$(5.56)$$

Similarly, for $i = m+1$, Equation (5.51) becomes

$$0 = -V_i \sum_{\substack{k=1 \\ \neq i}}^{n} V_k Y_{ik} \cos(\theta_i - \theta_k - \alpha_{ik}) \Delta\theta_{m+1}$$

$$+ \left[\frac{\partial Q_{L_i}(V_i)}{\partial V_i} - \sum_{k=1}^{n} V_k Y_{ik} \sin(\theta_i - \theta_k - \alpha_{ik}) \right] \Delta V_{m+1}$$

$$- V_i Y_{ik} \sin(\theta_i - \theta_k - \alpha_{ik}) \Delta V_1 - V_i Y_{ik} \sin(\theta_i - \theta_k - \alpha_{ik}) \Delta V_2 - \cdots$$

$$- V_i Y_{ik} \sin(\theta_i - \theta_k - \alpha_{ik}) \Delta V_m - V_i Y_{ik} \sin(\theta_i - \theta_k - \alpha_{ik}) \Delta V_{m+1} - \cdots$$

$$- V_i Y_{ik} \sin(\theta_i - \theta_k - \alpha_{ik}) \Delta V_n + V_i V_k Y_{ik} \cos(\theta_i - \theta_k - \alpha_{ik}) \Delta\theta_1$$

$$+ V_i V_k Y_{ik} \cos(\theta_i - \theta_k - \alpha_{ik}) \Delta\theta_2 + \cdots + V_i V_k Y_{ik} \cos(\theta_i - \theta_k - \alpha_{ik}) \Delta\theta_m$$

$$+ V_i V_k Y_{ik} \cos(\theta_i - \theta_k - \alpha_{ik}) \Delta\theta_{m+2} + \cdots + V_i V_k Y_{ik} \cos(\theta_i - \theta_k - \alpha_{ik}) \Delta\theta_n \quad (5.57)$$

Rearranging the terms containing the variables $\Delta V_{g_1} \ldots \Delta V_{g_m}$ and $\Delta V_{1_{m+1}} \ldots \Delta V_{1_n}$, Equation (5.57) can be written as

$$0 = [V_i V_k Y_{ik} \cos(\theta_i - \theta_k - \alpha_{ik}) \quad -V_i Y_{ik} \sin(\theta_i - \theta_k - \alpha_{ik})] \begin{bmatrix} \Delta\theta_1 \\ \Delta V_1 \end{bmatrix}$$

$$+ [V_i V_k Y_{ik} \cos(\theta_i - \theta_k - \alpha_{ik}) \quad -V_i Y_{ik} \sin(\theta_i - \theta_k - \alpha_{ik})] \begin{bmatrix} \Delta\theta_2 \\ \Delta V_2 \end{bmatrix} + \cdots$$

$$+ [V_i V_k Y_{ik} \cos(\theta_i - \theta_k - \alpha_{ik}) \quad -V_i Y_{ik} \sin(\theta_i - \theta_k - \alpha_{ik})] \begin{bmatrix} \Delta\theta_m \\ \Delta V_m \end{bmatrix}$$

$$+ \left[\left[-V_i \sum_{\substack{k=1 \\ \neq i}}^{n} V_k Y_{ik} \cos(\theta_i - \theta_k - \alpha_{ik}) \right] \quad \frac{\partial Q_{L_i}(V_i)}{\partial V_i} - \left[\sum_{k=1}^{n} V_k Y_{ik} \sin(\theta_i - \theta_k - \alpha_{ik}) \right] \right.$$

$$\left. - V_i Y_{ik} \sin(\theta_i - \theta_k - \alpha_{ik}) \right] \begin{bmatrix} \Delta\theta_{m+1} \\ \Delta V_{m+1} \end{bmatrix}$$

$$+ [V_i V_k Y_{ik} \cos(\theta_i - \theta_k - \alpha_{ik}) \quad -V_i Y_{ik} \sin(\theta_i - \theta_k - \alpha_{ik})] \begin{bmatrix} \Delta\theta_{m+2} \\ \Delta V_{m+2} \end{bmatrix} + \cdots$$

$$+ [V_i V_k Y_{ik} \cos(\theta_i - \theta_k - \alpha_{ik}) \quad -V_i Y_{ik} \sin(\theta_i - \theta_k - \alpha_{ik})] \begin{bmatrix} \Delta\theta_n \\ \Delta V_n \end{bmatrix}$$

$$(5.58)$$

Combining Equations (5.56) and (5.58) together in matrix form for $i = m+1$ results in

$$
0 = \begin{bmatrix} -V_i V_k Y_{ik}\sin(\theta_i - \theta_k - \alpha_{ik}) & -V_i Y_{ik}\cos(\theta_i - \theta_k - \alpha_{ik}) \\ V_i V_k Y_{ik}\cos(\theta_i - \theta_k - \alpha_{ik}) & -V_i Y_{ik}\sin(\theta_i - \theta_k - \alpha_{ik}) \end{bmatrix}\begin{bmatrix} \Delta\theta_1 \\ \Delta V_1 \end{bmatrix}
$$

$$
+ \begin{bmatrix} -V_i V_k Y_{ik}\sin(\theta_i - \theta_k - \alpha_{ik}) & -V_i Y_{ik}\cos(\theta_i - \theta_k - \alpha_{ik}) \\ V_i V_k Y_{ik}\cos(\theta_i - \theta_k - \alpha_{ik}) & -V_i Y_{ik}\sin(\theta_i - \theta_k - \alpha_{ik}) \end{bmatrix}\begin{bmatrix} \Delta\theta_2 \\ \Delta V_2 \end{bmatrix} + \cdots
$$

$$
+ \begin{bmatrix} -V_i V_k Y_{ik}\sin(\theta_i - \theta_k - \alpha_{ik}) & -V_i Y_{ik}\cos(\theta_i - \theta_k - \alpha_{ik}) \\ V_i V_k Y_{ik}\cos(\theta_i - \theta_k - \alpha_{ik}) & -V_i Y_{ik}\sin(\theta_i - \theta_k - \alpha_{ik}) \end{bmatrix}\begin{bmatrix} \Delta\theta_m \\ \Delta V_m \end{bmatrix}
$$

$$
+ \begin{bmatrix} \left[\sum_{\substack{k=1 \\ \neq i}}^{n} V_i V_k Y_{ik}\sin(\theta_i - \theta_k - \alpha_{ik}) \right] \\ \left[\sum_{\substack{k=1 \\ \neq i}}^{n} V_i V_k Y_{ik}\cos(\theta_i - \theta_k - \alpha_{ik}) \right] \end{bmatrix}
$$

$$
\begin{bmatrix} \dfrac{\partial P_{L_i}(V_i)}{\partial V_i} - \left[\sum_{k=1}^{n} V_k Y_{ik}\cos(\theta_i - \theta_k - \alpha_{ik}) \right] - V_i Y_{ik}\cos(\theta_i - \theta_k - \alpha_{ik}) \\ \dfrac{\partial Q_{L_i}(V_i)}{\partial V_i} - \left[\sum_{k=1}^{n} V_k Y_{ik}\sin(\theta_i - \theta_k - \alpha_{ik}) \right] - V_i Y_{ik}\sin(\theta_i - \theta_k - \alpha_{ik}) \end{bmatrix}\begin{bmatrix} \Delta\theta_{m+1} \\ \Delta V_{m+1} \end{bmatrix} + \cdots
$$

$$
+ \begin{bmatrix} -V_i V_k Y_{ik}\sin(\theta_i - \theta_k - \alpha_{ik}) & -V_i Y_{ik}\cos(\theta_i - \theta_k - \alpha_{ik}) \\ V_i V_k Y_{ik}\cos(\theta_i - \theta_k - \alpha_{ik}) & -V_i Y_{ik}\sin(\theta_i - \theta_k - \alpha_{ik}) \end{bmatrix}\begin{bmatrix} \Delta\theta_n \\ \Delta V_n \end{bmatrix}
$$

$$(5.59)$$

Therefore, for $i = m+1$ and $k = 1, 2, \ldots, m$, the submatrix of the *first row* of the D_6 matrix can be obtained from Equation (5.59). Again, for $i = m+1$ and $k = m+1$, $m+2, \ldots, n$, other submatrices of the *first row* of the D_7 matrix can be calculated from Equation (5.59).

For $i = n$, Equation (5.50) gives

$$
0 = \left[\sum_{\substack{k=1 \\ \neq i}}^{n} V_i V_k Y_{ik}\sin(\theta_i - \theta_k - \alpha_{ik}) \right]\Delta\theta_n
$$

$$
+ \left(\dfrac{\partial P_{L_n}(V_n)}{\partial V_n} - \left[\sum_{k=1}^{n} V_k Y_{ik}\cos(\theta_i - \theta_k - \alpha_{ik}) \right] \right)\Delta V_n
$$

$$
- V_i Y_{ik}\cos(\theta_i - \theta_k - \alpha_{ik})\Delta V_1 - V_i Y_{ik}\cos(\theta_i - \theta_k - \alpha_{ik})\Delta V_2 - \cdots
$$

$$
- V_i Y_{ik}\cos(\theta_i - \theta_k - \alpha_{ik})\Delta V_m - V_i Y_{ik}\cos(\theta_i - \theta_k - \alpha_{ik})\Delta V_{m+1} - \cdots
$$

$$
- V_i Y_{ik}\cos(\theta_i - \theta_k - \alpha_{ik})\Delta V_n - V_i V_k Y_{ik}\sin(\theta_i - \theta_k - \alpha_{ik})\Delta\theta_1
$$

$$
- V_i V_k Y_{ik}\sin(\theta_i - \theta_k - \alpha_{ik})\Delta\theta_2 - \cdots - V_i V_k Y_{ik}\sin(\theta_i - \theta_k - \alpha_{ik})\Delta\theta_m
$$

$$
- V_i V_k Y_{ik}\sin(\theta_i - \theta_k - \alpha_{ik})\Delta\theta_{m+1} - \cdots - V_i V_k Y_{ik}\sin(\theta_i - \theta_k - \alpha_{ik})\Delta\theta_{n-1} \quad (5.60)
$$

Rearranging the terms containing the variables $\Delta V_{g_1} \ldots \Delta V_{g_m}$ and $\Delta V_{1_{m+1}} \ldots \Delta V_{1_n}$, Equation (5.60) can be written as

$$
\begin{aligned}
0 = &[-V_i V_k Y_{ik} \sin(\theta_i - \theta_k - \alpha_{ik}) \quad -V_i Y_{ik} \cos(\theta_i - \theta_k - \alpha_{ik})] \begin{bmatrix} \Delta\theta_1 \\ \Delta V_1 \end{bmatrix} \\
&+[-V_i V_k Y_{ik} \sin(\theta_i - \theta_k - \alpha_{ik}) \quad -V_i Y_{ik} \cos(\theta_i - \theta_k - \alpha_{ik})] \begin{bmatrix} \Delta\theta_2 \\ \Delta V_2 \end{bmatrix} + \cdots \\
&+[-V_i V_k Y_{ik} \sin(\theta_i - \theta_k - \alpha_{ik}) \quad -V_i Y_{ik} \cos(\theta_i - \theta_k - \alpha_{ik})] \begin{bmatrix} \Delta\theta_m \\ \Delta V_m \end{bmatrix} \\
&+[-V_i V_k Y_{ik} \sin(\theta_i - \theta_k - \alpha_{ik}) \quad -V_i Y_{ik} \cos(\theta_i - \theta_k - \alpha_{ik})] \begin{bmatrix} \Delta\theta_{m+1} \\ \Delta V_{m+1} \end{bmatrix} + \cdots \\
&+\left[\sum_{\substack{k=1 \\ \neq i}}^{n} V_i V_k Y_{ik} \sin(\theta_i - \theta_k - \alpha_{ik}) \right. \\
&\left. \frac{\partial Q_{L_i}(V_i)}{\partial V_i} - \left[\sum_{k=1}^{n} V_k Y_{ik} \cos(\theta_i - \theta_k - \alpha_{ik}) \right] \right. \\
&\left. -V_1 Y_{ik} \cos(\theta_i - \theta_k - \alpha_{ik}) \right] \begin{bmatrix} \Delta\theta_n \\ \Delta V_n \end{bmatrix}
\end{aligned}
$$

$$(5.61)$$

Similarly, for $i = n$, Equation (5.51) becomes

$$
\begin{aligned}
0 = &-\sum_{\substack{k=1 \\ \neq i}}^{n} V_i V_k Y_{ik} \cos(\theta_i - \theta_k - \alpha_{ik}) \Delta\theta_n \\
&+\left[\frac{\partial Q_{L_n}(V_n)}{\partial V_n} - \sum_{k=1}^{n} V_k Y_{ik} \sin(\theta_i - \theta_k - \alpha_{ik}) \right] \Delta V_n \\
&-V_i Y_{ik} \sin(\theta_i - \theta_k - \alpha_{ik}) \Delta V_1 - V_i Y_{ik} \sin(\theta_i - \theta_k - \alpha_{ik}) \Delta V_2 - \cdots \\
&-V_i Y_{ik} \sin(\theta_i - \theta_k - \alpha_{ik}) \Delta V_m - V_i Y_{ik} \sin(\theta_i - \theta_k - \alpha_{ik}) \Delta V_{m+1} - \cdots \\
&-V_i Y_{ik} \sin(\theta_i - \theta_k - \alpha_{ik}) \Delta V_n + V_i V_k Y_{ik} \cos(\theta_i - \theta_k - \alpha_{ik}) \Delta\theta_1 \\
&+V_i V_k Y_{ik} \cos(\theta_i - \theta_k - \alpha_{ik}) \Delta\theta_2 + \cdots + V_i V_k Y_{ik} \cos(\theta_i - \theta_k - \alpha_{ik}) \Delta\theta_m \\
&+V_i V_k Y_{ik} \cos(\theta_i - \theta_k - \alpha_{ik}) \Delta\theta_{m+1} + \cdots + V_i V_k Y_{ik} \cos(\theta_i - \theta_k - \alpha_{ik}) \Delta\theta_{n-1}
\end{aligned}
$$

$$(5.62)$$

Rearranging the terms containing the variables $\Delta V_{g_1} \ldots \Delta V_{g_m}$ and $\Delta V_{1_{m+1}} \ldots \Delta V_{1_n}$, Equation (5.62) can be written as

$$0 = [V_i V_k Y_{ik} \cos(\theta_i - \theta_k - \alpha_{ik}) \quad -V_i Y_{ik} \sin(\theta_i - \theta_k - \alpha_{ik})] \begin{bmatrix} \Delta\theta_1 \\ \Delta V_1 \end{bmatrix}$$

$$+[V_i V_k Y_{ik} \cos(\theta_i - \theta_k - \alpha_{ik}) \quad -V_i Y_{ik} \sin(\theta_i - \theta_k - \alpha_{ik})] \begin{bmatrix} \Delta\theta_2 \\ \Delta V_2 \end{bmatrix} + \cdots$$

$$+[V_i V_k Y_{ik} \cos(\theta_i - \theta_k - \alpha_{ik}) \quad -V_i Y_{ik} \sin(\theta_i - \theta_k - \alpha_{ik})] \begin{bmatrix} \Delta\theta_m \\ \Delta V_m \end{bmatrix}$$

$$+[V_i V_k Y_{ik} \cos(\theta_i - \theta_k - \alpha_{ik}) \quad -V_i Y_{ik} \sin(\theta_i - \theta_k - \alpha_{ik})] \begin{bmatrix} \Delta\theta_{m+1} \\ \Delta V_{m+1} \end{bmatrix} + \cdots$$

$$+[V_i V_k Y_{ik} \cos(\theta_i - \theta_k - \alpha_{ik}) \quad -V_i Y_{ik} \sin(\theta_i - \theta_k - \alpha_{ik})] \begin{bmatrix} \Delta\theta_{m+2} \\ \Delta V_{m+2} \end{bmatrix} + \cdots$$

$$+ \left[-\left[V_i \sum_{\substack{k=1 \\ \neq i}}^{n} V_k Y_{ik} \cos(\theta_i - \theta_k - \alpha_{ik}) \right] \right.$$

$$\frac{\partial Q_{L_i}(V_i)}{\partial V_i} - \left[\sum_{k=1}^{n} V_k Y_{ik} \sin(\theta_i - \theta_k - \alpha_{ik}) \right]$$

$$\left. -V_i Y_{ik} \sin(\theta_i - \theta_k - \alpha_{ik}) \right] \begin{bmatrix} \Delta\theta_n \\ \Delta V_n \end{bmatrix} \tag{5.63}$$

Combining Equations (5.61) and (5.63) in matrix form for $i = n$ results in

$$0 = \begin{bmatrix} -V_i V_k Y_{ik} \sin(\theta_i - \theta_k - \alpha_{ik}) & -V_i Y_{ik} \cos(\theta_i - \theta_k - \alpha_{ik}) \\ V_i V_k Y_{ik} \cos(\theta_i - \theta_k - \alpha_{ik}) & -V_i Y_{ik} \sin(\theta_i - \theta_k - \alpha_{ik}) \end{bmatrix} \begin{bmatrix} \Delta\theta_1 \\ \Delta V_1 \end{bmatrix}$$

$$+ \begin{bmatrix} -V_i V_k Y_{ik} \sin(\theta_i - \theta_k - \alpha_{ik}) & -V_i Y_{ik} \cos(\theta_i - \theta_k - \alpha_{ik}) \\ V_i V_k Y_{ik} \cos(\theta_i - \theta_k - \alpha_{ik}) & -V_i Y_{ik} \sin(\theta_i - \theta_k - \alpha_{ik}) \end{bmatrix} \begin{bmatrix} \Delta\theta_2 \\ \Delta V_2 \end{bmatrix} + \cdots$$

$$+ \begin{bmatrix} -V_i V_k Y_{ik} \sin(\theta_i - \theta_k - \alpha_{ik}) & -V_i Y_{ik} \cos(\theta_i - \theta_k - \alpha_{ik}) \\ V_i V_k Y_{ik} \cos(\theta_i - \theta_k - \alpha_{ik}) & -V_i Y_{ik} \sin(\theta_i - \theta_k - \alpha_{ik}) \end{bmatrix} \begin{bmatrix} \Delta\theta_m \\ \Delta V_m \end{bmatrix}$$

$$+ \begin{bmatrix} -V_i V_k Y_{ik} \sin(\theta_i - \theta_k - \alpha_{ik}) & -V_i Y_{ik} \cos(\theta_i - \theta_k - \alpha_{ik}) \\ V_i V_k Y_{ik} \cos(\theta_i - \theta_k - \alpha_{ik}) & -V_i Y_{ik} \sin(\theta_i - \theta_k - \alpha_{ik}) \end{bmatrix} \begin{bmatrix} \Delta\theta_{m+1} \\ \Delta V_{m+1} \end{bmatrix}$$

$$+ \begin{bmatrix} \left[\sum_{\substack{k=1 \\ \neq i}}^{n} V_i V_k Y_{ik} \sin(\theta_i - \theta_k - \alpha_{ik}) \right] \\ \left[-\sum_{\substack{k=1 \\ \neq i}}^{n} V_i V_k Y_{ik} \cos(\theta_i - \theta_k - \alpha_{ik}) \right] \end{bmatrix}$$

$$\frac{\partial P_{L_i}(V_i)}{\partial V_i} - \left[\sum_{k=1}^{n} V_k Y_{ik} \cos(\theta_i - \theta_k - \alpha_{ik}) \right] - V_i Y_{ik} \cos(\theta_i - \theta_k - \alpha_{ik})$$

$$\frac{\partial Q_{L_i}(V_i)}{\partial V_i} - \left[\sum_{k=1}^{n} V_k Y_{ik} \sin(\theta_i - \theta_k - \alpha_{ik}) \right] - V_i Y_{ik} \sin(\theta_i - \theta_k - \alpha_{ik}) \end{bmatrix} \begin{bmatrix} \Delta\theta_n \\ \Delta V_n \end{bmatrix}$$

$$\tag{5.64}$$

Therefore, for $i=n$ and $k=1, 2,\ldots,m,$ the submatrix of the *last row* of the D_6 matrix can be obtained from Equation (5.64). For $i=n$ and $k=m+1, m+2,\ldots,n,$ other submatrices of the *last row* of the D_7 matrix can also be obtained from Equation (5.64). The submatrices of D_6 and D_7 for the intermediate rows corresponding to $i=m+2,\ldots,(n-1)$ can also be determined by the similar method.

Rewriting Equations (5.28), (5.33), (5.38), and (5.53) together,

$$\Delta\dot{X} = A_1\Delta X + B_1\Delta I_g + B_2\Delta V_g + E_1\Delta U \tag{5.65}$$

$$0 = C_1\Delta X + D_1\Delta I_g + D_2\Delta V_g \tag{5.66}$$

$$0 = C_2\Delta X + D_3\Delta I_g + D_4\Delta V_g + D_5\Delta V_1 \tag{5.67}$$

$$0 = D_6\Delta V_g + D_7\Delta V_1 \tag{5.68}$$

where $\quad X = \begin{bmatrix} X_1^T & X_2^T & \cdots & X_m^T \end{bmatrix}^T, \qquad X_i = \begin{bmatrix} \delta_i & \omega_i & E'_{q_i} & E'_{d_i} & E_{\mathrm{fd}_i} & V_{R_i} & R_{F_i} \end{bmatrix}^T,$

$I_g = \begin{bmatrix} I_{d_1} & I_{q_1} & I_{d_2} & I_{q_2} & \cdots & I_{d_m} & I_{q_m} \end{bmatrix}^T, \qquad V_g = \begin{bmatrix} \theta_1 & V_1 & \theta_2 & V_2 & \cdots & \theta_m & V_m \end{bmatrix}^T,$

$V_1 = \begin{bmatrix} \theta_{m+1} & V_{m+1} & \theta_{m+2} & V_{m+2} & \cdots & \theta_n & V_n \end{bmatrix}^T, \qquad U = \begin{bmatrix} U_1^T & U_2^T & \cdots & U_m^T \end{bmatrix}^T, \qquad$ and

$U_i = \begin{bmatrix} T_{M_i} & V_{\mathrm{ref}_i} \end{bmatrix}^T.$

This is the linearized differential-algebraic model for the multimachine system. This model is quite general and can easily be expanded to include frequency or voltage dependence at the load buses. The power system stabilizers (PSSs) and FACTS controllers can also be included easily.

In the aforementioned model, ΔI_g is not of interest and hence is eliminated from Equations (5.65) and (5.67) using Equation (5.66). Thus, from Equation (5.66),

$$\Delta I_g = -D_1^{-1}C_1\Delta X - D_1^{-1}D_2\Delta V_g \tag{5.69}$$

Substituting Equation (5.69) into Equation (5.65),

$$\Delta\dot{X} = A_1\Delta X + B_1\left\{-D_1^{-1}C_1\Delta X - D_1^{-1}D_2\Delta V_g\right\} + B_2\Delta V_g + E_1\Delta U \tag{5.70}$$

which after rearranging gives

$$\Delta\dot{X} = \left(A_1 - B_1 D_1^{-1} C_1\right)\Delta X + \left(B_2 - B_1 D_1^{-1} D_2\right)\Delta V_g + E_1\Delta U$$

Again, substitution for ΔI_g in Equation (5.67) results in

$$0 = C_2\Delta X + D_3\left(-D_1^{-1}C_1\Delta X - D_1^{-1}D_2\Delta V_g\right) + D_4\Delta V_g + D_5\Delta V_1 \tag{5.71}$$

or

$$0 = \left(C_2 - D_3 D_1^{-1} C_1\right)\Delta X + \left(D_4 - D_3 D_1^{-1} D_2\right)\Delta V_g + D_5\Delta V_1 \tag{5.72}$$

$$\therefore 0 = K_2\Delta X + K_1\Delta V_g + D_5\Delta V_1 \tag{5.73}$$

where $K_1 = \begin{bmatrix} D_4 - D_3 D_1^{-1} D_2 \end{bmatrix}$ and $K_2 = \begin{bmatrix} C_2 - D_3 D_1^{-1} C_1 \end{bmatrix}.$

Thus, the new overall differential-algebraic model becomes

$$\Delta \dot{X} = \left(A_1 - B_1 D_1^{-1} C_1\right) \Delta X + \left(B_2 - B_1 D_1^{-1} D_2\right) \Delta V_g + E_1 \Delta U \tag{5.74}$$

$$0 = K_2 \Delta X + K_1 \Delta V_g + D_5 \Delta V_1 \tag{5.75}$$

$$0 = D_6 \Delta V_g + D_7 \Delta V_1 \tag{5.76}$$

Writing Equations (5.74)–(5.76) in state-space representation,

$$\begin{bmatrix} \Delta \dot{X} \\ 0 \\ 0 \end{bmatrix} = \begin{bmatrix} A_1 - B_1 D_1^{-1} C_1 & B_2 - B_1 D_1^{-1} D_2 & 0 \\ K_2 & K_1 & D_5 \\ 0 & D_6 & D_7 \end{bmatrix} \begin{bmatrix} \Delta X \\ \Delta V_g \\ \Delta V_1 \end{bmatrix} + \begin{bmatrix} E_1 \\ 0 \\ 0 \end{bmatrix} \Delta U \tag{5.77}$$

which in more compact form can be written as

$$\begin{bmatrix} \Delta \dot{X} \\ 0 \end{bmatrix} = \begin{bmatrix} A' & B' \\ C' & D' \end{bmatrix} \begin{bmatrix} \Delta X \\ \Delta V_N \end{bmatrix} + \begin{bmatrix} E_1 \\ 0 \end{bmatrix} \Delta U \tag{5.78}$$

where $A' = A_1 - B_1 D_1^{-1} C_1$, $B' = \begin{bmatrix} B_2 - B_1 D_1^{-1} D_2 & 0 \end{bmatrix}$, $C' = \begin{bmatrix} K_2 \\ 0 \end{bmatrix}$, $D' = \begin{bmatrix} K_1 & D_5 \\ D_6 & D_7 \end{bmatrix}$,

and $\Delta V_N = \begin{bmatrix} \Delta V_g \\ \Delta V_1 \end{bmatrix}$. Defining $\Delta V_P = \begin{bmatrix} \Delta Y_C^T & \Delta Y_B^T \end{bmatrix}$,

$$= \begin{bmatrix} \Delta \theta_1 & \Delta V_1 & \dots & \Delta V_m & \vdots & \Delta \theta_2 & \dots & \Delta \theta_n & \Delta V_{m+1} & \dots & \Delta V_n \end{bmatrix}^T$$

where ΔY_B is the set of load-flow variables and ΔY_C is the set of other algebraic variables in the network equations; the differential-algebraic model can further be written as

$$\begin{bmatrix} \Delta \dot{X} \\ 0 \\ 0 \end{bmatrix} = \begin{bmatrix} A' & B_1' & B_2' \\ C_1' & D_{11}' & D_{12}' \\ C_2' & D_{21}' & D_{22}' \end{bmatrix} \begin{bmatrix} \Delta X \\ \Delta Y_C \\ \Delta Y_B \end{bmatrix} + \begin{bmatrix} E_1 \\ 0 \\ 0 \end{bmatrix} \Delta U \tag{5.79}$$

Now, D_{22}' is the load-flow Jacobian (J_{LF}) modified by the load representation and $\begin{bmatrix} D_{11}' & D_{12}' \\ D_{21}' & D_{22}' \end{bmatrix} = J_{AE}'$ is the network algebraic Jacobian. For voltage-dependent loads, only the appropriate diagonal elements of D_{11}' and D_{22}' will be affected. Now, the system matrix A_{sys} obtained from Equation (5.79) is

$$\Delta \dot{X} = A_{sys} \Delta X + E \Delta U \tag{5.80}$$

where $\left[A_{sys}\right]_{7m \times 7m} = [A'] - \begin{bmatrix} B_1' & B_2' \end{bmatrix} \left[J_{AE}'\right]^{-1} \begin{bmatrix} C_1' \\ C_2' \end{bmatrix}$

This model can be used to examine the effect of small-signal disturbance on the eigenvalues of the multimachine power system. When a PSS or any FACTS controllers are installed at any machine, the extra state variables corresponding to these controllers will be added with the system matrix.

5.2.3 Reduced-order flux-decay model

This model is widely used in eigenvalue analysis and PSS design. If the damper-winding constants are very small, then we can set them to zero, and from Equation (5.4), we have

$$0 = -E'_{d_i} + \left(X_{q_i} - X'_{q_i}\right)I_{q_i}, \quad i = 1, 2, \ldots, m \tag{5.81}$$

Using Equation (5.81), we can eliminate E'_{d_i} from Equations (5.2) and (5.8), and the synchronous machine dynamic circuit will be modified as in Figure 5.2. In flux-decay model, generally, a simplified exciter with one gain and one time constant is considered, which is shown in Figure 5.3.

- **The DAEs of generator with static exciter:**

$$\frac{d\delta_i}{dt} = \omega_i - \omega_s \tag{5.82}$$

$$\frac{d\omega_i}{dt} = \frac{T_{M_i}}{M_i} - \frac{E'_{q_i}I_{q_i}}{M_i} - \frac{\left(X'_{q_i} - X'_{d_i}\right)I_{d_i}I_{q_i}}{M_i} \tag{5.83}$$

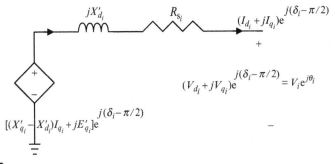

FIGURE 5.2

Dynamic circuit synchronous machine flux-decay model ($i = 1, 2, 3, \ldots, m$).

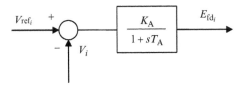

FIGURE 5.3

One-gain and one-time constant static exciter ($i = 1, 2, 3, \ldots, m$).

$$\frac{dE'_{q_i}}{dt} = -\frac{E'_{q_i}}{T'_{do_i}} - \frac{\left(X_{d_i} - X'_{d_i}\right)I_{d_i}}{T'_{do_i}} + \frac{E_{fd_i}}{T'_{do_i}} \tag{5.84}$$

$$\frac{dE_{fd_i}}{dt} = -\frac{E_{fd_i}}{T_{A_i}} + \frac{K_{A_i}}{T_{A_i}}(V_{ref_i} - V_i) \tag{5.85}$$

for $i = 1, 2, 3, \ldots, m$.

- **Stator algebraic equations:**

Assuming $R_s = 0$, and substituting for E'_{d_i} from Equation (5.81), we obtain the stator algebraic equations in polar form as

$$V_i \sin(\delta_i - \theta_i) - X_{q_i} I_{q_i} = 0 \tag{5.86}$$

$$V_i \cos(\delta_i - \theta_i) + X'_{d_i} I_{d_i} - E'_{q_i} = 0 \tag{5.87}$$

- **Network equations:**

The network equations for an n-bus system can be written in the same form as described for multimachine two-axis model represented by Equations (5.13)–(5.16).

5.3 COMPUTATION OF INITIAL CONDITIONS OF THE STATE VARIABLES

The initial conditions of the state variables for the model are computed systematically by solving the standard load-flow equations of the network, first, and then computing the other algebraic and state variables. The load-flow equations are part of the network equations, as shown in the succeeding text.

- **Load-flow formulation:**

The standard load flow is computed on the basis of constant PQ loads and has been the traditional mechanism for computing a proposed steady-state operating point.

The net power injected at a bus is defined as

$$P_i(\delta_i, I_{d_i}, I_{q_i}, V_i, \theta_i) + jQ_i(\delta_i, I_{d_i}, I_{q_i}, V_i, \theta_i) = $$
$$(P_{G_i} + P_{L_i}(V_i)) + j(Q_{G_i} + Q_{L_i}(V_i)), \quad \text{for } i = 1, 2, \ldots, m. \tag{5.88}$$

Thus, the real and reactive power-balance equations at the buses $1, 2, \ldots, n$ are

$$P_i(\delta_i, I_{d_i}, I_{q_i}, V_i, \theta_i) = \sum_{k=1}^{n} V_i V_k Y_{ik} \cos(\theta_i - \theta_k - \alpha_{ik}) = 0, \quad i = 1, 2, \ldots, m \tag{5.89}$$

$$P_{L_i}(V_i) = \sum_{k=1}^{n} V_i V_k Y_{ik} \cos(\theta_i - \theta_k - \alpha_{ik}), \quad i = m+1, m+2,\ldots, n \qquad (5.90)$$

$$Q_i(\delta_i, I_{d_i}, I_{q_i}, V_i, \theta_i) = \sum_{k=1}^{n} V_i V_k Y_{ik} \sin(\theta_i - \theta_k - \alpha_{ik}) = 0, \quad i = 1, 2,\ldots, m. \qquad (5.91)$$

$$Q_{L_i}(V_i) = \sum_{k=1}^{n} V_i V_k Y_{ik} \sin(\theta_i - \theta_k - \alpha_{ik}), \quad i = m+1, m+2,\ldots, n \qquad (5.92)$$

The standard load-flow equations result from Equations (5.89) to (5.92) and the chosen criterion:

1. Specify bus voltage magnitudes numbered 1 to m.
2. Specify bus voltage angle at bus number 1 (slack bus).
3. Specify net real power P_i injected at buses numbered 2 to m.
4. Specify load power P_{L_i} and Q_{L_i} at buses numbered $m+1$ to n.

Thus, we have

$$0 = -P_i + \sum_{k=1}^{n} V_i V_k Y_{ik} \cos(\theta_i - \theta_k - \alpha_{ik}) \quad i = 2,\ldots, m. \text{ PV buses} \qquad (5.93)$$

$$0 = -P_{L_i} + \sum_{k=1}^{n} V_i V_k Y_{ik} \cos(\theta_i - \theta_k - \alpha_{ik}) \quad i = m+1,\ldots, m. \text{ PQ buses} \qquad (5.94)$$

$$0 = -Q_i + \sum_{k=1}^{n} V_i V_k Y_{ik} \sin(\theta_i - \theta_k - \alpha_{ik}) \quad i = 1,2,\ldots, m. \text{ PV buses} \qquad (5.95)$$

$$0 = -Q_{L_i} + \sum_{k=1}^{n} V_i V_k Y_{ik} \sin(\theta_i - \theta_k - \alpha_{ik}) \quad i = m+1,\ldots, n. \text{ PQ buses} \qquad (5.96)$$

where P_i ($i = 2,\ldots,m$), V_i ($i = 2,\ldots,m$), P_{L_i} ($i = m+1,\ldots,n$), Q_{L_i} ($i = m+1,\ldots,n$), and θ_1 are specified numbers. The standard load-flow program solves Equations (5.93)–(5.96) for $\theta_2 \ldots \theta_n$ and $V_{m+1} \ldots V_n$. After the load-flow solution, the net power injected at the slack bus and the generator buses are computed.

The generator powers are given by $P_{G_i} = P_i - P_{L_i}$ and $Q_{G_i} = Q_i - Q_{L_i}$ ($i = 1,\ldots,m$). Again, from the synchronous machine dynamic circuit given in Figure 5.4, we have

$$P_{G_i} + jQ_{G_i} = \overline{V}_i \overline{I}_{G_i}^* = V_i e^{j\theta_i} (I_{d_i} - I_{q_i}) e^{-j(\delta_i - \pi/2)}$$
$$= V_i(\cos\theta_i + j\sin\theta_i)(I_{d_i} - jI_{q_i})(\sin\delta_i + j\cos\delta_i) \qquad (5.97)$$

Now, equating real and imaginary parts of Equation (5.97) gives

$$P_{G_i} = I_{d_i} V_i \sin(\delta_i - \theta_i) + I_{q_i} V_i \cos(\delta_i - \theta_i) \qquad (5.98)$$

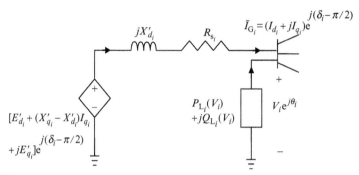

FIGURE 5.4

Dynamic circuit of synchronous machine.

$$Q_{G_i} = I_{d_i} V_i \cos(\delta_i - \theta_i) - I_{q_i} V_i \sin(\delta_i - \theta_i) \tag{5.99}$$

and

$$\bar{I}_{G_i} = I_{G_i} e^{j\gamma_i} = \left(I_{d_i} + I_{q_i}\right) e^{j(\delta_i - \pi/2)} \tag{5.100}$$

is the generator current injected at the generator bus.

In power system dynamic analysis, the initial values of all dynamic states and the fixed inputs T_{M_i} and V_{ref_i} $(i = 1, 2, \ldots, m)$ are found from the base case load-flow solution. To see how this is done, let us assume that a load-flow solution of Equations (5.93)–(5.96) has been found, and then, the first step in computing the initial conditions is to calculate the generator current from Equation (5.100):

- Step 1

$$\bar{I}_{G_i} = I_{G_i} e^{j\gamma_i} = \frac{(P_{G_i} - jQ_{G_i})}{\bar{V}_i^*}$$

Since $P_{G_i} = P_i - P_{L_i}$ and $Q_{G_i} = Q_i - Q_{L_i}$,

$$I_{G_i} e^{j\gamma_i} = \frac{((P_i - P_{L_i}) - j(Q_i - Q_{L_i}))}{V_i e^{-j\theta_i}}, \quad i = 1, 2, \ldots, m \tag{5.101}$$

This current is in the network reference frame and is equal to $\left(I_{d_i} + I_{q_i}\right) e^{j(\delta_i - \pi/2)}$

In steady state, all the derivatives are zero in the differential equations (5.1)–(5.7). The first step is to calculate the rotor angle δ_i at all the machines.

From Equation (5.4), setting $\dot{E}'_{d_i} = 0$,

$$E'_{d_i} = \left(X_{q_i} - X'_{q_i}\right) I_{q_i}, \quad i = 1, 2, \ldots, m. \tag{5.102}$$

Substituting Equation (5.102) into the complex stator algebraic equation (5.8), we have

$$0 = V_i e^{j\theta_i} + R_{s_i}\left(I_{d_i} + jI_{q_i}\right) e^{j(\delta_i - \pi/2)} + jX'_{d_i} I_{d_i} e^{j(\delta_i - \pi/2)} - X_{q_i} I_{q_i} e^{j(\delta_i - \pi/2)} - jE'_{q_i} e^{j(\delta_i - \pi/2)} \tag{5.103}$$

Adding and subtracting $jX_{q_i}I_{d_i}\, e^{j(\delta_i-\pi/2)}$ from left-hand side of Equation (5.103),

$$0 = V_i e^{j\theta_i} + \left(R_{s_i} + jX'_{q_i}\right)\left(I_{d_i} + jI_{q_i}\right)e^{j(\delta_i-\pi/2)}$$
$$-j\left[\left(X_{q_i} - X'_{d_i}\right)I_{d_i} + E'_{q_i}\right]e^{j(\delta_i-\pi/2)} \tag{5.104}$$

$$V_i e^{j\theta_i} + \left(R_{s_i} + jX'_{q_i}\right)I_{G_i}e^{j\gamma_i} = \left[\left(X_{q_i} - X'_{d_i}\right)I_{d_i} + E'_{q_i}\right]e^{j\delta_i}, \quad i = 1, 2, \ldots, m. \tag{5.105}$$

The right-hand side of Equation (5.105) is a voltage behind the impedance $\left(R_{s_i} + jX'_{q_i}\right)$ and has a magnitude $\left[\left(X_{q_i} - X'_{d_i}\right)I_{d_i} + E'_{q_i}\right]$ and an angle $\delta_i = $ angle of $\left(V_i e^{j\theta_i} + \left(R_{s_i} + jX'_{q_i}\right)I_{G_i}e^{j\gamma_i}\right)$.

- Step 2

  δ_i is computed as $\delta_i = $ angle of $\left(V_i e^{j\theta_i} + \left(R_{s_i} + jX'_{q_i}\right)I_{G_i}e^{j\gamma_i}\right)$.

- Step 3

  Computation of I_{d_i}, I_{q_i}, V_{d_i}, and V_{q_i} for each machine from the equations is

  $$I_{d_i} + jI_{q_i} = I_{G_i}e^{j(\gamma_i-\delta_i+\pi/2)}, \quad i = 1, 2, \ldots, m. \tag{5.106}$$

  $$V_{d_i} + jV_{q_i} = V_i e^{j(\theta_i-\delta_i+\pi/2)}, \quad i = 1, 2, \ldots, m. \tag{5.107}$$

- Step 4

  Computation of E'_{d_i} from Equation (5.9) is

  $$E'_{d_i} = V_{d_i} + R_{s_i}I_{q_i} - X'_{q_i}I_{q_i}$$
  $$= \left(X_{q_1} - X'_{q_1}\right)I_{q_1}, \quad i = 1, 2, \ldots, m. \tag{5.108}$$

- Step 5

  Computation of E'_{q_i} from Equation (5.10) is

  $$E'_{q_i} = V_{q_i} + R_{s_i}I_{q_i} + X'_{d_i}I_{d_i}, \quad i = 1, 2, \ldots, m \tag{5.109}$$

- Step 6

  Computation of E_{fd_i} from Equation (5.3) after setting derivative equal to zero is

  $$E_{fd_i} = E'_{q_i} + \left(X_{d_i} - X'_{d_i}\right)I_{d_i}, \quad i = 1, 2, \ldots, m. \tag{5.110}$$

- Step 7

With the known field voltage E_{fd_i}, the other variables R_{F_i}, V_{R_i}, and V_{ref_i} can be found from Equations (5.5) to (5.7) after setting the derivatives equal to zero:

$$V_{\text{R}_i} = (K_{\text{E}_i} + S_{\text{E}_i}(E_{\text{fd}_i}))E_{\text{fd}_i} \tag{5.111}$$

$$R_{\text{F}_i} = \frac{K_{\text{F}_i}}{T_{\text{F}_i}} E_{\text{fd}_i} \tag{5.112}$$

$$V_{\text{ref}_i} = V_i + \left(\frac{V_{\text{R}_i}}{K_{\text{A}_i}}\right), \quad \text{for } i = 1, 2, 3, \dots, m. \tag{5.113}$$

- Step 8

The mechanical states ω_i and T_{M_i} are found from Equations (5.1) and (5.2) after setting the derivative equal to zero:

$$\omega_i = \omega_s$$
$$T_{\text{M}_i} = E'_{d_i} I_{d_i} + E'_{q_i} I_{q_i} + \left(X'_{q_i} - X'_{d_i}\right)I_{q_i}I_{d_i} \tag{5.114}$$

This completes the computation of all dynamic-state initial conditions and fixed inputs.

5.3.1 An illustration

The initial conditions or the steady-state variables of all the three machines of the test system given in Figure 5.5 are computed here based on the solved load-flow data. The machine data, exciter data, and the load-flow results are given in Section B.3 of Appendix B. All results are in pu.

Machine 1
- Step 1

$$I_{\text{G}_1} e^{j\gamma_1} = \frac{(P_{\text{G}_1} - jQ_{\text{G}_1})}{V_1^*}$$

$$= \frac{(0.719 - j0.546)}{1.04\angle 0^\circ}$$

$$= 0.6913 - j0.5250$$

$$= 0.8681 \angle -37.231^\circ$$

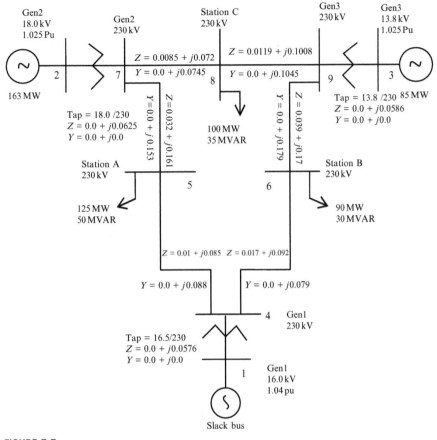

FIGURE 5.5

WSCC type 3-machine, 9-bus system; the value of Y is half the line charging.

- Step 2

  The machine rotor angle is

$$\delta_1 = \text{Angle of } \left\{ V_1 e^{j\theta_1} + \left(R_{s_1} + jX'_{q_1} \right) I_{G_1} e^{j\gamma_1} \right\}$$

$$= \text{Angle of } \left\{ 1.04\angle 0° + (0.089 + j0.0969) \times 0.8681 \angle -37.231° \right\}$$

$$= \text{Angle of } \left\{ 1.04 + (0.089 + j0.0969) \times 0.8681 \times (\cos 37.231° - j\sin 37.231°) \right\}$$

$$= \text{Angle of } \left\{ 1.04 + (0.089 + j0.0969) \times (0.6913 - j0.5250) \right\}$$

$$= 1.0080°$$

- Step 3

$$I_{d_1} + jI_{q_1} = I_{G_1} e^{j(\gamma_1 - \delta_1 + \pi/2)}$$

$$= 0.8681 \times e^{j(-37.231° - 1.0080° + 90°)}$$

$$= 0.8681 \times e^{51.761°}$$

$$= 0.8681 \times (\cos 51.761° + \sin 51.761°)$$

$$= 0.5376 + j0.6816$$

$$\therefore I_{d_1} = 0.5376 \quad \text{and} \quad I_{q_1} = 0.6816$$

$$V_{d_1} + jV_{q_1} = V_1 e^{j(\theta_1 - \delta_1 + \pi/2)}$$

$$= 1.04 \times e^{j(0° - 1.0080° + 90°)}$$

$$= 1.04 \times (\cos 88.992° + j\sin 88.992°)$$

$$= 0.0191 + j1.0398$$

$$\therefore V_{d_1} = 0.0191 \quad \text{and} \quad V_{q_1} = 0.0398$$

- Step 4

$$E'_{d_1} = \left(X_{q_1} - X'_{q_1} \right) I_{q_1}$$

$$= (0.0969 - 0.0969) \times 0.6816$$

$$= 0$$

- Step 5

$$E'_{q_1} = V_{q_1} + R_{s_1} I_{q_1} + X'_{d_1} I_{d_1}$$

$$= 0.398 + 0.089 \times 0.6816 + 0.0608 \times 0.5376$$

$$= 1.1332$$

- Step 6

$$E_{fd_1} = E'_{q_1} + \left(X_{d_1} - X'_{d_1} \right) I_{d_1}$$

$$= 1.1332 + (0.269 - 0.0608) \times 0.5376$$

$$= 1.2451$$

- Step 7

$$V_{R_1} = (K_{E_1} + S_{E_1}(E_{fd_1})) E_{fd_1}$$

$$= \left(K_{E_1} + 0.0039 e^{1.555 E_{fd_1}} \right) E_{fd_1}$$

$$= \left\{ 1.0 + 0.0039 \times e^{(1.555 \times 1.2451)} \right\} \times 1.2451$$

$$= 1.2788$$

$$R_{F_1} = \frac{K_{F_1}}{T_{F_1}}E_{fd_1} = \frac{0.063}{0.35} \times 1.2451 = 0.2241$$

$$V_{ref_1} = V_1 + \left(\frac{V_{R_1}}{K_{A_1}}\right) = 1.04 + \frac{1.2788}{35} = 1.0765$$

$$T_{M_1} = E'_{d_1}I_{d_1} + E'_{q_1}I_{q_1} + \left(X'_{q_1} - X'_{d_1}\right)I_{q_1}I_{d_1}$$
$$= 0 \times 0.5376 + 1.1332 \times 0.6816 + (0.0969 - 0.0608) \times 0.6816 \times 0.5376$$
$$= 0.7856$$

Machine 2
- Step 1

$$I_{G_2}e^{j\gamma_2} = \frac{(P_{G_2} - jQ_{G_2})}{V_2^*} = \frac{1.63 - j0.304}{1.025\angle 9.48°}$$

$$= 1.6174 - j0.00308$$

$$= 1.6177\angle - 1.0898°$$

- Step 2

  The machine rotor angle

$$\delta_2 = \text{Angle of } \left\{V_2e^{j\theta_2} + \left(R_{s_2} + jX'_{q_2}\right)I_{G_2}e^{j\gamma_2}\right\}$$
$$= \text{Angle of } \{1.025\angle 9.48° + (0.089 + j0.8645) \times 1.6177\angle - 1.0898°\}$$
$$= \text{Angle of } \{(1.0110 - j0.1687) + (0.089 + j0.8645) \times (1.6174 - j0.0308)\}$$
$$= \text{Angle of } 1.1815 + j1.2267$$
$$= 46.098°$$

- Step 3

$$I_{d_2} + jI_{q_2} = I_{G_2}e^{j(\gamma_2 - \delta_2 + \pi/2)}$$

$$= 1.6177 \times e^{j(-1.0898° - 46.098° + 90°)}$$

$$= 1.6177 \times e^{42.813°}$$

$$= 1.6177 \times (\cos 42.813° + \sin 42.813°)$$

$$= 1.1871 + j1.0989$$

$$\therefore I_{d_2} = 1.1871 \text{ and } I_{q_2} = 1.0989$$

$$V_{d_2} + jV_{q_2} = V_2 e^{j(\theta_2 - \delta_2 + \pi/2)}$$

$$= 1.025 \times e^{j(9.48° - 46.098° + 90°)}$$

$$= 1.025 \times (\cos 53.382° + j\sin 53.382°)$$

$$= 0.8457 + j0.5792$$

$$\therefore V_{d_2} = 0.8457 \quad \text{and} \quad V_{q_2} = 0.5792$$

- Step 4

$$E'_{d_2} = \left(X_{q_2} - X'_{q_2} \right) I_{q_2}$$

$$= (0.8645 - 0.8645) \times 1.0989 = 0$$

- Step 5

$$E'_{q_2} = V_{q_2} + R_{s_2} I_{q_2} + X'_{d_2} I_{d_2}$$

$$= 0.5792 + 0.089 \times 1.0989 + 0.1198 \times 1.1871$$

$$= 0.8192$$

- Step 6

$$E_{fd_2} = E'_{q_2} + \left(X_{d_2} - X'_{d_2} \right) I_{d_2}$$

$$= 0.8192 + (0.8958 - 0.1198) \times 1.1871$$

$$= 1.7404$$

- Step 7

$$V_{R_2} = (K_{E_2} + S_{E_2}(E_{fd_2})) E_{fd_2}$$

$$= \left\{ 1.0 + 0.0039 \times e^{(1.555 \times 1.7404)} \right\} \times 1.7404$$

$$= 1.8420$$

$$R_{F_2} = \frac{K_{F_2}}{T_{F_2}} E_{fd_2} = \frac{0.063}{0.35} \times 1.7404 = 0.3133$$

$$V_{ref_2} = V_2 + \left(\frac{V_{R_2}}{K_{A_2}} \right) = 1.025 + \frac{1.8420}{35} = 1.0776$$

$$T_{M_2} = E'_{d_2} I_{d_2} + E'_{q_2} I_{q_2} + \left(X'_{q_2} - X'_{d_2} \right) I_{q_2} I_{d_2}$$

$$= 0 \times 1.1871 + 0.8192 \times 1.0989 + (0.8645 - 0.1198) \times 1.0989 \times 1.1871$$

$$= 1.8717$$

Machine 3
- Step 1

$$I_{G_3}e^{j\gamma_2} = \frac{(P_{G_3} - jQ_{G_3})}{V_3^*} = \frac{0.85 - j0.142}{1.025\angle 4.77°}$$
$$= 0.8379 - j0.0691$$
$$= 0.8408\angle - 4.7190°$$

- Step 2

The machine rotor angle

$$\delta_3 = \text{Angle of} \left\{ V_3 e^{j\theta_3} + \left(R_{s_3} + jX'_{q_3} \right)I_{G_3}e^{j\gamma_3} \right\}$$
$$= \text{Angle of} \{1.025\angle 4.77° + (0.089 + j1.2578) \times 0.8408\angle - 4.7190°\}$$
$$= \text{Angle of} \{(1.0215 - j0.0852) + (0.089 + j1.2578) \times (0.8379 - j0.0691)\}$$
$$= \text{Angle of } 1.1830 + j0.9626$$
$$= 39.154°$$

- Step 3

$$I_{d_3} + jI_{q_3} = I_{G_3}e^{j(\gamma_3 - \delta_3 + \pi/2)}$$
$$= 0.8408 \times e^{j(-4.7190° - 39.154° + 90°)}$$
$$= 0.8408 \times e^{46.127°}$$
$$= 0.5830 + j0.6058$$
$$\therefore I_{d_3} = 0.5830 \text{ and } I_{q_3} = 0.6058$$

$$V_{d_3} + jV_{q_3} = V_3 e^{j(\theta_3 - \delta_3 + \pi/2)}$$
$$= 1.025 \times e^{j(4.77° - 39.154° + 90°)}$$
$$= 0.7114 + j0.7380$$
$$\therefore V_{d_3} = 0.7114 \text{ and } V_{q_3} = 0.7380$$

- Step 4

$$E'_{d_3} = \left(X_{q_3} - X'_{q_3} \right)I_{q_3} = (1.2578 - 1.2578) \times 0.6058 = 0$$

- Step 5

$$E'_{q_3} = V_{q_3} + R_{s_3}I_{q_3} + X'_{d_3}I_{d_3}$$
$$= 0.7380 + 0.089 \times 0.6058 + 0.1813 \times 0.5830$$
$$= 0.8976$$

- Step 6

$$E_{fd_3} = E'_{q_3} + \left(X_{d_3} - X'_{d_3}\right)I_{d_3}$$
$$= 0.8976 + (1.998 - 0.1813) \times 0.5830 = 1.9566$$

- Step 7

$$V_{R_3} = (K_{E_3} + S_{E_3}(E_{fd_3}))E_{fd_3}$$
$$= \left\{1.0 + 0.0039 \times e^{(1.555 \times 1.9566)}\right\} \times 1.9566 = 2.1166$$

$$R_{F_3} = \frac{K_{F_3}}{T_{F_3}}E_{fd_3} = \frac{0.063}{0.35} \times 1.9566 = 0.3522$$

$$V_{ref_3} = V_3 + \left(\frac{V_{R_3}}{K_{A_3}}\right) = 1.025 + \frac{2.1166}{35} = 1.0855$$

$$T_{M_3} = E'_{d_3}I_{d_3} + E'_{q_3}I_{q_3} + \left(X'_{q_3} - X'_{d_3}\right)I_{q_3}I_{d_3}$$
$$= 0 \times 0.5830 + 0.8976 \times 0.6058 + (1.2578 - 0.1813) \times 0.5830 \times 0.6058$$
$$= 0.9240$$

5.4 IDENTIFICATION OF ELECTROMECHANICAL SWING MODES

5.4.1 Participation factor analysis

Participation factor is a tool for identifying the state variables that have significant participation in a selected mode among many modes in a multigenerator power system [5]. It is natural to say that the significant state variables for an eigenvalue λ_p are those that correspond to large entries in the corresponding eigenvector ϕ_p. But the problem of using right and left eigenvector entries individually for identifying the relationship between the states and the modes is that the elements of the eigenvectors are dependent on dimension and scaling associated with the state variables. As a solution of this problem, a matrix called the participation matrix (P) is suggested in which the right and left eigenvectors entries are combined, and it is used as a measure of the association between the state variables and the modes:

$$P = [P_1 \quad P_2 \quad \cdots \quad P_r]$$

with

$$P_p = \begin{bmatrix} P_{1p} \\ P_{2p} \\ \vdots \\ P_{rp} \end{bmatrix} = \begin{bmatrix} \phi_{1p}\psi_{p1} \\ \phi_{2p}\psi_{p2} \\ \vdots \\ \phi_{rp}\psi_{pr} \end{bmatrix} \tag{5.115}$$

where $\phi_{\kappa p}$ is the element on the κth row and pth column of the modal matrix, Φ is the κth entry of the right eigenvector ϕ_p, $\psi_{p\kappa}$ is the element on the pth row and κth column of the modal matrix, and Ψ is the κth entry of the left eigenvector ψ_p.

The element $P_{\kappa p} = \phi_{\kappa p}\psi_{p\kappa}$ is termed the *participation factor*. It is a measure of the relative participation of the κth state variable in the pth mode, and vice versa.

Since $\phi_{\kappa p}$ measures the *activity* of the variable X_κ in the pth mode, and $\psi_{p\kappa}$ *weighs* the contribution of this activity to the mode, the product $P_{\kappa p}$ measures the *net participation*. The effect of multiplying the elements of the left and right eigenvectors makes the $P_{\kappa p}$ dimensionless. In view of the eigenvector normalization, the sum of the participation factors associated with any mode $\sum_{p=1}^{r} P_{\kappa p}$ or with any state variable $\left(\sum_{\kappa=1}^{r} P_{\kappa p}\right)$ is equal to 1. For a given autonomous linear system

$$\Delta \dot{X} = A_{sys}\Delta X \tag{5.116}$$

the participation factor is actually a measurement of sensitivity of the eigenvalue λ_p to the diagonal element $a_{\kappa\kappa}$ of the state matrix A. This is defined as

$$P_{\kappa p} = \frac{\partial \lambda_p}{\partial a_{\kappa\kappa}}, \quad \kappa = 1, 2, \ldots, r \tag{5.117}$$

The participation factor may also be defined by

$$P_{\kappa p} = \frac{\psi_{\kappa p}\phi_{\kappa p}}{\psi_p^T \phi_p} \tag{5.118}$$

where $\psi_{\kappa p}$ and $\phi_{\kappa p}$ are the κth entries in the left and right eigenvector associated with the pth eigenvalue.

EXAMPLE 5.1

Compute eigenvalues and the participation matrix (P) of the following system matrix:

$$A = \begin{bmatrix} -0.4 & 0 & -0.01 \\ 1 & 0 & 0 \\ -1.4 & 9.8 & -0.02 \end{bmatrix}$$

Solution

Eigenvalues are $\lambda_1 = -0.6565$; $\lambda_2, \lambda_3 = 0.1183 \pm j0.3678$
     The right eigenvectors are given by

$$\left(A - \lambda_p I\right)\phi = 0$$

Therefore,

$$\begin{bmatrix} -0.4 - \lambda_p & 0 & -0.01 \\ 1 & -\lambda_p & 0 \\ -1.4 & 9.8 & -0.02 - \lambda_p \end{bmatrix} \begin{bmatrix} \phi_{1p} \\ \phi_{2p} \\ \phi_{3p} \end{bmatrix} = 0$$

$$(-0.4 - \lambda_p)\phi_{1p} - 0.01\phi_{3p} = 0$$

$$\phi_{1p} - \lambda_p \phi_{2p} = 0$$

$$-1.4\phi_{1p} + 9.8\phi_{2p} + (-0.02 - \lambda_p)\phi_{3p} = 0$$

Solving this homogeneous equations for $p = 1, 2, 3$ for the eigenvalues λ_1, λ_2, and λ_3, respectively, the right eigenvectors are

$$\phi = \begin{bmatrix} -0.0389 & -0.0128 + j0.0091 & -0.0128 - j0.0091 \\ 0.0592 & 0.0123 + j0.0388 & 0.0123 - j0.0388 \\ -0.9975 & -0.9990 & -0.9990 \end{bmatrix}$$

The left eigenvectors are normalized so that $\phi\psi = I$ and are given by

$$\psi = \phi^{-1} = \frac{\text{adj}(\phi)}{|\phi|}$$

$$\therefore \psi = \begin{bmatrix} -14.6095 & -7.2933 - j13.4578 & -7.2933 + j13.4578 \\ 3.4262 & 1.7104 - j9.7317 & 1.7104 + j9.7317 \\ -0.2295 & 0.3859 - j0.0532 & 0.3859 + j0.0532 \end{bmatrix}$$

Therefore, the *participation matrix* is

$$P = \begin{bmatrix} \phi_{11}\psi_{11} & \phi_{12}\psi_{21} & \phi_{13}\psi_{31} \\ \phi_{21}\psi_{12} & \phi_{22}\psi_{22} & \phi_{23}\psi_{32} \\ \phi_{31}\psi_{13} & \phi_{32}\psi_{23} & \phi_{33}\psi_{33} \end{bmatrix}$$

i.e.,

$$P = \begin{bmatrix} 0.5681 & 0.2159 + j0.01062 & 0.2159 + j0.01062 \\ 0.2029 & 0.3985 - j0.0530 & 0.3985 + j0.0530 \\ 0.2289 & 0.3855 - j0.0532 & 0.3855 + j0.0532 \end{bmatrix}$$

Taking only the magnitudes,

$$P = \begin{bmatrix} 0.5681 & 0.2406 & 0.2406 \\ 0.2029 & 0.4020 & 0.4020 \\ 0.2289 & 0.3892 & 0.3892 \end{bmatrix} \begin{matrix} x_1 \\ x_2 \\ x_3 \end{matrix}$$
$$\quad\quad\; \lambda_1 \quad\; \lambda_2 \quad\; \lambda_3$$

5.4.2 **Swing mode and participation ratio**

The swing mode of a power system can be identified by the criterion proposed here as follows. Where authors used a swing mode identification index is called the swing-loop *participation ratio* [6]. The participation ratio was originally introduced based on the concept of participation factor. The characteristics of the swing modes are that they are closely related to the electromechanical swing-loops associated with the relevant state variables like rotor angle ($\Delta\delta$) and machine speed ($\Delta\omega$). The swing-loop participation ratio (ρ) for the pth mode is defined by

$$\rho_p = \left| \frac{\sum_{\kappa=1}^{\ell} P_{\kappa p}}{\sum_{\kappa=\ell+1}^{r} P_{\kappa p}} \right| \tag{5.119}$$

where $P_{\kappa p}$ is the participation factor of the κth state variable for the pth electromechanical mode. Here, "ℓ" represents the number of relevant states belonging to the state variable set $[\Delta\delta, \Delta\omega]$, and "$r$" is the total number of states (relevant and nonrelevant) belonging and not belonging to the state variable set $[\Delta\delta, \Delta\omega]$. As the sum of the participation factor for a particular mode corresponding to all relevant and nonrelevant states is equal to 1, Equation (5.119) can also be written as

$$\rho_p = \left| \frac{\sum_{\kappa=1}^{\ell} P_{\kappa p}}{1 - \sum_{\kappa=1}^{\ell} P_{\kappa p}} \right| \tag{5.120}$$

The proposed criterion states that for the swing modes, the oscillation frequency is in the range of 0.2-2.5 Hz. and its swing-loop participation ratio (ρ)$\gg 1$.

5.5 **AN ILLUSTRATION: A TEST CASE**

The small-signal stability of a power system may be analyzed using any of the methods applicable to linear systems. However, the modal analysis approach using eigenvalue and swing-mode computation techniques is a very fundamental tool. The poorly damped electromechanical oscillations have negative impact on the power transfer capability in a power system, and in some cases, it induces stress in the mechanical shaft. In this section, eigenvalues and swing modes of a multimachine system have been computed prior to application of PSS, and the critical swing mode is identified from them. In the following chapters, the improvement of damping of this critical mode has been observed with the application of PSS and FACTS devices.

- **Calculation of eigenvalue and swing mode:**

  A WSCC type 3-machine, 9-bus system (Figure 5.5) has been considered as a test system. The base is 100 MVA, and system frequency is 50 Hz. The machine #1 is treated as slack bus. Uniform damping has been assumed in all three machines. The converged load-flow data are obtained by running the standard load-flow program given in Table B.2 (Appendix B). The constant power loads are treated as

Table 5.1 Eigenvalues and Swing Modes of the Study System

#	Eigenvalue (λ)	Frequency (f) (Hz)	Damping Ratio (ζ)	Swing-Loop Participation Ratio ($|\rho|$)
1	$-2.4892 \pm j10.8650$	1.7290	0.2233	10.1575
2	$-5.1617 \pm j11.2755$	1.7943	0.4162	12.4678
3	$-5.3063 \pm j10.3299$	1.6438	0.4569	0.0406
4	$-5.6837 \pm j10.3601$	1.6486	0.4810	0.2146
5	$-5.5957 \pm j10.3330$	1.6443	0.4762	0.0102
6	-2.5226	0	1.0000	2.1054
7	0.0000	0	1.0000	∞
8	$-0.4087 \pm j0.8293$	0.1320	0.4421	0.0625
9	$-0.4759 \pm j0.5616$	0.0894	0.6465	0.0933
10	$-0.4164 \pm j0.6618i$	0.1053	0.5325	0.0536
11	-3.2258	0	1.0000	0
12	-1.8692	0	1.0000	0
13	-1.6667	0	1.0000	0

injected into the buses. Steady-state variables and initial conditions are obtained by the procedure described in Section 5.3. A MATLAB program has been developed to obtain the system matrix (A_{sys}) and the eigenvalues and the electromechanical modes. The eigenvalues and the electromechanical modes of the system are listed in Table 5.1. The frequencies (f) and damping ratios (ζ) of the electromechanical modes are calculated by assuming that $\lambda = \delta + j\omega_d$ and $\zeta = \frac{\sigma}{\sqrt{\sigma^2 + \omega_d^2}}$, where $\omega_d = 2\pi f$ (rad/s).

It has been observed from the Table 5.1 that among 21 eigenvalues, one has zero magnitude, 4 are real, and the rest (16) are complex conjugate. The real parts contribute damping, and the imaginary part is responsible for electromechanical oscillation to the system. It is found from the 4th column of the Table 5.1 that the damping ratio (ζ) of the electromechanical mode #1 (λ_1) is the smallest and therefore the behavior of this mode is important to study the small-signal stability of the system. This mode has been referred to as the *critical swing mode*. The mode frequency and the participation factor analysis suggest that the nature of this critical mode is a local mode and is strongly associated with the machine 2 and the system states ($\Delta\delta, \Delta\omega$).

The swing-loop participation ratios for each electromechanical mode have been shown in column 5 of Table 5.1, which interprets that the mode #1 and mode #2 are satisfying the criterion for the swing modes (Section 5.4.2) and among which, mode #1 is the most *critical swing mode*.

Hence, the *PSS should be placed at an optimum location, so that it can yield maximum damping to the electromechanical oscillation of the critical swing mode (#1).* The application of PSS and its optimum location have been discussed in the following chapters.

EXERCISES

5.1 Write down the (i) DAEs and the (ii) stator algebraic equations of a synchronous machine for two-axis (*d–q*) model. Obtain the linearized state-space
equations from them.

5.2 In a multimachine system, write down the network equations pertaining to the
generator (*P–V*) buses and the load (*P–Q*) bus in power-balance and in current
balance forms. Formulate the load-flow equations in real power and reactive
power-balance form.

5.3 Explain the participation factor and participation ratio. Explain how swing
modes of any multimachine system can be identified applying participation
ratio analysis.

5.4 Find the participation factors of the eigenvalues for the following system
matrices, where $\dot{x} = Ax + Bu$

(**a**) $A = \begin{bmatrix} 3 & 7 \\ 2 & 4 \end{bmatrix}$:

(**b**) $A = \begin{bmatrix} 2 & 1 & 1 \\ 0 & 3 & 1 \\ 0 & 4 & -1 \end{bmatrix}$

5.5 The single-line diagram of two-area multimachine power system is given in
Figure 5.6. The system comprises four generators, 15 transmission lines,
and 6 numbers of load buses. Using two-axis model for the generator and constant power load representation, obtain
(**a**) the initial conditions of the state variables of the system,
(**b**) eigenvalues and the swing modes of the system.

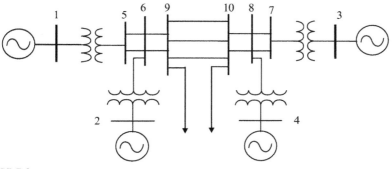

FIGURE 5.6

Two-area power system.

The transmission line data, machine data, excitation system data, and load-flow data are given in Section B.3 of Appendix B.

5.6 Repeat the problem given in 5.5 for 15% load increase in bus 9 and for one tie-line outage between buses 9 and 10.

References

[1] P.W. Sauer, M.A. Pai, Power System Dynamics and Stability, Pearson Education Pte. Ltd., Singapore, 1998.

[2] M.A. Pai, D.P. Sengupta, K.R. Padiyar, Small Signal Analysis of Power Systems, Narosa Publishing House, India, 2004.

[3] Y. Obata, S. Takeda, H. Suzuki, An efficient eigenvalue estimation technique for multi-machine power system dynamic stability analysis, IEEE Trans. Power Apparatus Syst. PAS-100 (1) (1981) 259–263.

[4] N. Martins, Efficient eigenvalue and frequency response methods applied to power system small-signal stability studies, IEEE Trans. Power Syst. PWRS-1 (1) (1986) 217–224.

[5] C.L. Chen, Y.Y. Hu, An efficient algorithm for design of decentralized output feedback power system stabilizer, IEEE Trans. Power Syst. 3 (3) (1988) 999–1004.

[6] E.Z. Zhou, O.P. Malik, G.S. Hope, A reduced-order iterative method for swing mode computation, IEEE Trans. Power Syst. 6 (3) (1991) 1224–1230.

Mitigation of Small-Signal Stability Problem Employing Power System Stabilizer

6

6.1 INTRODUCTION

In Chapter 4, it has been discussed the need of installing a power system stabilizer (PSS) in a power system in order to introduce additional damping to the rotor oscillations of the synchronous machine. The enhancement of damping in power systems by means of a PSS has been a subject of great attention in the past three decades [1–4]. It is much more significant today when many large and complex power systems frequently operate close to their stability limits.

In this chapter, the problem of small-signal stability has been investigated by applying the conventional PSS. A speed input single-stage PSS has been applied in the linearized model of a SMIB power system, and then, the application of PSS has been extended in a multimachine network. In both cases, investigation is carried out by studying the behavior of the critical eigenvalue or the critical swing mode.

A PSS is a lead-lag compensator; it uses any auxiliary stabilizing signal, like the machine speed, as input and compensates the phase lag introduced by the machine and excitation system and must produce a component of electric torque in phase with the rotor speed change so as to increase the damping of the rotor oscillations. Though there is a common perception that the application of PSS is almost a mandatory requirement on all generators in modern multimachine power network, but in developing countries, where power networks are mostly longitudinal in nature, with constrained economic limits, the use of a costly PSS with each and every generator is not done. In view of this requirement, this chapter also discussed the method of selection of optimal location of a PSS in a multimachine power system. There are several methods of PSS location selection that are available in the literatures [5–9], among which participation factor (PF) analysis, sensitivity of PSS effect (SPE), and the optimum PSS location index (OPLI) are discussed. A WSCC type 3-machine, 9-bus test system has been taken as a test case.

6.2 THE APPLICATION OF PSS IN AN SMIB SYSTEM

A simple single-machine infinite bus (SMIB) system has been shown in Figure 6.1. It is assumed that the machine is equipped with a fast exciter. In order to improve small-signal oscillations, a PSS is incorporated in this system.

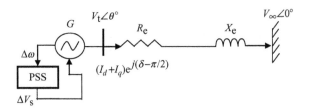

FIGURE 6.1

Single-machine infinite bus system with PSS.

6.2.1 Combined model of SMIB system with PSS

A PSS is a lead-lag compensator, which produces a component of electric torque to damp generator rotor oscillations by controlling its excitation. The basic block diagram of a speed input single-stage PSS, which acts through excitation system, is depicted in Figure 6.2.

Neglecting washout stage, the linearized Heffron–Phillips model of the SMIB system, including PSS dynamics, can be represented by the following state-space equations:

$$\Delta \dot{E}'_q = -\frac{1}{K_3 T'_{do}} \Delta E'_q - \frac{K_4}{T'_{do}} \Delta \delta + \frac{1}{T'_{do}} \Delta E_{fd} \tag{6.1}$$

$$\Delta \dot{\delta} = \omega_s \Delta v \tag{6.2}$$

$$\Delta \dot{v} = -\frac{K_2}{2H} \Delta E'_q - \frac{K_1}{2H} \Delta \delta - \frac{D \omega_s}{2H} \Delta v + \frac{1}{2H} \Delta T_M \tag{6.3}$$

$$\Delta \dot{E}_{fd} = -\frac{1}{T_A} \Delta E_{fd} - \frac{K_A K_5}{T_A} \Delta \delta - \frac{K_A K_6}{T_A} \Delta E'_q + \frac{K_A}{T_A} \Delta V_{ref} \tag{6.4}$$

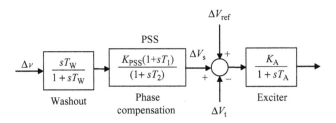

FIGURE 6.2

Exciter with PSS.

$$\Delta \dot{V}_s = -\frac{1}{T_2}\Delta V_s - \frac{K_{PSS}T_1}{T_2}\frac{K_2}{2H}\Delta E'_q - \frac{K_{PSS}T_1}{T_2}\frac{K_1}{2H}\Delta \delta$$

$$+ \left(\frac{K_{PSS}}{T_2} - \frac{K_{PSS}T_1}{T_2}\frac{D\omega_s}{2H} \right)\Delta v \qquad (6.5)$$

where $K_2 = \dfrac{\partial P_e}{\partial E'_q}$ and $K_1 = \dfrac{\partial P_e}{\partial \delta}$. Assuming the stator resistance $R_s = 0$, the electric

power $P_e = \dfrac{E'_q V_\infty}{X_T}\sin\delta$, where $X_T = X'_d + X_e$.

Here, Equation (6.5) is added to the general equations (6.1)–(6.4) of the SMIB system because of the installation of a PSS. The system matrix (A_PSS) of this combined model has been presented in Equation (6.6). The system matrix without PSS can be easily obtained by excluding the PSS output state (V_s):

$$A\_PSS = \begin{bmatrix} -\dfrac{1}{K_3 T'_{do}} & -\dfrac{K_4}{T'_{do}} & 0 & \dfrac{1}{T'_{do}} & 0 \\[2ex] 0 & 0 & \omega_s & 0 & 0 \\[2ex] -\dfrac{K_2}{2H} & -\dfrac{K_1}{2H} & -\dfrac{D\omega_s}{2H} & 0 & 0 \\[2ex] -\dfrac{K_A K_6}{T_A} & -\dfrac{K_A K_5}{T_A} & 0 & -\dfrac{1}{T_A} & \dfrac{K_A}{T_A} \\[2ex] -\dfrac{K_2 T_1}{T_2}\left(\dfrac{K_{PSS}}{2H}\right) & -\dfrac{K_1 T_1}{T_2}\left(\dfrac{K_{PSS}}{2H}\right) & \left(\dfrac{K_{PSS}}{T_2} - \dfrac{K_{PSS}T_1}{T_2}\dfrac{D\omega_s}{2H}\right) & 0 & \dfrac{1}{T_2} \end{bmatrix}$$

$$(6.6)$$

The washout filter stage is neglected here, since its objective is to offset the dc steady-state error and not have any effect on phase shift or gain at the oscillating frequency. The application of washout stage is not a critical task. Its dynamics can be included easily with suitable choice of the parameter T_W. The value of T_W is generally set within 10-20 s.

6.2.2 Results and discussion

- **Eigenvalue analysis**

In this section, eigenvalues and the electromechanical swing modes of a SMIB system are computed in MATLAB from the system matrix, A_PSS, presented in Equation (6.6). The machine and exciter data are given in Section B.1 of Appendix B. The eigenvalues of the system without and with PSS are listed in Table 6.1. It is evident that the damping ratio of the electromechanical swing mode #2 (second row, third column) is small compared to the other mode; therefore, the behavior of this mode is more important to study the small-signal stability problem of this system and this mode has been referred to as the *critical mode*. When a PSS is installed in the system, the damping ratio of this critical mode #2 is enhanced significantly. The value of the damping ratio with PSS has been shown in the second row, column six of Table 6.1.

Table 6.1 Eigenvalues Without and With PSS

	Before Application of PSS			After Application of PSS	
#	Eigenvalue	Damping Ratio	PSS Parameters	Eigenvalue	Damping Ratio
1	$-2.6626 \pm j15.136$	0.1733	$K_{PSS} = 1.0$	$-2.0541 \pm j15.3253$	0.1328
2	$-0.05265 \pm j7.3428$	0.0072	$T_1 = 0.5$ s	$-0.4116 \pm j7.1110$	0.0578
3	–	–	$T_2 = 0.1$ s	-10.4989	1.0

- **Time domain analysis**

The small-signal stability response of this system has been examined further by plotting the rotor angle deviation under different values of the PSS gain (K_{PSS}) for a unit change in mechanical step power input (ΔT_M) with a reasonable simulation time of 600 s. It has been observed that the application of PSS introduces significant improvement in damping on the rotor angle oscillations and is shown in Figure 6.3a. It has been further observed that the better enhancement of damping and settling time can be achieved with the increase of the PSS gain and the corresponding plot is presented in Figure 6.3b. Thus, it may be reasonable to remark that the installation of PSS in a SMIB system not only damps the rotor angle oscillations effectively but also enhances its performance with increasing PSS gain.

6.3 MULTIMACHINE SMALL-SIGNAL STABILITY IMPROVEMENT

The small-signal model of a multimachine system with IEEE Type I exciter has been described in Chapter 5 (Section 5.2.1). All equations relating to the performance of the machine with exciter and network power flow were linearized around the nominal operating condition to obtain the dynamic model of the system for eigenvalue analysis and are represented by the following state-space equations:

$$\Delta \dot{X} = A_1 \Delta X + B_1 \Delta I_g + B_2 \Delta V_g + E_1 \Delta U \tag{6.7}$$

$$0 = C_1 \Delta X + D_1 \Delta I_g + D_2 \Delta V_g \tag{6.8}$$

$$0 = C_2 \Delta X + D_3 \Delta I_g + D_4 \Delta V_g + D_5 \Delta V_1 \tag{6.9}$$

$$0 = D_6 \Delta V_g + D_7 \Delta V_1 \tag{6.10}$$

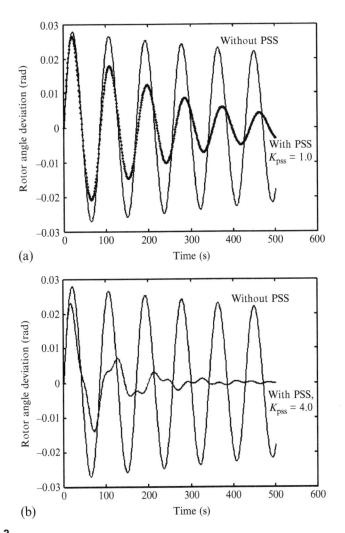

FIGURE 6.3

Response of rotor angle deviation with different PSS gain (a) with $K_{PSS}=1$ and (b) with $K_{PSS}=4$.

6.3.1 **Multimachine model with PSS**

The multimachine model with a PSS has been formulated by adding the state variable (V_s) associated with the PSS in Equations (6.7)–(6.9). Therefore, the order of the system matrix and the number of eigenvalues of the system will be increased by one.

The added state equation corresponding to the PSS can be obtained from Chapter 3, Section 3.8 as

$$\Delta \dot{V}_{s_i} = -\frac{1}{T_2}\Delta V_{s_i} + \frac{K_{PSS}}{T_2}\frac{\Delta \omega_i}{\omega_s} + \frac{K_{PSS}T_1}{T_2}\frac{\Delta \dot{\omega}_i}{\omega_s} \tag{6.11}$$

Substituting the expression of $\Delta \dot{\omega}_i$ from Equation (5.21) (Chapter 5), the previously mentioned Equation (6.11) becomes

$$\Delta \dot{V}_{s_i} = -\frac{1}{T_2}\Delta V_{s_i} + \left(\frac{K_{PSS}}{T_2\omega_s} - \frac{K_{PSS}T_1D_i}{T_2\omega_sM_i}\right)\Delta\omega_i$$

$$+ \left(-\frac{K_{PSS}T_1E'_{q_i}}{T_2\omega_sM_i} + \frac{K_{PSS}T_1X'_{d_i}I_{d_i}}{T_2\omega_sM_i} - \frac{K_{PSS}T_1X'_{q_i}I_{d_i}}{T_2\omega_sM_i}\right)\Delta I_{q_i}$$

$$+ \left(-\frac{K_{PSS}T_1E'_{d_i}}{T_2\omega_sM_i} + \frac{K_{PSS}T_1X'_{d_i}I_{q_i}}{T_2\omega_sM_i} - \frac{K_{PSS}T_1X'_{q_i}I_{q_i}}{T_2\omega_sM_i}\right)\Delta I_{d_i}$$

$$- \frac{K_{PSS}T_1I_{q_i}}{T_2\omega_sM_i}\Delta E'_{q_i} - \frac{K_{PSS}T_1I_{d_i}}{T_2\omega_sM_i}\Delta E'_{d_i} + \frac{K_{PSS}T_1}{T_2\omega_sM_i}\Delta T_{M_i} \tag{6.12}$$

The state variables are then modified as

$$\Delta X = \begin{bmatrix} \Delta X_1^T & \Delta X_2^T & \cdots & \Delta X_m^T \end{bmatrix}^T$$

where

$$\Delta X_i = \begin{bmatrix} \Delta\delta_i & \Delta\omega_i & \Delta E'_{q_i} & \Delta E'_{d_i} & \Delta E_{fd_i} & \Delta V_{R_i} & \Delta R_{F_i} & \Delta V_{s_i} \end{bmatrix}^T$$
$$\Delta I_g = \begin{bmatrix} \Delta I_{d_1} & \Delta I_{q_1} & \Delta I_{d_2} & \Delta I_{q_2} & \cdots & \Delta I_{d_m} & \Delta I_{q_m} \end{bmatrix}^T$$
$$\Delta V_g = \begin{bmatrix} \Delta\theta_1 & \Delta V_1 & \Delta\theta_2 & \Delta V_2 & \cdots & \Delta\theta_m & \Delta V_m \end{bmatrix}^T$$
$$\Delta V_l = \begin{bmatrix} \Delta\theta_{m+1} & \Delta V_{m+1} & \Delta\theta_{m+2} & \Delta V_{m+2} & \cdots & \Delta\theta_n & \Delta V_n \end{bmatrix}^T$$
$$\Delta U = \begin{bmatrix} \Delta U_1^T & \Delta U_2^T & \cdots & \Delta U_m^T \end{bmatrix}^T$$
$$\Delta U_i = \begin{bmatrix} \Delta T_{M_i} & \Delta V_{ref_i} \end{bmatrix}^T$$

for $i = 1, 2, \ldots, m$ (the number of PV buses) and $i = m+1, m+2, \ldots, n$ (the number of PQ buses). Here, in ΔX_i, the state variable of the PSS ΔV_{s_i} is included.

Eliminating ΔI_g from the respective equations (6.7)–(6.9), the overall state-space model is

$$\begin{bmatrix} \Delta \dot{X} \\ 0 \end{bmatrix} = \begin{bmatrix} A' & B' \\ C' & D' \end{bmatrix}\begin{bmatrix} \Delta X \\ \Delta V_N \end{bmatrix} + \begin{bmatrix} E_1 \\ 0 \end{bmatrix}\Delta U \tag{6.13}$$

where $\Delta V_N = \begin{bmatrix} \Delta V_g & \Delta V_l \end{bmatrix}^T$; $A' = [A_1 - B_1D_1^{-1}C_1]$; $B' = \begin{bmatrix} B_2 - B_1D_1^{-1}D_2 & 0 \end{bmatrix}$; $C' = \begin{bmatrix} K_2 \\ 0 \end{bmatrix}$;

and $D' = \begin{bmatrix} K_1 & D_5 \\ D_6 & D_7 \end{bmatrix}$ with $K_1 = \begin{bmatrix} D_4 - D_3D_1^{-1}D_2 \end{bmatrix}$ and $K_2 = \begin{bmatrix} C_2 - D_3D_1^{-1}C_1 \end{bmatrix}$.

Therefore, the system matrix A_PSS for a multimachine system with a PSS can be obtained as

$$[A\_PSS]_{(7m+1)\times(7m+1)} = [A'] - [B'][D']^{-1}[C'] \qquad (6.14)$$

This model has been used in the following section to study the effect of small-signal disturbance on the eigenvalues of a multimachine power system and their improvement with the application of a PSS. The system matrix without PSS can be obtained excluding the state-space variable ΔV_{s_i}.

6.3.2 An illustration-computation of eigenvalues and swing modes

In Chapter 5, the computation of system matrix (A_{sys}) and the electromechanical swing modes without PSS has been illustrated for a WSCC type 3-machine, 9-bus test system and results are presented in Table 5.1. In this section, the PSS has been applied to this proposed test system in order to improve the damping ratio of the critical swing mode #1. The parameters of the PSS are assumed as $K_{PSS} = 20$, $T_1 = 0.15$, and $T_2 = 0.11$. Even though the critical swing mode has a 22% damping without PSS, further enhancement of the system stability is achieved by the application of PSS via shifting the critical swing mode to a more desirable position in the s-plane. The PSS is installed with each machine separately and the eigenvalues, frequency, and its damping ratios are listed in Tables 6.2, 6.3, and 6.4, respectively.

It can be seen that the swing modes get affected with the installation of the PSS at any of the three machines. However, with the response of the critical swing mode being of prime concern, it has been observed that the improvement in the critical swing mode is of highest degree (Table 6.3) if the PSS is installed in machine 2.

Table 6.2 Eigenvalue, Frequency, and Corresponding Damping Ratio with PSS Installed at Machine 1

Mode #	Eigenvalue (λ)	Frequency (f) (Hz)	Damping Ratio (ζ)
1	$-2.5291 \pm j10.8920$	1.7333	0.2262
2	-9.1046	0	1.0000
3	$-5.1630 \pm j11.3145$	1.8005	0.4151
4	$-5.2715 \pm j10.2838$	1.6365	0.4562
5	$-5.6843 \pm j10.3443$	1.6461	0.4816
6	$-5.5913 \pm j10.3325i$	1.6443	0.4759
7	-2.5070	0	1.0000
8	0.0000	0	1.0000
9	$-0.5562 \pm j0.5387$	0.0857	0.7183
10	$-0.3346 \pm j0.7015$	0.1116	0.4306
11	$-0.4086 \pm j0.6217$	0.0989	0.5493
12	-3.2258	0	1.0000
13	-1.8692	0	1.0000
14	-1.6667	0	1.0000

Table 6.3 Eigenvalue, Frequency, and Corresponding Damping Ratio with PSS Installed at Machine 2

Mode #	Eigenvalue (λ)	Frequency (f) (Hz)	Damping Ratio (ζ)
1	**−3.5586 ± j10.8354**	**1.7243**	**0.3120**
2	−9.2182	0	1.0000
3	−4.9144 ± j 11.2478i	1.7899	0.4004
4	−4.7038 ± j 10.3468i	1.6465	0.4139
5	−5.6886 ± j 10.3285i	1.6436	0.4824
6	−5.3437 ± j 10.2399i	1.6295	0.4626
7	−2.5030	0	1.0000
8	0.0000	0	1.0000
9	−0.4198 ± j 0.8407i	0.1338	0.4467
10	−0.5257 ± j 0.5070i	0.0807	0.7198
11	−0.3291 ± j 0.5646i	0.0898	0.5036
12	−3.2258	0	1.0000
13	−1.8692	0	1.0000
14	−1.6667	0	1.0000

Table 6.4 Eigenvalue, Frequency, and Corresponding Damping Ratio with PSS Installed at Machine 3

Mode #	Eigenvalue (λ)	Frequency (f) (Hz)	Damping Ratio (ζ)
1	**−2.4834 ± j 10.8865**	**1.7324**	**0.2224**
2	−5.4393 ± j 12.4858i	1.9869	0.3994
3	−9.2964	0	1.0000
4	−5.3085 ± j 8.9013i	1.4165	0.5122
5	−5.3364 ± j 10.3270i	1.6434	0.4591
6	−5.5996 ± j 10.3330i	1.6443	0.4765
7	−2.5030	0	1.0000
8	0.0000	0	1.0000
9	−0.4160 ± j 0.8356i	0.1330	0.4457
10	−0.4609 ± j 0.4637i	0.0738	0.7050
11	−0.4005 ± j 0.6596i	0.1050	0.5189
12	−3.2258	0	1.0000
13	−1.8692	0	1.0000
14	−1.6667	0	1.0000

The value of damping ratio (ζ) for the installation of PSS in machines 1, 2, and 3 are, respectively, obtained as $\zeta=0.2262$, 0.3120, and 0.2224.

A study of root locus of the critical swing mode with variation of PSS gain has also been investigated. The root-locus plots for the installation of PSS at three machines are shown in Figure 6.4a, b, and c, respectively.

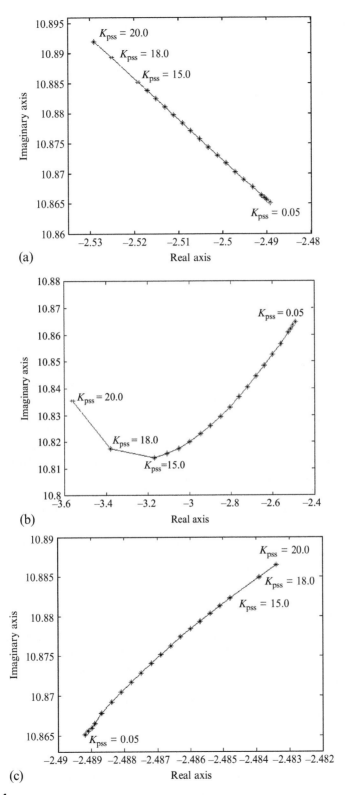

FIGURE 6.4

Root locus of the critical swing mode when PSS is installed at (a) machine 1, (b) machine 2, and (c) machine 3.

In these figures, the real axis and the imaginary axis represent the real part $(-\sigma)$ and the imaginary part $(+j\omega)$ of the critical swing mode (at different values of the PSS gain K_{PSS}), respectively. The negative imaginary part $(-j\omega)$ can be obtained from symmetry. It has been found that with the installation of a PSS in machine 2, the damping of the critical swing mode increases and simultaneously oscillation decreases, unlike the case for machine 1 where with the increase of PSS gain, damping improves, but oscillation also increases. In the application of PSS in machine 3, the critical mode moves towards instability with marginal increase in the gain of PSS. In view of this result, it may be concluded that machine 2 should be the possible choice of installation of the PSS.

6.4 DEVELOPMENT OF A LOCATION SELECTION INDICATOR OF PSS

During the application of PSS to a multimachine power system to achieve the largest improvement in damping, first, the best location of PSS must be identified among a number of interconnected machines. Study reveals that the PSS displaces the swing mode from its critical position to a more desirable position, changing the response of the excitation system. Based on the change of the exciter transfer function, a simple and easy indicator called OPLI [9] has been introduced in Section 6.4.3 to identify the best location of the PSS in a multimachine system. The results of OPLI method have also been compared with the existing SPE method and the PF analysis method. The special emphasis has been given on the comparison between OPLI and SPE methods as it has been reported in [8] that SPE method is more accurate than PF analysis method. It appears that the proposed OPLI method gives identical prediction with the existing methods (SPE and PF) on the selection of optimum location of PSS. It has been further observed that both OPLI and SPE bear almost similar characteristics with variation of PSS gain. The computation of OPLI and its comparison with SPE have been illustrated for a 3-machine, 9-bus test system, and its validity has also been tested in an IEEE type 14-bus test system.

6.4.1 Participation factor

The concept of PF has been discussed in Chapter 5, Section 5.4. Here, this PF analysis method has been applied to find the best location of a PSS in a multimachine power system. The PF is a quantitative measure of how a particular mode (eigenvalue) is affected by the various state variables in the system. In a multimachine system, the PF (PF_i) for the ith machine considering speed deviation ($\Delta\omega_i$) as the respective state variable is defined by

$$PF_i = \phi_{j,\Delta\omega_i}\psi_{j,\Delta\omega_i} \tag{6.15}$$

where $\phi_{j,\Delta\omega_i}$ is the right-eigenvector entry and $\psi_{j,\Delta\omega_i}$ is the left-eigenvector entry of the jth electromechanical swing mode corresponding to the state variable $\Delta\omega_i$ of the *i*th machine. The machine having the highest PF for the most poorly damped swing mode signifies the most effective location of stabilizer application.

6.4.2 Sensitivity of PSS effect

The PSS installed on a machine in a power system is a closed-loop controller. If a machine is selected for installation of PSS, for best effect, first, the amplitude of PSS input that is measured by the right eigenvector corresponding to speed change $\Delta\omega$ should be relatively large, and second, the control effect of PSS measured by the coefficient S_{ji} should be strong.

The control effect of PSS on the system (by the PSS output state ΔV_{s_i} and the system mode λ_j) can be calculated by $S_{ji} = \psi_{j,\Delta E_{fd_i}}$, where $\psi_{j,\Delta E_{fd_i}}$ is the left-eigenvector entry of the *j*th mode (λ_j) corresponding to the state variable ΔE_{fd_i} of the *i*th machine. In order to take into consideration the effect of both the PSS input and the PSS control in selecting the PSS location, SPE for the *i*th machine has been considered as

$$\text{SPE}_i = \phi_{j,\Delta\omega_i}\psi_{j,\Delta E_{fd_i}} \tag{6.16}$$

for $i = 1, 2, \ldots, m$ (the number of machines) where $\phi_{j,\Delta\omega_i}$ is the right-eigenvector entry and $\psi_{j,\Delta E_{fd_i}}$ is the left-eigenvector entry of the *j*th mode corresponding to the state $\Delta\omega_i$ and ΔE_{fd_i} of the *i*th machine. SPE measures both the activity of PSS input ($\Delta\omega_i$) participating in a certain oscillatory mode and the control effect of PSS, on this mode. The larger the magnitude of the SPE, the better is the overall performance of the PSS. In a multimachine power system, there may be several swing modes that are of interest, and for each mode, a set $\{\text{SPE}_i, i = 1, 2, \ldots, m\}$ can be calculated by Equation (6.16). The SPE with the largest magnitude of any *i*th machine identifies the best location of PSS.

Algorithms of calculation of SPE
1. Compute the initial conditions of the state variables.
2. Install the PSS at any *i*th machine and obtain the system matrix $[A_{\text{sys}}]$ and eigenvalues.
3. Identify the critical swing mode.
4. Derive the right-eigenvector matrix $[\phi]$ and a left-eigenvector matrix $[\psi]$.
5. The right-eigenvector entry $\phi_{j,\Delta\omega_i}$ corresponding to the state variable $\Delta\omega_i$ (PSS input) and the left-eigenvector entry $\psi_{j,\Delta E_{fd_i}}$ corresponding to the state variable ΔE_{fd_i} (PSS control effect) are noted for the *j*th critical swing mode for the *i*th machine.
6. Calculate $|\text{SPE}_i|$ for the *i*th machine using Equation (6.16).
7. Repeat steps 1-6 for each machine.

6.4.3 Optimum PSS location index

The newly proposed concept of OPLI is based on the change of exciter transfer function with respect to the PSS transfer function in a certain swing mode. The PSS on a machine is a closed-loop controller and that considers usually the machine speed or power as its input and introduces a damping so that the system moves from a less stable region to a more stable region. As the PSS acts through the excitation system, the effect of displacement of swing modes due to the installation of PSS will change the response of the excitation system. The response of the excitation system at a swing mode λ' can be obtained by replacing λ' for "s" in its transfer function $G_{ex}(s)$.

The change of response of the excitation system with respect to the PSS response for a swing mode λ' is determined by the proposed index OPLI and is defined by

$$|OPLI_i| = \frac{|(G_{ex_i}(\lambda') - G_{ex_i}(\lambda^\circ))|}{|G_{PSS}(\lambda')|} \qquad (6.17)$$

for $i = 1, 2, \ldots, m$ (the number of machines).

Here, λ° and λ' are the critical swing modes before and after the installation of PSS, respectively. The magnitude of OPLI measures the effect of PSS on the exciter response in a swing mode λ' of interest. The larger the value of the OPLI, the larger is the control effect of PSS on the exciter and the better is the overall performance of PSS in the power system.

Algorithms of calculation of OPLI
1. Derive the transfer function of the excitation system $G_{ex}(s)$.
2. Calculate the $G_{ex}(\lambda^\circ)$; here, λ° is the most critical swing mode, before the application of PSS.
3. Install the PSS at any machine. Here, $G_{PSS}(s) = \dfrac{K_{PSS}(T_1+1)}{(sT_2+1)}$.
4. Compute the system matrix A_PSS and eigenvalues after the application of PSS.
5. Note the critical swing mode λ' to obtain $G_{ex}(\lambda')$ and $G_{PSS}(\lambda')$.
6. Calculate the OPLI applying Equation (6.17).
7. Repeat steps 1-6 for each machine.

The transfer function of the exciter for the ith machine $G_{ex_i}(s) = \dfrac{\Delta E_{fd_i}(s)}{\Delta V_{s_i}(s)}$ can be obtained following Equations (2.113) and (2.114) (Chapter 2).

6.4.4 An illustration

The PSS location selection indicators PF, SPE, and OPLI are calculated using Equations (6.15)–(6.17) for a WSCC type 3-machine, 9-bus test system. The machine data and exciter data are taken from Appendix B. The magnitudes of OPLI, SPE, and PF are listed in Tables 6.5, 6.6, and 6.7, respectively.

Table 6.5 PF When PSS is Installed in Individual Machine

PSS Installed in	Right Eigenvector of Critical Swing Mode #1 $|(\phi_{1,\Delta\omega_l})|$	Left Eigenvector of Critical Swing Mode #1 $|(\psi_{1,\Delta\omega_l})|$	$|(PF_i)|$ ($i=1, 2, 3$)
Machine #1	0.15345	0.6269	0.0962
Machine #2	0.26739	3.3132	0.8856
Machine #3	0.15627	0.0525	0.0084

Table 6.6 SPE When PSS is Installed in Individual Machine

PSS Installed in	Right Eigenvector of Critical Swing Mode #1 $\left\|\left(\phi_{1,\Delta\omega_i}\right)\right\|$	Left-Eigenvector of Critical Swing Mode #1 $\left\|\left(\psi_{1,\Delta E_{fd_i}}\right)\right\|$	$\|(SPE_i)\|$ $(i=1, 2, 3)$
Machine #1	0.15345	0.07926	0.01216
Machine #2	0.26739	1.8630	0.49814
Machine #3	0.15627	0.0270	0.00421

Table 6.7 OPLI When PSS is Installed in Individual Machine

Swing Mode ($\lambda°$) Before Installation of PSS	PSS Installed in	Swing Mode (λ') After Installation of PSS	$\|OPLI\| = \frac{\|(G_{ex}(\lambda') - G_{ex}(\lambda°))\|}{\|G_{PSS}(\lambda')\|}$
$-2.4892 \pm j10.8650$	Machine #1	$-2.5291 \pm j10.8920$	0.00215
	Machine #2	$-3.5586 + j10.8354$	0.05174
	Machine #3	$-2.4834 \pm j10.8865$	0.00096

In Table 6.5, for the critical swing mode #1 ($j=1$), the right-eigenvector entry $\phi_{j,\Delta\omega_i}$ and the left-eigenvector entry $\psi_{j,\Delta\omega_i}$ corresponding to the state $\Delta\omega_i$ for the installation of PSS in the ith machine are computed from the right-eigenvector matrix $[\phi]$ and the left-eigenvector matrix $[\psi]$, and $|PF_i|$ are calculated correspondingly.

In Table 6.6, for the critical swing mode #1 ($j=1$), the right-eigenvector entry $\phi_{1,\Delta\omega_i}$ corresponding to the state $\Delta\omega_i$ and the left-eigenvector entry $\psi_{1,\Delta E_{fd_i}}$ corresponding to the state ΔE_{fd_i} after the installation of PSS in the ith machine are computed from the right-eigenvector matrix $[\phi]$ and the left-eigenvector matrix $[\psi]$, and I SPE$_i$| are calculated for each machine for $i=1, 2, 3$.

Therefore, $G_{ex_i}(s)$ for $i=1, 2, 3$ for the study system are computed in MATLAB and are obtained as

$$G_{ex_1} = \frac{318.4713s + 909.9181}{s^3 + 11.2452s^2 + 98.2312s + 48.4012}$$

$$G_{ex_2} = \frac{318.4713s + 909.9181}{s^3 + 11.9388s^2 + 103.6805s + 58.309}$$

and

$$G_{ex_3} = \frac{318.4713s + 909.9181}{s^3 + 11.4333s^2 + 99.709s + 51.088}$$

respectively.

Calculation of OPLI for application of PSS at machine 1

The critical swing mode $\lambda' = -2.5291 + j10.8920$.

$$G_{\text{PSS}}(\lambda') = \frac{20(0.15\lambda' + 1)}{(0.11\lambda' + 1)}$$

$$G_{\text{ex}_1}(\lambda^\circ) = \frac{(318.4713\lambda^\circ + 909.9181)}{\lambda^{\circ 3} + 11.2452\lambda^{\circ 2} + 98.2312\lambda^\circ + 48.4012}$$

and

$$G_{\text{ex}_1}(\lambda') = \frac{(318.4713\lambda' + 909.9181)}{\lambda'^3 + 11.2452\lambda'^2 + 98.2312\lambda' + 48.4012}$$

$$|\text{OPLI}_1| = \frac{|(G_{\text{ex}_1}(\lambda') - G_{\text{ex}_1}(\lambda^\circ))|}{|G_{\text{PSS}}(\lambda')|} = 0.00215.$$

Calculation of OPLI for application of PSS at machine 2

The critical swing mode $\lambda' = -3.5586 + j10.8354$.

$$|\text{OPLI}_2| = \frac{|(G_{\text{ex}_2}(\lambda') - G_{\text{ex}_2}(\lambda^\circ))|}{|G_{\text{PSS}}(\lambda')|} = 0.05174.$$

Calculation of OPLI for application of PSS at machine 3

The critical swing mode $\lambda' = -2.4834 + j10.8865$.

$$|\text{OPLI}_3| = \frac{|(G_{\text{ex}_3}(\lambda') - G_{\text{ex}_3}(\lambda^\circ))|}{|G_{\text{PSS}}(\lambda')|} = 0.00096.$$

It has been observed from these tables (fourth column, second row) that the values of OPLI, SPE, and PF are large for the installation of PSS in machine 2 compared to the values for the other two locations. *In view of root locus of the critical swing mode and the magnitudes of these three indicators, it is possible to conclude that machine 2 should be the best location of PSS.*

6.4.5 Implication of PSS gain in SPE and OPLI characteristics

The characteristics of OPLI with variation of PSS gain have been described and it is compared with the characteristics of SPE. With PSS installed in machines 1 and 2, both the SPE and OPLI characteristics show increment with increase in PSS gain. For machine 3, both of these sensitivity parameters exhibit decrement with increasing PSS gain (Figures 6.5c and 6.6c). It has been further observed that the slopes of the profile of SPE and OPLI both are high for the optimum location of the PSS (Figures 6.5b and 6.6b). Thus, it appears that the proposed index OPLI bears similar characteristics as SPE and can be effectively used instead of SPE to predict the optimum location of PSS.

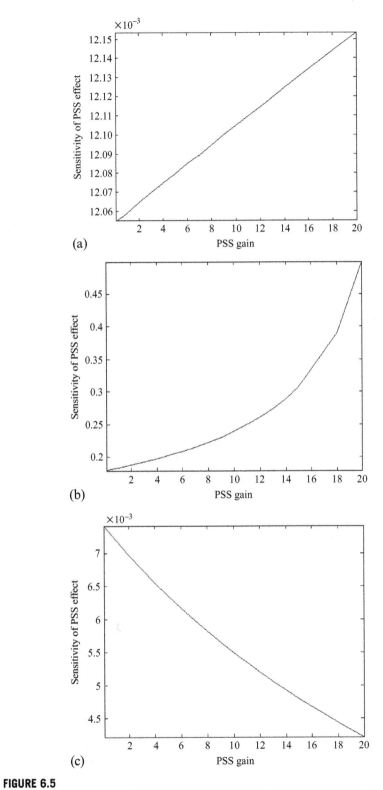

FIGURE 6.5

SPE versus PSS gain when PSS is installed in (a) machine 1, (b) machine 2, and (c) machine 3.

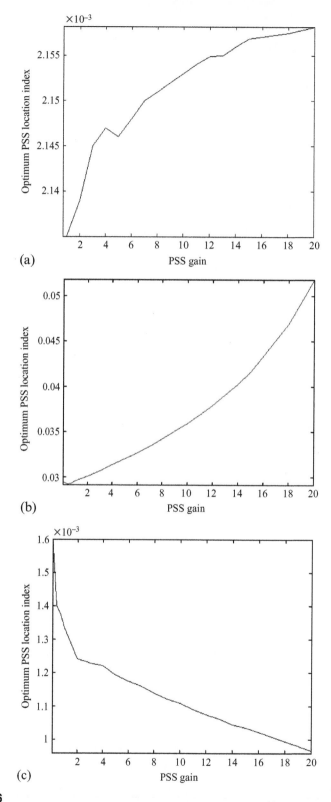

FIGURE 6.6

OPLI versus PSS gain when PSS is installed in (a) machine 1, (b) machine 2, and (c) machine 3.

The validation of the OPLI characteristics has also been tested separately in an IEEE type 14-bus test system where there are 2 generators and 3 synchronous compensators. This bus system is also suitable for small-signal stability analysis [10]. In our study, the investigation has been made on generator 1. The magnitudes of SPE and OPLI for the installation of PSS in generator 1 are, respectively, plotted in Figure 6.7a and b with increasing PSS gain. These plots also interpret that the

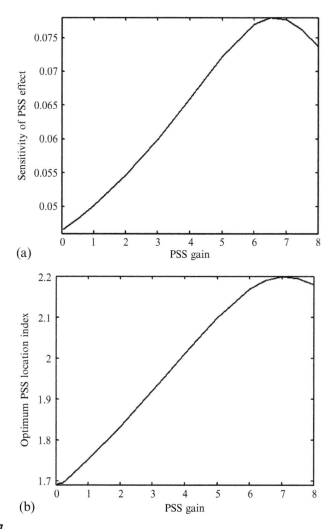

FIGURE 6.7

(a) SPE versus PSS gain when PSS is installed at generator and (b) OPLI versus PSS gain when PSS is installed at generator 1.

SPE and OPLI bear almost similar characteristics and are applicable to indentify the best location of PSS in this system.

6.5 EFFECT OF LOAD

The nonlinear equations of different types of load and its linearized model have been described in Chapter 3, Section 3.7, which is represented here for the purpose of small-signal stability analysis.

For constant power-type load, we have

$$\Delta P_{L_i} = 0 \tag{6.18}$$

$$\Delta Q_{L_i} = 0 \tag{6.19}$$

and for constant current-type characteristics, we have

$$\Delta P_{L_i} = \frac{P_{L_{io}}}{V_{io}} \Delta V_i \tag{6.20}$$

$$\Delta Q_{L_i} = \frac{Q_{L_{io}}}{V_{io}} \Delta V_i \tag{6.21}$$

and when load is considered a constant impedance type, we have

$$\Delta P_{L_i} = 2 \frac{P_{L_{io}}}{(V_{io})^2} V_i \Delta V_i \tag{6.22}$$

$$\Delta Q_{L_i} = 2 \frac{Q_{L_{io}}}{(V_{io})^2} V_i \Delta V_i \tag{6.23}$$

In order to study the effect of the different types of load on small-signal stability, Equations (6.18)–(6.23) are to be incorporated into the linearized network equations (6.9)–(6.10) of the dynamic model of the multimachine power system. It is to be noted that the small-signal stability problem of the 3-machine, 9-bus test system has been studied in Section 6.3.2, based on eigenvalue analysis, and their assumed load is of the constant power type. In this section, this issue has been investigated further in the said multimachine system when load is being of the constant current and constant impedance type.

6.5.1 Effect of type of load

- **Constant power-type load at buses 5, 6, and 8**

In this case, loads in buses 5, 6, and 8 are treated as constant power type. Therefore, Equations (6.18) and (6.19) are incorporated into the linearized network equations for the corresponding buses 5, 6, and 8, and the eigenvalues of the study system are computed from the system matrix.

- **Constant current-type load at buses 5, 6, and 8**

In this study, loads in buses 5, 6, and 8 are assumed of the constant current type. Therefore, Equations (6.20) and (6.21) are incorporated into the linearized network equations of the corresponding buses 5, 6, and 8, and the eigenvalues of the study system are computed from the system matrix.

- **Constant Impedance Type Load at Buses 5, 6, and 8**

Here, the linearized load model given by Equations (6.22) and (6.23) for the constant impedance-type load is included in the linearized network equations of the multimachine system for the respective buses 5, 6, and 8.

The eigenvalues for various kinds of load are shown in Tables 6.8 and 6.9. It is evident from this table that for a nominal load ($P_{L5}, Q_{L5} = 1.25, 0.5$ pu; P_{L6}, $Q_{L6} = 0.9, 0.3$ pu; and $P_{L8}, Q_{L8} = 1.00, 0.35$ pu) at buses 5, 6, and 8, the system is stable for all types of load and it has been observed that the damping ratio of the critical swing mode #1 is comparatively greater for the case of the constant impedance-type load than the constant current- and constant power-type load.

In case of constant power-type load, it has been further observed that an increase of load at bus 5 ($P_{L5}, Q_{L5} = 4.5, 0.5$ pu), one of the eigenvalue moves to the right half of the s-plane, which makes the system dynamically unstable, whereas for the other two types of loads, all the eigenvalues are stable.

Thus, in view of these results, it is possible to conclude that the proposed multimachine system is relatively more stable in constant impedance-type load rather than in constant current- and constant power-type load. It can be further concluded that the system is suitable for mixed loading only at nominal load because in this situation, the system is stable for all kinds of load and with the increase of load, the small-signal stability of the system deteriorates because of the effect of constant power-type component and finally becomes unstable at very high magnitude of load.

The method of load modeling and the effect of various kinds of load presented in this section can be applied to any multimachine system suitable for small-signal stability analysis.

6.5.2 Effect of load on critical swing mode

The real or reactive load (constant power type) at a particular bus is increased in the steps for the said 3-machine, 9-bus test system.

> *Case 1*: The real load P_L is increased at load bus 5 (heaviest load bus) from a base load of 1.25 to 3.5 pu, at constant reactive load $Q_L = 0.5$ pu.
> *Case 2*: The reactive load Q_L is increased at load bus 5 (heaviest load bus) from a base load of 0.5 to 1.5 pu, at constant real load $P_L = 1.25$ pu.

For each step, the execution of load flow program followed by the computation of initial conditions of the state variables is carried out to get the system matrix and eigenvalues. The critical eigenvalue is noted for each case without and with

Table 6.8 Eigenvalues for Different Types of Load at Buses 5, 6, and 8 When $P_{L5}, Q_{L5} = 1.25, 0.5$ pu; $P_{L6}, Q_{L6} = 0.9, 0.3$ pu; and $P_{L8}, Q_{L8} = 1.00, 0.35$ pu

Constant Power		Constant Current		Constant Impedance	
Eigenvalue (λ)	Damping Ratio (ζ)	Eigenvalue (λ)	Damping Ratio (ζ)	Eigenvalue (λ)	Damping Ratio (ζ)
-2.4892 ± j10.8650	**0.2233**	**-1.1236 ± j7.6409**	**0.1454**	**-2.8153 ± j11.072**	**0.24642**
-5.1617 ± j11.2755	0.4162	-7.4798 ± j10.262	0.5890	-7.3738 ± j10.242	0.58429
-5.3063 ± j10.3299	0.4569	-5.1195 ± j10.174	0.4495	-4.6961 ± j10.464	0.40945
-5.6837 ± j10.3601	0.4810	-5.7054 ± j10.319	0.4838	-5.6739 ± j10.332	0.48135
-5.5957 ± j10.3330	0.4762	-3.1812 ± j11.630	0.2638	-2.9029 ± j5.9838	0.43647
-2.5226	1.0000	-0.6811 ± j1.7112	0.3698	-8.1798	1
-0.0000	1.0000	0.000	1	0.000	1
-0.4087 ± j 0.8293	0.4421	-0.4099 ± j0.9254	0.4050	-1.3574 ± j1.5898	0.64934
-0.4759 ± j 0.5616	0.6465	-0.4667 ± j0.5912	0.6196	-0.3973 ± j0.9244	0.39489
-0.4164 ± j 0.6618	0.5325	-2.7557	1	-0.4582 ± j0.5823	0.61839
-3.2258	1.0000	-3.2258	1	-3.2258	1
-1.8692	1.0000	-1.870	1	-1.8692	1
-1.6667	1.0000	-1.666	1	-1.6667	1

Table 6.9 Eigenvalues for Different Types of Load at Buses 5, 6, and 8 When P_{L5}, $Q_{L5}=4.5, 0.5$ pu; P_{L6}, $Q_{L6}=0.9, 0.3$ pu; and P_{L8}, $Q_{L8}=1.00, 0.35$ pu

Constant Power		Constant Current		Constant Impedance	
Eigenvalue (λ)	Damping Ratio (ζ)	Eigenvalue (λ)	Damping Ratio (ζ)	Eigenvalue (λ)	Damping Ratio (ζ)
-2.1644 ± j5.3949	**0.3723**	**-1.8474 ± j7.5445**	**0.2378**	**-1.8166 ± j7.6093**	**0.2322**
-13.881 ± j10.913	0.7861	-12.226 ± j8.7109	0.8144	-12.217 ± j8.6820	0.8151
-5.8667 ± j10.212	0.4981	-7.4156 ± j10.487	0.5773	-7.3793 ± j10.4490	0.5768
-3.1500 ± j10.424	0.2892	-3.4284 ± j10.576	0.3083	-3.4298 ± j10.5870	0.3082
-9.4512 ± j 9.889	0.6909	-5.526 ± j10.369	0.4703	-5.5003 ± j10.3590	0.4689
3.7189	1	-1.2701 ± j0.5462	0.9186	-1.317	1
0.000	1	0.0000	1	0.000	1
-3.1906 ± j0.9831	0.9556	-0.3993 ± j0.8301	0.4334	-0.35857 ± j0.8503	0.3885
-0.4632 ± j0.8270	0.4886	-0.2787 ± j0.2040	0.8069	-0.92432 ± j0.4193	0.9106
-0.6504 ± j0.1968	0.9571	-0.8469	1	-0.4407 ± j0.4273	0.7178
-3.2258	1	-3.2258	1	-3.2258	1
-1.8692	1	-1.8692	1	-1.8692	1
-1.6667	1	-1.6667	1	-1.6667	1

Table 6.10 Effect of Load on Critical Swing Mode

#	Real Load (P_L) (pu)	Reactive Load (Q_L) (pu)	Critical Swing Mode (λ^o) Before the Installation of PSS	Damping Ratio	Critical Swing Mode (λ') for Optimum Location of PSS (Machine #2)	Damping Ratio
Base load	1.25	0.5	$-2.4892 \pm j10.8650$	0.2233	$-3.5586 \pm j10.8354$	0.3120
Case #1	1.5	0.5	$-2.4745 \pm j10.9692$	0.2201	$-3.3502 \pm j10.9232$	0.3932
	2.5	0.5	$-2.4031 \pm j11.3400$	0.2073	$-3.0547 \pm j11.3793$	0.2593
	3.5	0.5	$-2.3074 \pm j11.6323$	0.1946	$-2.7576 \pm j11.6532$	0.2303
Case #2	1.25	1.0	$-2.4468 \pm j10.6290$	0.2243	$-3.4368 \pm j10.4549$	0.3123
	1.25	1.5	$-2.4210 \pm j10.8862$	0.2171	$-3.2897 \pm j10.8578$	0.3030

installation of PSS, and when PSS is installed at the optimum location, i.e., machine 2, the obtained eigenvalues and corresponding damping ratios are presented in Table 6.10. This table illustrates that with an increase of load (real or reactive) without PSS, the damping ratio of the critical swing mode decreases, and with the installation of PSS, the damping ratio of this mode improves significantly. It has also been confirmed in this study that the relative improvement of stability at the selected optimum location of PSS is more in comparison to the other two locations (machines 1 and 3).

6.5.3 Effect on PSS location indicators

The effect of load on SPE and OPLI has also been investigated in this section. The magnitudes of SPE and OPLI for the previously mentioned load variations are computed in Table 6.11 employing algorithms given in the previous sections. It has been observed that even with increasing load, both the sensitivity parameters are reasonably accurate. The present study also reveals that the proposed index, OPLI, is suitable for the application of PSS even during heavy loading condition and till the system approaches its critical operating limit.

Table 6.11 Effect of Load on PSS Location Indicators

#	Real Load (P_L) (pu)	Reactive Load (Q_L) (pu)	SPE at Optimum Location of PSS	OPLI at Optimum Location of PSS
Base load	1.25	0.5	0.4981	0.0517
Case #1	1.5	0.5	0.3258	0.0395
	2.5	0.5	0.2076	0.0239
	3.5	0.5	0.1334	0.0145
Case #2	1.25	1.0	0.4176	0.0537
	1.25	1.5	0.3198	0.0395

EXERCISES

5.1. Write down the Heffron–Philips state-space model of a SMIB power system without and with PSS. Obtain the system matrix with and without PSS.

5.2. In a SMIB system, the machine and exciter parameters are given as follows:

$$R_e = 0, \ X_e = 0.5\,\text{pu}, \ V_t \angle\theta° = 1\angle15°\,\text{pu}, \ V_\infty \angle\theta° = 1.05\angle0°\,\text{pu},$$
$$H = 3.2\,\text{s}, \ T'_{do} = 9.6\,\text{s}, \ K_A = 400, \ T_A = 0.2\,\text{s}, \ R_s = 0.0\,\text{pu},$$
$$X_q = 2.1\,\text{pu}, \ X_d = 2.5\,\text{pu}, \ X'_d = 0.39\,\text{pu}, \ D = 0, \ \text{and} \ \omega_s = 314\,\text{rad/s}.$$

(i) Calculate the system matrix and eigenvalues of the system without and with PSS.

   (ii) Identify the critical swing mode and explain the effect of PSS on improving small-signal stability of this system.

   Assume the PSS parameters are $K_{PSS} = 10$; $T_1 = 1.0$; and $T_2 = 0.5$.

5.3. Discuss the methods and algorithms for the calculation of the following PSS location selection indicators:

   (i) SPE and (ii) OPLI. Explain their significance.

5.4. The transfer function of a multistage PSS is given by the equation

$$G_{PSS} = K_{PSS} \frac{sT_W}{(1+sT_W)} \left(\frac{1+sT_1}{1+sT_2} \right)^n$$

where K_{PSS} is the PSS gain, T_W is the washout time constant, and T_1 and T_2 are the lead-lag time constants. For $n=2$, derive the state-space equation of the PSS for application in a multimachine power system.

5.5. The machine data and exciter data of an IEEE type 14-bus test system are given in Appendix B. Find the optimal location of PSS applying the method of SPE.

References

[1] E.V. Larsen, D.A. Swann, Applying power system stabilizer, part I: general concept, part II: performance objective and tuning concept, part III: practical considerations, IEEE Trans. Power Apparatus Syst. 100 (12) (1981) 3017–3046.

[2] S. Abe, A. Doi, A new power system stabilizer synthesis in multimachine power systems, IEEE Trans. Power Apparatus Syst. 102 (12) (1983) 3910–3918.

[3] P. Kundur, M. Klein, G.J. Rogers, M.S. Zywno, Application of power system stabilizers for enhancement of overall system stability, IEEE Trans. Power Syst. 4 (2) (1989) 614–626.

[4] G.J. Rogers, The application of power system stabilizers to a multigenerator plant, IEEE Trans. Power Syst. 15 (1) (2000) 350–355.

[5] T. Hiyama, Coherency-based identification of optimum site for stabilizer applications, Proc. IEEE 130 (Pt. C, 2) (1983) 71–74.

[6] J.L. Chiang, J.S. Thorp, Identification of optimum site for power system stabilizer applications, IEEE Trans. Power Syst. 5 (4) (1990) 1302–1308.

[7] N. Martins, L.T.G. Lima, Determination of suitable locations for power system stabilizers and static VAR compensators for damping electromechanical oscillations in large scale power system, IEEE Trans. Power Syst. 5 (4) (1990) 1455–1469.

[8] E.Z. Zhou, O.P. Malik, G.S. Hope, Theory and method for selection of power system stabilizer location, IEEE Trans. Energy Convers. 6 (1) (1991) 170–176.

[9] D. Mondal, A. Chakrabarti, A. Sengupta, A unique method for selection of optimum location of power system stabilizer to mitigate small-signal stability problem, Journal of Institution of Engineers (India) 91 (2010) 14–20, pt: EL/2.

[10] S.K.M. Kodsi, C.A. Canizares, Modelling and simulation of IEEE 14 bus systems with FACTS controllers, Technical Report, University of Waterloo, 2003, pp. 1–46.

Application of FACTS Controller

7.1 INTRODUCTION

The potential benefits of using PSS to damp low-frequency electromechanical oscillations for enhancing power system stability are discussed in Chapter 6. In recent times, the development of power electronics devices, like FACTS [1], has generated rapid interest from the researchers to mitigate various problems in the operation and control of modern power systems. In addition to the small-signal stability problem, they have various roles such as scheduling power flow, reducing net loss, providing voltage support, limiting short-circuit currents, and mitigating subsynchronous resonance. FACTS controllers are capable of controlling network conditions in a fast manner, and this feature of FACTS can be exploited to improve several power system problems [2,3]. Though PSS associated with generators are mandatory requirements for damping of oscillations in the power system, the performance of the power network still gets affected by changes in network configurations, load variations, etc., and is liable to cause small-signal oscillations. Hence, the installation of FACTS device has been suggested in several literatures [4,5] to achieve appreciable damping of system oscillations.

Depending on the power electronic devices used in the control, the FACTS controllers can be categorized as (a) variable impedance type (b) voltage source converter (VSC) type.

The variable impedance-type controllers include the followings:

(i) Static Var compensator (SVC) (shunt-connected)
(ii) Thyristor-controlled series capacitor or compensator (TCSC) (series-connected)
(iii) Thyristor-controlled phase shifting transformer (combined shunt and series)

The VSC-based FACTS controllers are as follows:

(i) Static synchronous compensator (STATCOM) (shunt-connected)
(ii) Static synchronous series compensator (series-connected)
(iii) Unified power flow controller (combined shunt and series)

Among the variable impedance-type FACTS devices, SVC and TCSC are proven to be robust and effective in mitigating power system operation problems. In this chapter, these two prominent FACTS devices along with one VSC-based FACTS device (STATCOM) are simulated for application in single-machine infinite-bus power (SMIB) as well as in multimachine power systems. The performance of SVC, TCSC, and STATCOM controllers is compared with the performance of PSS. It has been revealed that these FACTS controllers are more effective than PSS in mitigating small-signal oscillations. It has been further revealed that the performance of both STATCOM and TCSC controllers is reasonably better than the SVC controller.

Finally, in this chapter, the small-signal stability problem of a regional longitudinal power system (24-machine, 203-bus system) has been investigated in the face of three major types of power system contingencies, e.g., load increase, generation drop, and transmission line outage.

7.2 FACTS TECHNOLOGY [6]

The FACTS controller is defined as "a power electronic based system and other static equipment that provide control of one or more ac transmission system parameters." FACTS could be connected in series with the power systems (*series compensation*) and in shunt with the power systems (*shunt compensation*) or both in series and shunt with the power systems.

7.2.1 Series compensation

In series compensation, the FACTS devices are connected in series with the power system transmission line and work as a controllable voltage source. In long transmission lines, series inductance occurs that causes a large voltage drop when a large current flows through it. To compensate it, series capacitors (X_C) are connected in transmission network (Figure 7.1a), which modify line reactance (X) by decreasing its value so as to increase the transmittable active power. However, more reactive

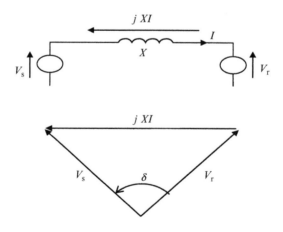

FIGURE 7.1a

Transmission network and phasor diagram.

power must be provided. The expression for active and reactive power flow through the transmission line is given by

$$P = \frac{V^2}{X - X_C} \sin \delta \qquad (7.1)$$

or

$$P = \frac{V^2}{X(1 - K_c)} \sin \delta \qquad (7.2)$$

where $K_c = \dfrac{X_C}{X}$; $0 \le K_c \le 1$ is the degree of compensation of the transmission line reactance. It is to be noted that sending end voltage $(V_s) =$ receiving end voltage $(V_r) = V$ (say) for or a loss less line.

The reactive power supplied by the series capacitor is given by $Q_c = I^2 X_c$. From the phasor diagram (Figure 7.1b), the expression for line current is

$$I = \frac{2V \sin \frac{\delta}{2}}{X(1 - K_c)} \qquad (7.3)$$

Therefore,

$$Q_c = I^2 X_c = \frac{4V^2 \sin^2 \frac{\delta}{2}}{X^2 (1 - K_c)^2} K_c X \qquad (7.4)$$

After simplification, this gives

$$Q_c = \frac{2V^2}{X} \frac{K_c}{(1 - K_c)^2} (1 - \cos \delta) \qquad (7.5)$$

As the series capacitor is used to cancel part of the reactance of the line, it increases the maximum power, reduces the transmission angle at a given level of power transfer, and increases the virtual natural load. Since effective line reactance is reduced,

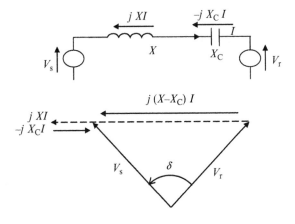

FIGURE 7.1b

Series compensation.

it absorbs less of the line-charging reactive power, so shunt reactors may be needed. A line with 100% series compensation would have a resonant frequency equal to the power frequency, and since the damping in power systems is low, such a system would be hypersensitive to small changes. For this reason, the degree of series compensation is limited in practice to about 80%.

7.2.2 Shunt compensation

In shunt compensation, FACTS are connected in parallel with the power system transmission line. It works as a controllable current source. A reactive current is injected into the line to maintain constant voltage magnitude by varying shunt impedance. Therefore, the transmittable active power is increased but at the expense of increasing the reactive power demand. There are two methods of shunt compensations:

(i) *Shunt capacitive compensation.*

This method is used improve the power factor. Whenever an inductive load is connected to the transmission line, power factor lags because of lagging load current. To compensate it, a shunt capacitor is connected, which draws current leading to the source voltage. The net result is improvement in power factor.

(ii) *Shunt inductive compensation.*

This method is used either when charging the transmission line or when there is very low load at the receiving end. Due to very low or no load, a very low current flows through the transmission line. Shunt capacitance in the transmission line causes voltage amplification (*Ferranti effect*). The receiving end voltage (V_r) may become double the sending end voltage (V_s) (generally in case of very long transmission lines). To compensate it, shunt inductors are connected across the transmission line.

Figure 7.2a and b show the arrangement of the ideal midpoint shunt compensator, which maintains a voltage, V_c, equal to the bus bar voltage such that $V_s = V_r = V_c = V$. Each half of the line is represented by a π equivalent circuit. The synchronous machines at the ends are assumed to supply or absorb the reactive power for the leftmost and rightmost half sections, leaving the compensator to supply or absorb only the reactive power for the central half of the line. It can be seen that the compensator does not consume real power since the compensator voltage, V_c and its current, I_c, are in quadrature. If the compensator can vary its admittance continuously in such a way as to maintain midpoint voltage $V_c = V$, then in the steady state, the line is sectioned into two independent halves. The power (P) transferred from the sending end to the midpoint is equal to the power transferred from the midpoint to the receiving end and is given by

$$P = \frac{V^2 \sin\frac{\delta}{2}}{X/2} = 2\frac{V^2}{X}\sin\frac{\delta}{2} \tag{7.6}$$

The maximum transmissible power is $\frac{2V^2}{X}$, twice the steady-state limit of the uncompensated line. It is reached when $\delta/2 = \pi/2$, that is, with a transmission angle δ or 90°

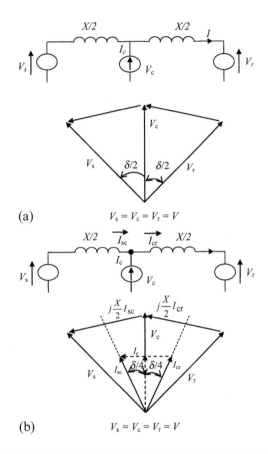

FIGURE 7.2

Shunt compensation (a) Voltage phasor, (b) Voltage and current phasor.

across each half of the line and a total transmission angle of 180° across the whole line. The reactive power generated by the compensator, (Q_p), is generally given by the relation $Q_p = I_c V_c = I_c V$. From the phasor diagram, I_c can be expressed in terms of I_{sc} and δ:

$$I_c = 2 I_{sc} \sin \frac{\delta}{4} \tag{7.7}$$

Therefore,

$$Q_p = I_c V_c = 2 \frac{2V \sin \frac{\delta}{4}}{X/2} \sin \frac{\delta}{4} V \tag{7.8}$$

This can be rearranged as

$$Q_p = \frac{4V^2}{X} \left(1 - \cos \frac{\delta}{2}\right) \tag{7.9}$$

7.3 APPLICATION OF SVC IN SMALL-SIGNAL STABILITY IMPROVEMENT

SVC is a FACTS device, used for shunt compensation, to maintain bus voltage magnitude and hence power system stability. SVC regulates the bus voltage to compensate continuously the change of reactive power loading. The configuration of the basic SVC module and the block diagram of an SVC controller were shown in figures 3.11 and 3.13, respectively, in Chapter 3. The machine speed, $\Delta v (=\Delta\omega/\omega_s)$, is taken as the control input to the auxiliary controller. ΔV_0 is the output signal generated by the auxiliary controller. The firing angle (α) of the thyristor determines the value of susceptance (ΔB_{svc}) to be included in the network in order to maintain bus voltage magnitude.

7.3.1 Model of SMIB system with SVC

For small-signal stability studies of an SMIB power system, the linear model of Heffron-Phillips has been used for many years, providing reliable results. The general Heffron-Phillips model of the SMIB system is given in Section 4.2 (Chapter 4). The inclusion of an SVC controller in this system results in an addition of state variables, $\Delta x_{svc} = [\Delta V_0 \ \Delta\alpha \ \Delta B_{svc}]^T$ corresponding to the SVC controller. Therefore, state-space equations of the SMIB system are modified with additional state-space equations of the SVC controller. The single line diagram of an SMIB system with an SVC controller is shown in Figure. 7.3.

The combined state-space representation of an SMIB system with an SVC controller is then given by

$$\Delta\dot{E}'_q = -\frac{1}{K_3 T'_{do}}\Delta E'_q - \frac{K_4}{T'_{do}}\Delta\delta + \frac{1}{T'_{do}}\Delta E_{fd} \tag{7.10}$$

$$\Delta\dot{\delta} = \omega_s \Delta v \tag{7.11}$$

$$\Delta\dot{v} = -\frac{K_2}{2H}\Delta E'_q - \frac{K_1}{2H}\Delta\delta - \frac{K_\alpha}{2H}\Delta\alpha - \frac{D\omega_s}{2H}\Delta v + \frac{1}{2H}\Delta T_M \tag{7.12}$$

$$\Delta\dot{E}_{fd} = -\frac{1}{T_A}\Delta E_{fd} - \frac{K_A K_5}{T_A}\Delta\delta - \frac{K_A K_6}{T_A}\Delta E'_q + \frac{K_A}{T_A}\Delta V_{ref} \tag{7.13}$$

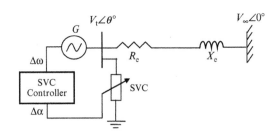

FIGURE 7.3

Single line diagram of an SMIB system with SVC.

$$\Delta \dot{V}_0 = -\frac{K_{\text{SVC}}T_1}{T_2}\frac{K_2}{2H}\Delta E'_q - \frac{K_{\text{SVC}}T_1}{T_2}\frac{K_1}{2H}\Delta \delta$$

$$+ \left(\frac{K_{\text{SVC}}}{T_2} - \frac{K_{\text{SVC}}T_1}{T_2}\frac{D\omega_s}{2H}\right)\Delta v - \frac{1}{T_2}\Delta V_0$$

$$- \frac{K_{\text{SVC}}T_1 K_\alpha}{2HT_2}\Delta \alpha + \frac{K_{\text{SVC}}T_1}{2HT_2}\Delta T_M \tag{7.14}$$

$$\Delta \dot{\alpha} = -K_1 \Delta V_0 + K_1 \Delta V_{\text{svc}} - K_1 \Delta V_{\text{ref}} \tag{7.15}$$

$$\Delta \dot{B}_{\text{svc}} = -\frac{1}{T_{\text{svc}}}\Delta \alpha - \frac{1}{T_{\text{svc}}}\Delta B_{\text{svc}} \tag{7.16}$$

Here, Equations (7.14)–(7.16) are added due to the installation of SVC in the Heffron-Phillips model. Assuming stator resistance $R_s = 0$, the electrical power $P_e = \dfrac{E'_q V_\infty \sin \delta}{X_T}$, where $X_T = X'_d + X_{\text{eff}}$ and $X_{\text{eff}} = X_e + X_{\text{svc}}(\alpha)$. K_2, K_1, and K_α are obtained from the following equations: $K_2 = \dfrac{\partial P_e}{\partial E'_q} = \dfrac{V_\infty \sin \delta}{X_T}$,

$K_1 = \dfrac{\partial P_e}{\partial \delta} = \dfrac{E'_q V_\infty \cos \delta}{X_T}$, and $K_\alpha = \dfrac{\partial P_e}{\partial \alpha} = E'_q V_\infty \sin \delta \dfrac{\partial B_{\text{svc}}}{\partial \alpha}$.

The system matrix (A_SVC) of this model has been presented in Equation (7.17).

7.3.2 An illustration: Simulation result

In this section, the system matrix and the eigenvalues of the SMIB power system with the SVC controller have been computed using Equation (7.17) and then compared the performance of the SVC controller with that of the PSS.

$$A\_SVC = \begin{bmatrix} -\dfrac{1}{K_3 T'_{\text{do}}} & -\dfrac{K_4}{T'_{\text{do}}} & 0 & \dfrac{1}{T'_{\text{do}}} & 0 & 0 & 0 \\[2mm] 0 & 0 & \omega_s & 0 & 0 & 0 & 0 \\[2mm] -\dfrac{K_2}{2H} & -\dfrac{K_1}{2H} & -\dfrac{D\omega_s}{2H} & 0 & 0 & \dfrac{K_\alpha}{2H} & 0 \\[2mm] -\dfrac{K_A K_6}{T_A} & -\dfrac{K_A K_5}{T_A} & 0 & -\dfrac{1}{T_A} & 0 & 0 & 0 \\[2mm] -\dfrac{K_2 T_1}{T_2}\left(\dfrac{K_{\text{SVC}}}{2H}\right) & -\dfrac{K_1 T_1}{T_2}\left(\dfrac{K_{\text{SVC}}}{2H}\right) & \left(\dfrac{K_{\text{SVC}}}{T_2} - \dfrac{K_{\text{SVC}}T_1}{T_2}\dfrac{D\omega_s}{2H}\right) & 0 & -\dfrac{1}{T_2} & -\dfrac{K_{\text{SVC}}T_1 K_\alpha}{2HT_2} & 0 \\[2mm] 0 & 0 & 0 & 0 & K_1 & 0 & 0 \\[2mm] 0 & 0 & 0 & 0 & 0 & -\dfrac{1}{T_{\text{svc}}} & -\dfrac{1}{T_{\text{svc}}} \end{bmatrix}$$

$$\tag{7.17}$$

The parameters of the SMIB system and SVC are given in Sections B.1 and B.1.1 of Appendix B, respectively. Simulation studies of the two cases are carried out in MATLAB: SMIB system without SVC and with SVC. The computed eigenvalues are shown in Table 7.1. It is observed that the electromechanical mode #2 is the *critical mode*. The damping ratio of this mode is enhanced substantially (more than 78%) with the application of an SVC controller compared to the PSS. The time domain analysis presented in Figure 7.4 also reveals that the application of SVC imparts significantly better settling time (less than 50%) compared to the PSS.

7.4 APPLICATION OF A TCSC CONTROLLER IN AN SMIB SYSTEM

This section describes the application of a TCSC controller in an SMIB system in order to improve small-signal stability through series compensation. Simulation results established the superiority of the TCSC controller over PSS and SVC controllers. With changes in the firing angle of the thyristors, the TCSC can change its apparent reactance smoothly and rapidly. Because of its rapid and flexible regulation ability, it can improve transient stability and dynamic performance and is capable of providing positive damping effect to the electromechanical oscillation modes of the power systems.

7.4.1 Model of an SMIB system with a TCSC controller

A simple SMIB system with TCSC controller has been shown in Figure 7.5. The small-signal model of the TCSC controller has been described in Section 3.9.3 (Chapter 3).

The state-space model of SMIB system with a TCSC controller can be formulated by adding the state variables $\Delta x_{\text{tcsc}} = [\Delta \alpha \ \Delta X_{\text{TCSC}}]^{\text{T}}$ corresponding to the TCSC controller in the general Heffron-Phillips model of SMIB system. Therefore, combined state-space model of the SMIB system with a TCSC controller can be represented by the following equations:

$$\Delta \dot{E}'_q = -\frac{1}{K_3 T'_{\text{do}}} \Delta E'_q - \frac{K_4}{T'_{\text{do}}} \Delta \delta + \frac{1}{T'_{\text{do}}} \Delta E_{\text{fd}} \qquad (7.18)$$

$$\Delta \dot{\delta} = \omega_{\text{s}} \Delta v \qquad (7.19)$$

$$\Delta \dot{v} = -\frac{K_2}{2H} \Delta E'_q - \frac{K_1}{2H} \Delta \delta + \frac{K_\alpha}{2H} \Delta \alpha - \frac{D \omega_{\text{s}}}{2H} \Delta v + \frac{1}{2H} \Delta T_{\text{M}} \qquad (7.20)$$

$$\Delta \dot{E}_{\text{fd}} = -\frac{1}{T_{\text{A}}} \Delta E_{\text{fd}} - \frac{K_{\text{A}} K_5}{T_{\text{A}}} \Delta \delta - \frac{K_{\text{A}} K_6}{T_{\text{A}}} \Delta E'_q + \frac{K_{\text{A}}}{T_{\text{A}}} \Delta V_{\text{ref}} \qquad (7.21)$$

Table 7.1 Eigenvalues Without and with the Application of PSS and SVC

#	Eigenvalues Without Controller	Damping Ratio	Eigenvalues with PSS	Damping Ratio	Eigenvalues with SVC	Damping Ratio
1	$-2.6626 \pm j15.136$	0.1733	$-2.0541 \pm j15.325$	0.1328	$-2.2268 \pm j16.949$	0.1303
2	$-0.05265 \pm j7.3428$	**0.0072**	$-\mathbf{0.4116} \pm j\mathbf{7.1110}$	**0.0578**	$-\mathbf{1.4422} \pm j\mathbf{13.887}$	**0.1033**
3	–	–	-10.498	1.0	-0.9531	1.0
4	–	–	–	–	-5.6543	1.0
5	–	–	–	–	-50.063	1.0

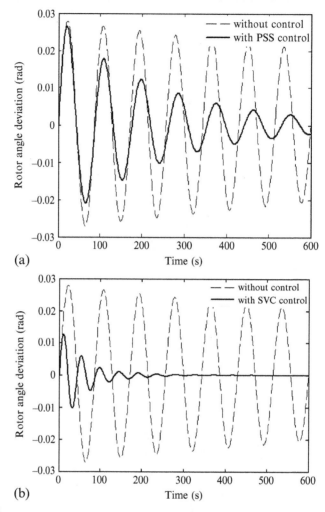

(a)

(b)

FIGURE 7.4

Rotor angle response (a) with and without PSS control and (b) with and without SVC control.

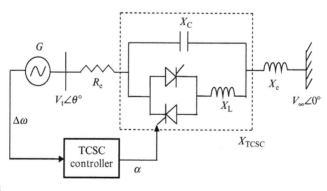

FIGURE 7.5

SMIB system with TCSC.

$$\Delta \dot{\alpha} = \frac{K_{\text{TCSC}} T_1}{T_2} \frac{K_2}{2H} \Delta E'_q + \frac{K_{\text{TCSC}} T_1}{T_2} \frac{K_1}{2H} \Delta \delta$$

$$+ \left(-\frac{K_{\text{TCSC}}}{T_2} + \frac{K_{\text{TCSC}} T_1}{T_2} \frac{D \omega_s}{2H} \right) \Delta v$$

$$+ \left(-\frac{1}{T_2} + \left(\frac{-K_{\text{TCSC}} T_1 K_\alpha}{2H T_2} \right) \right) \Delta \alpha - \frac{K_{\text{TCSC}} T_1}{2H T_2} \Delta T_{\text{M}} \tag{7.22}$$

$$\Delta \dot{X}_{\text{TCSC}} = -\frac{1}{T_{\text{TCSC}}} \Delta \alpha - \frac{1}{T_{\text{TCSC}}} \Delta X_{\text{TCSC}} \tag{7.23}$$

Equations (7.22) and (7.23) are added due to the installation of the TCSC. Here, $K_2 = \frac{\partial P_e}{\partial E'_q}$, $K_1 = \frac{\partial P_e}{\partial \delta}$, and $K_\alpha = \frac{\partial P_e}{\partial \alpha}$. The electrical power ($P_e$) assuming the stator resistance $R_s = 0$ is $P_e = \frac{E'_q V_\infty}{X_{\text{T}}} \sin \delta$, where $X_{\text{T}} = X'_d + X_{\text{eff}}$ and $X_{\text{eff}} = X_e - X_{\text{TCSC}}(\alpha)$. "$\alpha$" is the firing angle of the thyristors. The system matrix (A_TCSC) of the corresponding model can be obtained as

$$A\_TCSC = \begin{bmatrix} -\dfrac{1}{K_3 T'_{\text{do}}} & -\dfrac{K_4}{T'_{\text{do}}} & 0 & \dfrac{1}{T'_{\text{do}}} & 0 & 0 \\[2ex] 0 & 0 & \omega_s & 0 & 0 & 0 \\[2ex] -\dfrac{K_2}{2H} & -\dfrac{K_1}{2H} & -\dfrac{D\omega_s}{2H} & \dfrac{K_\alpha}{2H} & 0 & 0 \\[2ex] -\dfrac{K_A K_6}{T_A} & -\dfrac{K_A K_5}{T_A} & 0 & -\dfrac{1}{T_A} & \dfrac{K_A}{T_A} & 0 \\[2ex] \dfrac{K_2 T_1}{T_2}\left(\dfrac{K_{\text{TCSC}}}{2H}\right) & \dfrac{K_1 T_1}{T_2}\left(\dfrac{K_{\text{TCSC}}}{2H}\right) & \left(-\dfrac{K_{\text{TCSC}}}{T_2} + \dfrac{K_{\text{TCSC}} T_1}{T_2}\dfrac{D\omega_s}{2H}\right) & 0 & -\dfrac{1}{T_2} + \left(\dfrac{-K_{\text{TCSC}} T_1 K_\alpha}{2H T_2}\right) & 0 \\[2ex] 0 & 0 & 0 & 0 & -\dfrac{1}{T_{\text{TCSC}}} & -\dfrac{1}{T_{\text{TCSC}}} \end{bmatrix}$$

$$\tag{7.24}$$

7.4.2 An illustration: Eigenvalue computation and performance analysis

The system matrix of an SMIB system with a TCSC controller is simulated in MATLAB following Equation (7.24). The eigenvalues of the system are computed without and with the TCSC and PSS. It is found that the damping ratio of the critical mode #2 (Table 7.2, second row and third column) is improved satisfactorily with the application of both the PSS and the TCSC controllers, but the improvement is reasonably more with the application of the latter one.

A comparison between the performances of the PSS and the TCSC controllers is also made by increasing in steps the TCSC conduction angle ($\sigma = \pi - \alpha$) for different values of the line compensations (Table 7.3). The TCSC has been designed with resonance at firing

Table 7.2 Eigenvalues Without and with PSS and TCSC

#	Eigenvalues Without Controller	Damping Ratio	Eigenvalues with PSS	Damping Ratio	Eigenvalues with TCSC	Damping Ratio
1	$-2.6626 \pm j15.136$	0.1733	$-2.0541 \pm j15.325$	0.1328	$-2.9411 \pm j10.655$	0.2661
2	$-0.05265 \pm j7.3428$	0.0072	$-0.4116 \pm j7.1110$	0.0578	$-1.6463 \pm j5.9836$	0.2653
3	–	–	-10.498	1.0	-40.0000	1.0
4	–	–	–	–	-24.2581	1.0

Table 7.3 Comparison of the Performance of PSS and TCSC Controllers

PSS and TCSC Controller Gain	Critical Eigenvalues with PSS	Damping Ratio	TCSC Conduction Angle (σ) and % of Line Compensation	Critical Eigenvalues with TCSC	Damping Ratio
0.5	$-0.2401 \pm j7.2342$	0.0332	5° (8.93%)	$-0.1139 \pm j7.4245$	0.0153
1.0	$-0.4116 \pm j7.1110$	0.0578	10° (14.16%)	$-0.1922 \pm j7.4334$	0.0258
2.0	$-0.7008 \pm j6.8372$	0.1020	15° (20.27%)	$-0.3968 \pm j7.3285$	0.0541
3.0	$-0.9205 \pm j6.5510$	0.1391	20° (28.26%)	$-0.7851 \pm j6.9840$	0.1117
4.0	$-1.0820 \pm j6.2719$	0.1725	25° (40.75%)	$-1.4200 \pm j5.6944$	0.2420
5.0	$-1.1991 \pm j6.0096$	0.1957	30° (67.30%)	$-1.3046 \pm j2.7470$	0.4290

angle ($\alpha_r = 140°$). The gain for the PSS and the TCSC controllers is kept identical for each step. The operating region of the TCSC has been kept in the capacitive zone. It has been observed from Table 7.3 that with the variation of conduction angle up to $\sigma = 20°$, away from the resonant point, damping effect of the TCSC controller is less compared to the PSS. It has been further observed that with conduction angle near the resonant point ($\sigma \geq 25°$), enhancement of damping of the critical swing mode is appreciable with respect to the damping introduced with PSS (Table 7.3, row fifth and row sixth). This establishes that it is possible to achieve better enhancement of damping with TCSC controller compared to the PSS when operating region of the TCSC is kept near its resonant point.

The small-signal behavior of the system is next examined in time domain when it is being subjected to a change of unit mechanical step power input (ΔT_M). The time response analysis program is simulated in MATLAB.

The rotor angle deviations without and with TCSC controller are plotted in Figure 7.6a and b for line compensations around 40% and 67%, respectively. It has been found that the oscillations damp faster with the application of TCSC controller compared to that of the PSS (Figure 7.2a). Another important observation is that the impact of TCSC is relatively better with higher value of line compensation. In view of these results, it may be reasonable to conclude that TCSC is a more effective controller than PSS in mitigating small-signal oscillations.

Figure 7.6a and b employing TCSC with different line compensations is also compared with Figure 7.4b, which was obtained with the application of an SVC controller (Section 7.3.2). It is evident that the TCSC controller can provide better settling time and damping in rotor angle oscillations compared to an SVC controller.

7.5 MULTIMACHINE APPLICATION OF SVC
7.5.1 Multimachine model with SVC

An SVC is here included in a multimachine system. The linearized small-signal model of a general multimachine system combining exciter, PSS, and network equations has been described in Section 6.3 (Chapter 6). This model has been used here for the installation of an SVC, and respective equations are rewritten for convenience:

$$\Delta \dot{X} = A_1 \Delta X + B_1 \Delta I_g + B_2 \Delta V_g + E_1 \Delta U \tag{7.25}$$

$$0 = C_1 \Delta X + D_1 \Delta I_g + D_2 \Delta V_g \tag{7.26}$$

$$0 = C_2 \Delta X + D_3 \Delta I_g + D_4 \Delta V_g + D_5 \Delta V_1 \tag{7.27}$$

$$0 = D_6 \Delta V_g + D_7 \Delta V_1 \tag{7.28}$$

The installation of an SVC in this model results in an addition of state variables, $\Delta x_{svc} = [\Delta V_0 \ \Delta\alpha \ \Delta B_{svc}]^T$, corresponding to the SVC controller in Equations (7.25)–(7.27) and the SVC power flow equations in the network Equation (7.28).

Referring to equation (3.80) from Chapter 3, the linearized SVC power flow equations at bus n are given by

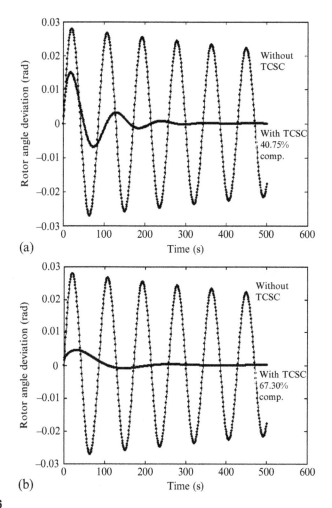

(a)

(b)

FIGURE 7.6

Rotor angle responses (a) without and with TCSC (40% line compensation) (b) without and with TCSC (67% line compensation).

$$\begin{bmatrix} \Delta P_n \\ \Delta Q_n \end{bmatrix} = \begin{bmatrix} 0 & 0 & 0 \\ 0 & -2V_n B_{\text{svc}} & \dfrac{2V_n^2(1-\cos 2\alpha)}{X_L} \end{bmatrix} \begin{bmatrix} \Delta\theta_n \\ \Delta V_n \\ \Delta\alpha \end{bmatrix} \qquad (7.29)$$

Therefore, state variables of the general multimachine model with PSS and SVC controllers are modified as follows:

$$\Delta X = \begin{bmatrix} \Delta X_1^{\text{T}} & \Delta X_2^{\text{T}} & \cdots & \Delta X_m^{\text{T}} \end{bmatrix}^{\text{T}},$$

where $\Delta X_i = \begin{bmatrix} \Delta\delta_i & \Delta\omega_i & \Delta E'_{q_i} & \Delta E'_{d_i} & \Delta E_{\text{fd}_i} & \Delta V_{\text{R}_i} & \Delta R_{\text{F}_i} & \Delta V_{\text{s}_i} & \Delta V_{0_i} & \Delta\alpha_i & \Delta B_{\text{SVC}_i} \end{bmatrix}^{\text{T}}$ for the ith generator at which SVC control unit is connected and $\Delta X_i = \begin{bmatrix} \Delta\delta_i & \Delta\omega_i & \Delta E'_{q_i} & \Delta E'_{d_i} & \Delta E_{\text{fd}_i} & \Delta V_{\text{R}_i} & \Delta R_{\text{F}_i} & \Delta V_{\text{s}_i} \end{bmatrix}^{\text{T}}$ for the remaining $i = 1$, 2, 3, ..., $(m-1)$ generators. ΔV_{si} is the state variable of the PSS.

Eliminating ΔI_{g} from the respective Equations (7.25)–(7.27), the overall modified m-machine system matrix with SVC controller can be obtained as

$$[A\_SVC]_{(8m+3)\times(8m+3)} = [A'] - [B'][D']^{-1}[C'] \qquad (7.30)$$

The structures of matrices $[A']$, $[B']$, $[C']$, and $[D']$ without controller are described in Section 5.2.2 (Chapter 5). Dimensions and elements of these matrices will be modified correspondingly in addition to SVC equations.

In the following section, this model has been used for eigenvalue computation and small-signal stability analysis.

7.5.2 An illustration

The WSCC-type 3-machine, 9-bus system described in Chapter 6 has been reconsidered here as a test system (Figure 7.7). For each generator, a speed-input PSS is equipped, and the SVC is installed in load bus 8. It is to be noted that the location of the SVC is not selected optimally; instead, it is placed arbitrarily at bus 8 in order to study the small-signal performance and to compare it with that of the PSS. The voltage input to the SVC is taken from bus 8 where it is connected. The size as well as the parameters of the SVC controller is mentioned in Section B.2.1 of Appendix B, following literatures [7]. During simulation, the initial value of the firing angle of the SVC is assigned in the capacitive zone ($\alpha = 136°$). The initial value of the firing angle is generally set within $\pm 10 - 15°$ of the resonant point where the susceptance variation of the SVC with firing angle is sharp. The choice of firing angle determines the amount of susceptance to be included with the SVC bus. The eigenvalues of the system with PSS and SVC controllers are computed using Equation (7.30) in MATLAB.

There are 21 eigenvalues of the system (refer to table 5.1, Chapter 5) excluding the PSS and SVC dynamics, among which 8 are complex conjugate eigenvalues that are separately listed in Table 7.4. There are two numbers of electromechanical modes of this multimachine system, which are mode #1 and mode #2, respectively. The electromechanical mode #1 is identified as the *critical swing mode* following swing mode identification criterion [9], which is stated in Section 5.4 of Chapter 5. The study of this mode is of our prime interest in mitigating small-signal oscillations. As this mode is found strongly associated with machine 2, the auxiliary control input ($\Delta\omega$) should be selected here from this machine. After application of an SVC, it may be observed from the first row of Table 7.4 that though the PSS introduces adequate damping to this critical swing mode #1, additional improvement can be achieved with the installation of the SVC controller. The improvement of another swing mode, i.e., mode #2, is also found satisfactory.

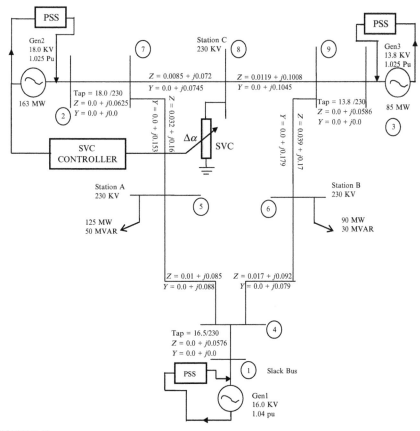

FIGURE 7.7

3-Machine, 9-bus system with PSS and SVC controllers.

In the following section, the small-signal performance of the said multimachine system has been investigated further applying a TCSC controller. The purpose is to study the impact of TCSC controller in comparison with SVC when it is installed in addition to PSS in a multimachine system. The combined model of a multimachine system with PSS and TCSC controller and eigenvalue analysis is described as follows.

7.6 APPLICATION OF TCSC IN A MULTIMACHINE POWER SYSTEM

The linearized model of a multimachine system referred in the previous article has been taken further here for the installation of a TCSC. A schematic block-diagram representation of a power system with PSS and TCSC controllers is depicted in Figure 7.8. The TCSC controller comprises the basic building blocks of a damping

Table 7.4 Eigenvalue and Damping Ratio Without and with PSS and SVC

#	Eigenvalue Without Control	Damping Ratio	Eigenvalue with PSS Only	Damping Ratio	Eigenvalue with PSS and SVC	Damping Ratio
1	**−2.4892 ± j10.8650**	**0.2233**	**−2.829 ± j6.7010**	**0.38894**	**−3.1655 ± j6.6907**	**0.42767**
2	**−5.1617 ± j11.2755**	**0.4162**	**−5.0991 ± j9.9843**	**0.45483**	**−5.1139 ± j9.8357**	**0.46131**
3	−5.3063 ± j10.3299	0.4569	−5.0464 ± j8.2216	0.52312	−4.8477 ± j8.1447	0.51146
4	−5.6837 ± j10.3601	0.4810	−5.2357 ± j7.7298	0.5608	−5.2384 ± j7.7716	0.55893
5	−5.5957 ± j10.3330	0.4762	−4.8846 ± j6.7741	0.58487	−4.9226 ± j6.9043	0.58053
6	−0.4087 ± j0.8293	0.4421	−0.5069 ± j1.3179	0.35898	−0.43508 ± j1.1899	0.3434
7	−0.4759 ± j0.5616	0.6465	−0.50382 ± j0.6433	0.61654	−0.44368 ± j0.5984	0.59556
8	−0.4164 ± j0.6618	0.5325	−0.46775 ± j0.8376	0.48755	−0.41218 ± j0.7784	0.46796

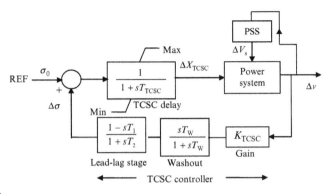

FIGURE 7.8

Block diagram: Power system with PSS and TCSC controller.

controller (controller gain, lead-lag time constants, and washout stage) and the TCSC internal delay. Therefore, the structure of the TCSC controller can be considered similar to the PSS except for the TCSC delay. The model of a TCSC controller is described in Section 3.9.3 of Chapter 3.

7.6.1 Multimachine model with TCSC

The installation of a TCSC results in an addition of state variables corresponding to the TCSC controller $(\Delta x_{\text{tcsc}} = [\Delta\alpha \ \ \Delta X_{\text{TCSC}}]^{\text{T}})$ in Equations (7.25)–(7.27) and the TCSC power flow equations in the network Equation (7.28). Therefore, the modified state variables with PSS and TCSC controllers are obtained as

$$\Delta X = \begin{bmatrix} \Delta X_1{}^{\text{T}} & \Delta X_2{}^{\text{T}} & \cdots & \Delta X_m{}^{\text{T}} \end{bmatrix}^{\text{T}},$$

where $\Delta X_i = \begin{bmatrix} \Delta\delta_i & \Delta\omega_i & \Delta E'_{q_i} & \Delta E'_{d_i} & \Delta E_{\text{fd}_i} & \Delta V_{\text{R}_i} & \Delta R_{\text{F}_i} & \Delta V_{\text{s}_i} & \Delta\alpha_i & \Delta X_{\text{TCSC}_i} \end{bmatrix}^{\text{T}}$ for the ith generator from which TCSC control unit receives auxiliary input signal (Δv) and $\Delta X_i = \begin{bmatrix} \Delta\delta_i & \Delta\omega_i & \Delta E'_{q_i} & \Delta E'_{d_i} & \Delta E_{\text{fd}_i} & \Delta V_{\text{R}_i} & \Delta R_{\text{F}_i} & \Delta V_{\text{s}_i} \end{bmatrix}^{\text{T}}$ for the remaining $i = 1, 2, 3, \ldots, (m-1)$ generators.

It is to be noted that in ΔX_i, the state variable $\Delta V_{\text{s}i}$ is corresponding to the PSS, $\Delta\alpha_i$ and $\Delta X_{\text{TCSC}i}$ are the state variables of the TCSC controller, and the other variables are already described in general multimachine model (Section 5.2.2, Chapter 5).

The linearized TCSC power flow equations at nodes 's' and 't' (figure 3.17, Chapter 3) are obtained referring to equations (3.127) and (3.128) (Chapter 3) and are given by

$$0 = \begin{bmatrix} \dfrac{\partial P_{\text{s}}}{\partial\theta_{\text{s}}} & \dfrac{\partial P_{\text{s}}}{\partial V_{\text{s}}} & \dfrac{\partial P_{\text{s}}}{\partial\alpha} \\[3mm] \dfrac{\partial Q_{\text{s}}}{\partial\theta_{\text{s}}} & \dfrac{\partial Q_{\text{s}}}{\partial V_{\text{s}}} & \dfrac{\partial Q_{\text{s}}}{\partial\alpha} \\[3mm] \dfrac{\partial P_{\text{st}}}{\partial\theta_{\text{s}}} & \dfrac{\partial P_{\text{st}}}{\partial V_{\text{s}}} & \dfrac{\partial P_{\text{st}}}{\partial\alpha} \end{bmatrix} \begin{bmatrix} \Delta\theta_{\text{s}} \\[3mm] \Delta V_{\text{s}} \\[3mm] \Delta\alpha \end{bmatrix} \qquad (7.31)$$

$$0 = \begin{bmatrix} \dfrac{\partial P_t}{\partial \theta_t} & \dfrac{\partial P_t}{\partial V_t} & \dfrac{\partial P_t}{\partial \alpha} \\[2mm] \dfrac{\partial Q_t}{\partial \theta_t} & \dfrac{\partial Q_t}{\partial V_t} & \dfrac{\partial Q_s}{\partial \alpha} \\[2mm] \dfrac{\partial P_{st}}{\partial \theta_t} & \dfrac{\partial P_{st}}{\partial V_t} & \dfrac{\partial P_{st}}{\partial \alpha} \end{bmatrix} \begin{bmatrix} \Delta \theta_t \\[2mm] \Delta V_t \\[2mm] \Delta \alpha \end{bmatrix} \tag{7.32}$$

where

$$P_{st} = V_s^2 g_{st} - V_s V_t (g_{st} \cos \theta_{st} + b_{st} \sin \theta_{st}) \tag{7.33}$$

$$Q_{st} = -V_s^2 b_{st} - V_s V_t (g_{st} \sin \theta_{st} - b_{st} \cos \theta_{st}) \tag{7.34}$$

with

$$Y_{st}^* = \frac{1}{R_{st} + j(X_{st} - X_{TCSC})} = \frac{R_{st} - j(X_{st} - X_{TCSC})}{R_{st}^2 + (X_{st} - X_{TCSC})^2}$$

$$= g_{st} - jb_{st} \tag{7.35}$$

where R_{st} and X_{st} are the resistance and reactance of the transmission line, which connect a TCSC between nodes s and t.

These linearized TCSC power flow equations are incorporated in the network Equation (7.28). Therefore, the overall system matrix with a PSS and a TCSC for an m-machine system can be obtained as

$$[A_{TCSC}]_{(8m+2) \times (8m+2)} = [A'] - [B'][D']^{-1}[C'] \tag{7.36}$$

where dimensions and elements of the matrices $[A']$, $[B']$, $[C']$, and $[D']$ will be modified corresponding to the equations of the TCSC controller. The system matrix without PSS and only with TCSC controller can be obtained excluding the state of the PSS (ΔV_{s_i}) from the state variable matrix ΔX_i. In the following section, this model will be used for eigenvalue computation and small-signal stability analysis of the said test system.

7.6.2 An illustration: Study of small-signal stability

In order to study the small-signal performance of the said 3-machine, 9-bus test system with TCSC controller, the TCSC module has been placed in branch 5-7 (Figure 7.9) associated with the highest load bus 5, and for each generator, a speed-input PSS has been equipped mandatorily. The size of the TCSC module and the parameters of the TCSC controller are given in Section B.2.2 of Appendix B. The initial value of the TCSC firing angle has been set at $\alpha = 146.5°$ with capacitive compensation around 60%. It is already mentioned that (Section 7.5.2) the initial value of the firing angle is generally set within $\pm 10 - 15°$ of the resonant point. Any particular value of the firing angle within this range can be decided depending upon the series compensation needed. The eigenvalues of the system with PSS and TCSC controllers are computed in MATLAB using Equation (7.36) and are listed in Table 7.5.

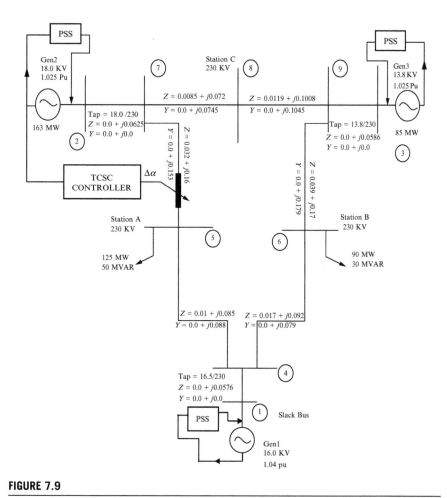

FIGURE 7.9

3-Machine, 9-bus system with PSS and TCSC.

It has been found from the first row of Table 7.5 that the damping ratio of the *critical swing mode* #1 is enhanced substantially in the presence of the TCSC controller in this multimachine system. The small-signal performance of the system has been once again investigated in the time domain. The mode frequency and right eigenvector analysis suggests that this critical swing mode is a local mode and strongly involved with the generator 2. Therefore, analysis of angular speed deviation response of generator 2 is particularly important in this study. The angular speed deviation response of generator 2 has been plotted in Figures 7.10 and 7.11 for different modes of control with simulation time of 10 s. It appears that

Table 7.5 Eigenvalue Analysis Without and with PSS and TCSC

#	Eigenvalue Without Control	Damping Ratio	Eigenvalue with PSS Only	Damping Ratio	Eigenvalue with PSS and TCSC	Damping Ratio
1	$-2.4892 \pm j10.8650$	0.2233	$-2.829 \pm j6.7010$	0.38894	$-2.9357 \pm j4.9231$	0.51216
2	$-5.1617 \pm j11.2755$	0.4162	$-5.0991 \pm j9.9843$	0.45483	$-5.5884 \pm j10.427$	0.47238
3	$-5.3063 \pm j10.3299$	0.4569	$-5.0464 \pm j8.2216$	0.52312	$-5.5244 \pm j8.4601$	0.54675
4	$-5.6837 \pm j10.3601$	0.4810	$-5.2357 \pm j7.7298$	0.5608	$-5.2931 \pm j7.9360$	0.55488
5	$-5.5957 \pm j10.3330$	0.4762	$-4.8846 \pm j6.7741$	0.58487	$-5.3071 \pm j6.7796$	0.61641
6	$-0.4087 \pm j0.8293$	0.4421	$-0.5069 \pm j1.3179$	0.35898	$-0.40502 \pm j0.8956$	0.41205
7	$-0.4759 \pm j0.5616$	0.6465	$-0.50382 \pm j0.6433$	0.61654	$-0.5184 \pm j0.5893$	0.66044
8	$-0.4164 \pm j0.6618$	0.5325	$-0.46775 \pm j0.8376$	0.48755	-0.54097 ± 0.9696	0.48722

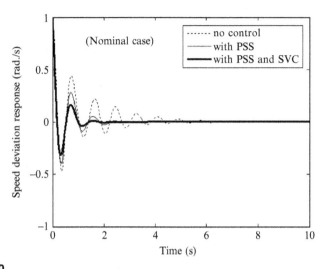

FIGURE 7.10

Angular speed response without and with PSS and SVC for generator 2.

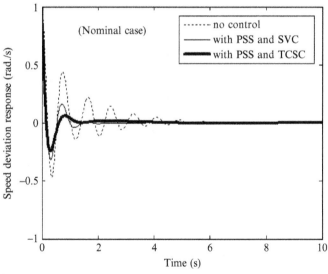

FIGURE 7.11

Angular speed response without and with SVC and TCSC for generator 2.

the installation of the SVC controller in addition to PSS introduces better damping in rotor speed deviation response compared to that of the PSS only. Again, plots of Figure 7.11 confirm that the installation of TCSC not only reduces peak overshoot but also introduces better settling time compared to the installation of SVC and PSS.

Thus, in this study, it has been established that TCSC in addition to PSS helps to damp out transient oscillations faster compared to the SVC with PSS in this multi-machine system.

7.7 VOLTAGE SOURCE CONVERTER-BASED FACTS DEVICE (STATCOM)

The STATCOM resembles in many respects a synchronous compensator but without the inertia. The basic electronic block of a STATCOM is the VSC, which in general converts an input dc voltage into a three-phase output voltage at fundamental frequency, with rapidly controllable amplitude and phase angle. In addition to this, the controller has a coupling transformer and a dc capacitor (figure 3.14, Chapter 3). The control system can be designed to maintain the magnitude of the bus voltage constant by controlling the magnitude and/or phase shift of the VSC output voltage.

The STATCOM is modeled here using the model described in [9], which is a fundamental frequency model of the controller that accurately represents the active and reactive power flows from and to the VSC. The model is basically a controllable voltage source behind impedance with the representation of the charging and discharging dynamics, of the dc capacitor, as well as the STATCOM ac and dc losses.

7.7.1 SMIB system with the STATCOM controller

Figure 7.12 is a SMIB system installed with a STATCOM. A voltage control strategy is assumed for control of the STATCOM bus voltage, and additional control block and signals are added for oscillation damping, given in figure 3.16 (Chapter 3). The

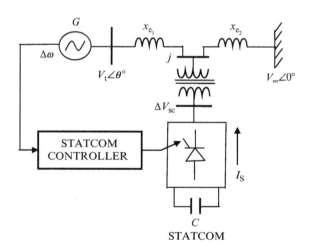

FIGURE 7.12

SMIB system with STATCOM.

linearized state-space model of an SMIB system with the STATCOM controller can be formulated by the following equations:

$$\Delta \dot{E}'_q = -\frac{1}{K_3 T'_{do}} \Delta E'_q - \frac{K_4}{T'_{do}} \Delta \delta + \frac{1}{T'_{do}} \Delta E_{fd} \qquad (7.37)$$

$$\Delta \dot{\delta} = \omega_s \Delta v \qquad (7.38)$$

$$\Delta \dot{v} = -\frac{K_2}{2H} \Delta E'_q - \frac{K_1}{2H} \Delta \delta - \frac{K_{V_{sc}}}{2H} \Delta V_{sc} - \frac{D\omega_s}{2H} \Delta v + \frac{1}{2H} \Delta T_M \qquad (7.39)$$

$$\Delta \dot{E}_{fd} = -\frac{1}{T_A} \Delta E_{fd} - \frac{K_A K_5}{T_A} \Delta \delta - \frac{K_A K_6}{T_A} \Delta E'_q + \frac{K_A}{T_A} \Delta V_{ref} \qquad (7.40)$$

$$\Delta \dot{X}_{s_2} = -\frac{1}{T_m} \Delta X_{s_2} + \frac{K_\omega}{T_m} \Delta \omega - \frac{1}{T_m} \Delta V_{meas} \qquad (7.41)$$

$$\Delta \dot{X}_{s_3} = \left(-\frac{K_P}{T_m} + K_I \right) \Delta X_{s_2} + \frac{K_P K_\omega}{T_m} \Delta \omega - \frac{K_P}{T_m} \Delta V_{meas} \qquad (7.42)$$

$$\Delta \dot{V}_{sc} = -\frac{1}{T_2} \Delta V_{sc} + \frac{1}{T_2} \Delta X_{s_3} + \frac{T_1}{T_2} \left(-\frac{K_P}{T_m} + K_I \right) \Delta X_{s_2} + \frac{T_1 K_P K_\omega}{T_2 T_m} \Delta \omega - \frac{T_1 K_P}{T_2 T_m} \Delta V_{meas} \qquad (7.43)$$

where Equations (7.37)–(7.40) correspond to the SMIB power system and additional Equations (7.41)–(7.43) are included for the STATCOM controller.

The coefficients K_1, K_2, and $K_{V_{sc}}$ can be derived from the following relations: $K_1 = \frac{\partial P_e}{\partial \delta}$, $K_2 = \frac{\partial P_e}{\partial E'_q}$, and $K_{V_{sc}} = \frac{\partial P_e}{\partial V_{sc}}$, where P_e is the generator power and V_{sc} is the STATCOM bus voltage. The generator power (P_e) and the STATCOM bus voltage (V_{sc}) are related by the following equations [10]:

$$P_e = \frac{(x_{e_1} + x'_d) x_{e_2} E'_q V_\infty I_s \sin \delta}{(x_{e_1} + x_{e_2} + x'_d) G_1} + \frac{E'_q V_\infty \sin \delta}{(x_{e_1} + x_{e_2} + x'_d)} \qquad (7.44)$$

$$V_{sc} = \frac{G_1}{x_{e_1} + x_{e_2} + x'_d} + \frac{(x_{e_1} + x'_d) x_{e_2}}{x_{e_1} + x_{e_2} + x'_d} I_s \qquad (7.45)$$

where

$$G_1 = \sqrt{\left(x_{e_2} E'_q \right)^2 + \left((x_{e_1} + x'_d) V_\infty \right)^2 + 2 (x_{e_1} + x'_d) x_{e_2} E'_q V_\infty \cos \delta}$$

Eliminating STATCOM current I_s from Equation (7.44), the equation for P_e becomes

$$P_e = \frac{E'_q V_\infty V_{sc} \sin \delta}{G_1} \qquad (7.46)$$

The system matrix ($A\_STATCOM$) of the corresponding model can be obtained as

$$
A\_STATCOM = \begin{bmatrix}
-\dfrac{1}{K_3 T'_{do}} & -\dfrac{K_4}{T'_{do}} & 0 & -\dfrac{1}{T'_{do}} & 0 & 0 & 0 \\[2ex]
0 & 0 & 1 & 0 & 0 & 0 & 0 \\[2ex]
-\dfrac{K_2 \omega_s}{2H} & -\dfrac{K_1 \omega_s}{2H} & -\dfrac{D\omega_s}{2H} & 0 & 0 & 0 & -\dfrac{K_{V_{sc}} \omega_s}{2H} \\[2ex]
-\dfrac{K_A K_6}{T_A} & -\dfrac{K_A K_5}{T_A} & 0 & -\dfrac{1}{T_A} & 0 & 0 & 0 \\[2ex]
0 & 0 & \dfrac{K_\omega}{Tm} & 0 & -\dfrac{1}{Tm} & 0 & 0 \\[2ex]
0 & 0 & \dfrac{K_P K_\omega}{Tm} & 0 & \left(-\dfrac{K_P}{Tm}+K_I\right) & 0 & 0 \\[2ex]
0 & 0 & \dfrac{T_1 K_P K_\omega}{T_2 Tm} & 0 & \dfrac{T_1}{T_2}\left(-\dfrac{K_P}{Tm}+K_I\right) & \dfrac{1}{T_2} & -\dfrac{1}{T_2}
\end{bmatrix}
$$

$$(7.47)$$

7.7.2 An illustration

The eigenvalues of an SMIB power system with the STATCOM controller have been computed in Table 7.6. The performance of this controller in comparison with SVC has also been illustrated. It has been observed that the damping of the *critical swing mode* #2 is enhanced more than 98% with the application of the STATCOM controller compared to the SVC controller. The deviation of generator speed response is also demonstrated in Figure 7.13 for the three cases: (i) without control, (ii) with SVC control, and (iii) STATCOM control. It is found that both SVC and STATCOM controllers provide satisfactory settling time for a step change in input with simulation time of 300 s, but the contribution of the latter is significantly better. Thus, it is possible to conclude that STATCOM is a superior controller than SVC in mitigating small-signal oscillations problem.

Table 7.6 Eigenvalues Without and with SVC and STATCOM

#	Eigenvalues with SVC	Damping Ratio	STATCOM Controller Parameters	Eigenvalues with STATCOM	Damping Ratio
1		0.1733	$K_P=0.8$	$-6.5881 \pm j19.275$	0.3234
2	**$-1.4422 \pm j13.887$**	**0.1033**	$K_I=0.0$	**$-3.0243 \pm j14.422$**	**0.2052**
3	-0.9531	1.0	$T_1=0.2$ s	$-9.0279 \pm j4.7604$	0.8845
4	-5.6543	1.0	$T_2=0.1$ s	0.00	—
5	-50.063	1.0	$K_\omega=10$	—	—
			$T_m=0.1$		

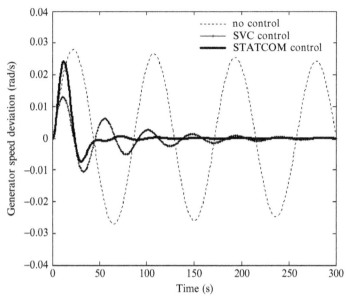

FIGURE 7.13

Generator speed response: (i) without control, (ii) with SVC, and (iii) STATCOM control.

7.7.3 Multimachine model with STATCOM

The linearized model of a multimachine system combining basic power system components, has been reconsidered here for installation of an STATCOM controller. The installation of a STATCOM in a multimachine system results in an addition of state variables of the STATCOM controller, $\Delta X_{\mathrm{STATCOM}} = [\Delta X_{s_2} \ \Delta X_{s_3} \ \Delta V_{\mathrm{sc}}]^{\mathrm{T}}$, with the machine DAE equations and the STATCOM power flow equations in the network equation. The state variables of a multimachine system with PSS and STATCOM are now then modified as follows:

$$\Delta X = \left[\Delta X_1^{\mathrm{T}} \ \Delta X_2^{\mathrm{T}} \ \cdots \ \Delta X_m^{\mathrm{T}}\right]^{\mathrm{T}},$$

where $\Delta X_i = \left[\Delta \delta_i \ \Delta \omega_i \ \Delta E'_{q_i} \ \Delta E'_{d_i} \ \Delta E_{\mathrm{fd}_i} \ \Delta V_{\mathrm{R}_i} \ \Delta R_{\mathrm{F}_i} \ \Delta V_{s_i} \ \Delta X_{s_2} \ \Delta X_{s_3} \ \Delta V_{\mathrm{sc}}\right]^{\mathrm{T}}$ for the ith generator from which the STATCOM controller receives auxiliary input signal $(\Delta \omega)$ and the state variable $\Delta X_i = \left[\Delta \delta_i \ \Delta \omega_i \ \Delta E'_{q_i} \ \Delta E'_{d_i} \ \Delta E_{\mathrm{fd}_i} \ \Delta V_{\mathrm{R}_i} \ \Delta R_{\mathrm{F}_i} \ \Delta V_{s_i}\right]^{\mathrm{T}}$ for the remaining $i = 1$, 2, 3, ..., $(m-1)$ generators.

It is assumed that the STATCOM is connected to any jth load bus, the power flow equations of a STATCOM can then be obtained as (refer to Section 3.9.2, Chapter 3)

$$P_{\mathrm{sc}} = |V_{\mathrm{sc}}|^2 G_{\mathrm{sc}} - |V_{\mathrm{sc}}||V_j|\left[G_{\mathrm{sc}} \cos\left(\delta_{\mathrm{sc}} - \theta_j\right) + B_{\mathrm{sc}} \sin\left(\delta_{\mathrm{sc}} - \theta_j\right)\right] \qquad (7.48)$$

$$Q_{sc} = -|V_{sc}|^2 B_{sc} - |V_{sc}||V_j| [G_{sc} \sin(\delta_{sc} - \theta_j) - B_{sc} \cos(\delta_{sc} - \theta_j)] \qquad (7.49)$$

where $V_{sc} \angle \delta_{sc}$ is the inverted voltage (ac) at the output of STATCOM and $Y_{sc} = G_{sc} + jB_{sc}$. G_{sc} and B_{sc} are the conductance and susceptance of the line between the jth load bus and the STATCOM. Therefore, linearized real and reactive power flow equations of the jth load bus can be represented by the Equation (7.50). Here, voltage magnitude V_{sc} and phase angle δ_{sc} are taken to be the state variables. The power flow equations for the other $i = m+1, m+2, \ldots, (n-1)$ load buses remain unaffected:

$$
\begin{bmatrix} \Delta P_j \\ \Delta Q_j \\ \Delta P_{sc} \\ \Delta Q_{sc} \end{bmatrix} =
\begin{bmatrix}
\dfrac{\partial P_j}{\partial \theta_j} & \dfrac{\partial P_j}{\partial V_j} & \dfrac{\partial P_j}{\partial \delta_{sc}} & \dfrac{\partial P_j}{\partial V_{sc}} \\[2mm]
\dfrac{\partial Q_j}{\partial \theta_j} & \dfrac{\partial Q_j}{\partial V_j} & \dfrac{\partial Q_j}{\partial \delta_{sc}} & \dfrac{\partial P_j}{\partial V_{sc}} \\[2mm]
\dfrac{\partial P_{sc}}{\partial \theta_j} & \dfrac{\partial P_{sc}}{\partial V_j} & \dfrac{\partial P_{sc}}{\partial \delta_{sc}} & \dfrac{\partial P_{sc}}{\partial V_{sc}} \\[2mm]
\dfrac{\partial Q_{sc}}{\partial \theta_j} & \dfrac{\partial Q_{sc}}{\partial \theta_j} & \dfrac{\partial Q_{sc}}{\partial \delta_{sc}} & \dfrac{\partial Q_{sc}}{\partial V_{sc}}
\end{bmatrix}
\begin{bmatrix} \Delta \theta_j \\ \Delta V_j \\ \Delta \delta_{sc} \\ \Delta V_{sc} \end{bmatrix}
\qquad (7.50)
$$

The system matrix with the STATCOM controller for an m-machine system can then be obtained as

$$[A\_STATCOM]_{(8m+3) \times (8m+3)} = [A'] - [B'][D']^{-1}[C'] \qquad (7.51)$$

Dimensions and elements of these matrices will be modified with the addition of the state variables corresponding to the STATCOM controller and the STATCOM power flow equations. In the following section, this model will be employed for eigenvalue computation and small-signal stability analysis in a multimachine power system.

7.7.4 Small-signal performance analysis

The performance of the WSCC-type 3-machine, 9-bus system will be investigated here with the installation of a STATCOM controller. For each generator, a speed-input PSS is equipped, and the STATCOM is installed with the load bus 5 (Figure 7.14). The STATCOM is placed arbitrarily at load bus 5 in order to study the small-signal performance of the system and to compare it with the performance of the SVC. The power flow result indicates that the STATCOM generates 43.8 MVAR in order to keep the nodal voltage magnitude at 1.00 p.u. The STATCOM parameters associated with this reactive power generation are $V_{sc} = 1.025$ pu and $\delta_{sc} = -20.5°$. $G_{sc} = 0.0253$ pu and $B_{sc} = 5.0250$ pu are, respectively, the conductance and susceptance of the line between the jth load bus and the STATCOM. The eigenvalues of the system with PSS and STATCOM controllers are computed using Equation (7.51).

It has already been described in Chapter 5 that the multimachine system under consideration has 21 eigenvalues (refer to table 5.1, Chapter 5) at nominal operating conditions. There are two numbers of electromechanical swing modes. The electromechanical mode #1 is identified as the *critical swing mode*, which is of our prime

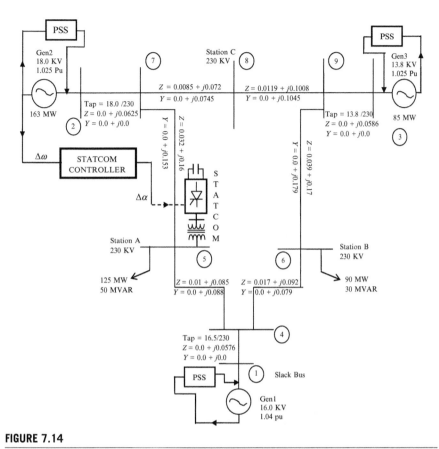

FIGURE 7.14

3-Machine, 9-bus system with PSS and STATCOM.

interest in mitigating small-signal oscillations. As this mode is found strongly associated with machine 2, the auxiliary control input ($\Delta\omega$) for STATCOM is selected from machine 2. In the presence of PSS and STATCOM controllers in this system results in 27 eigenvalues, among which 16 are complex conjugate, 9 are real, and 2 are zero magnitude. Only complex conjugate eigenvalues are listed in Table 7.7, and the effect of STATCOM on swing modes #1 and 2 is highlighted. A comparison with SVC and STATCOM is also simultaneously presented. It may be observed from the first row of Table 7.7 that though the PSS with SVC introduces adequate damping to this critical swing mode #1, additional improvement around 24% more can be achieved with the installation of a STATCOM controller, improvement of another electromechanical swing mode, that is, mode #2, is also found satisfactory. Following these results, it is possible to conclude that STATCOM is a more effective controller than SVC in mitigating power system small-signal stability problem.

Table 7.7 Eigenvalue Analysis Without and with SVC and STATCOM

#	Eigenvalue Without Control	Damping Ratio	Eigenvalue with PSS and SVC	Damping Ratio	Eigenvalue with PSS and STATCOM	Damping Ratio
1	$-2.4892 \pm j10.8650$	0.2233	$-3.1655 \pm j6.6907$	0.4276	$-4.9697 \pm j7.9606$	0.5296
2	$-5.1617 \pm j11.2755$	0.4162	$-5.1139 \pm j9.8357$	0.4613	$-5.2673 \pm j9.4063$	0.4886
3	$-5.3063 \pm j10.3299$	0.4569	$-4.8477 \pm j8.1447$	0.5114	$-5.6035 \pm j7.2264$	0.6128
4	$-5.6837 \pm j10.3601$	0.4810	$-5.2384 \pm j7.7716$	0.5589	$-2.3273 \pm j7.0746$	0.3125
5	$-5.5957 \pm j10.3330$	0.4762	$-4.9226 \pm j6.9043$	0.5805	$-10.841 \pm j2.3575$	0.9772
6	$-0.4087 \pm j0.8293$	0.4421	$-0.43508 \pm j1.1899$	0.3434	$-0.35903 \pm j1.3694$	0.2536
7	$-0.4759 \pm j0.5616$	0.6465	$-0.44368 \pm j0.5984$	0.5955	$-0.54119 \pm j0.57209$	0.6872
8	$-0.4164 \pm j0.6618$	0.5325	$-0.41218 \pm j0.7784$	0.4679	$-3.6922 \pm j0.7971$	0.9775

Table 7.8 Comparison Among PSS, SVC, STATCOM, and TCSC

Mode of Control	Critical Swing Mode	Damping Ratio
Without control	$-2.4892 \pm j10.8650$	0.22330
With PSS only	$-2.8290 \pm j6.7010$	0.38894
With PSS and SVC controllers	$-3.1655 \pm j6.6907$	0.42767
With PSS and TCSC controllers	$-2.9357 \pm j4.9231$	0.51216
With PSS and STATCOM	$-4.9697 \pm j7.9606$	0.5296

The effect of different control modes on the WSCC-type 3-machine, 9-bus system is separately illustrated in Table 7.8. It is clear that by applying STATCOM or TCSC in addition to PSS, it is possible to achieve higher level of damping compared to the application of SVC with PSS or application of PSS only in this multimachine system.

7.8 APPLICATION OF TCSC IN A LONGITUDINAL POWER SYSTEM

Small-signal stability of power systems using a TCSC controller has been studied in literatures mostly on standard test systems or sample systems only rather than on real power systems. Furthermore, simulations have been carried out using reduced-order models considering generator buses only. In this section, a real longitudinal multimachine power system has been taken as a test case, and simulations are carried out considering the full-order linearized model including all types of network buses. In addition to this, based on critical eigenvalue variations, a new indicator being termed as *small-signal stability rank* (*SSSR*) is proposed in Section 7.8.3 for the assessment of effectiveness of the TCSC controller in three commonly occurring contingencies, e.g., load increase, generation drop, and transmission line outage.

7.8.1 Description of the test system and base case study

The power system under consideration is one of the largest power networks of Eastern India. The whole power network has been configured as a 14-area, 24-machine system which consisting of 203 buses with 266 branches. It has 108 numbers 132 kV lines, 30 numbers 220 kV lines, 15 numbers 400 kV lines, and 6 numbers 66 kV lines. The whole network includes 35 numbers 3-winding line transformers and 37 numbers 2-winding load transformers. The actual tap positions of the transformers are included during simulation. Bus 1 is treated as a slack bus. There are 6 generators (1, 2, 3, 5, 17, and 20) having high capacity (540-600 MW), while 8 generators (4, 6, 7, 10, 11, 12, 13, and 19) have medium capacity (150-380 MW), and the rest (10 generators) are of low capacity (20-90 MW). All machines are assumed to be equipped with the IEEE Type I excitation system. For each machine, a speed-input power system stabilizer has been incorporated to ensure adequate damping of its local modes. The nodal voltage magnitudes and angles were solved by the conventional N-R load flow, while a separate subprogram was solved at the end of each iteration to update the state variables for the FACTS in order to meet the specified line-flow criteria.

The eigenvalues of this system are calculated following multimachine model with TCSC controller. The proposed system has a total of 168 eigenvalues for the base case. Among which 23 are identified as swing modes and are listed in Table 7.9. The frequency and damping ratio corresponding to each swing mode are also given in this table. It is evident that the damping ratio of the swing mode #16 is smallest compared to other swing modes and is referred to as the *critical swing mode* (λ). Therefore, the behavior of this mode is of prime concern for the study of small-signal oscillation problem in this system. The mode frequency and right eigenvector analysis suggests that the mode #16 is an interarea mode involved with almost

Table 7.9 Base Case Swing Modes Without PSS and TCSC

#	Swing Modes	Frequency (f)	Damping Ratio (ζ)	Remarks
1	$-3.0845 \pm j12.882$	2.0503	0.23286	Swing modes #1–15
2	$-3.0668 \pm j12.151$	1.9339	0.24471	
3	$-3.0733 + j11.660$	1.8558	0.25487	
4	$-3.1172 \pm j10.963$	1.7449	0.27348	
5	$-3.3891 \pm j8.1479$	1.2968	0.38405	
6	$-3.0920 \pm j8.1006$	1.2892	0.35661	
7	$-3.3347 \pm j7.9905$	1.2717	0.38514	
8	$-3.1006 \pm j7.7754$	1.2375	0.37040	
9	$-3.1915 \pm j7.7091$	1.2269	0.38251	
10	$-3.1702 \pm j7.4950$	1.1929	0.38956	
11	$-3.1783 \pm j6.8383$	1.0883	0.42148	
12	$-3.2243 \pm j6.7030$	1.0668	0.43348	
13	$-3.0431 \pm j6.0689$	0.9659	0.44823	
14	$-3.3906 \pm j5.8905$	0.9375	0.49886	
15	$-3.3215 \pm j5.7429$	0.9140	0.50066	
16	$\mathbf{-1.0363 \pm j4.3800}$	**0.6971**	**0.23023**	**Critical swing mode #16**
17	$-3.4131 \pm j5.0370$	0.8016	0.56095	Swing modes #17–23
18	$-3.2044 \pm j4.9464$	0.7872	0.54370	
19	$-3.8535 \pm j4.0865$	0.6503	0.68607	
20	$-3.2680 \pm j4.5548$	0.7249	0.58296	
21	$-3.3024 \pm j4.3813$	0.6973	0.60191	
22	$-3.3031 \pm j4.4201$	0.7034	0.59860	
23	$-2.9858 \pm j2.5168$	0.4005	0.76460	

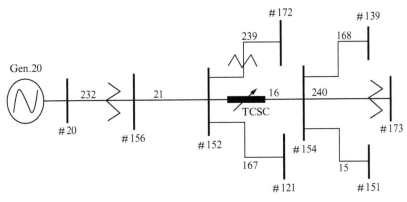

FIGURE 7.15

Part of the 14-area, 24-machine, 203-bus system with TCSC.

all machines and in particular it has a strong association with machine 20. The part of the study system associated with the machine 20 is shown in Figure 7.15.

The network branches associated with the machine 20 (lines 15, 16, 21, 167, 168, 239, and 240) between two load buses (Figure 7.15) are selected for probable installing locations of the TCSC module. The TCSC has been placed in network branch 16. The compensations (X_{TCSC}/X_{line}) of the each selected lines were kept to be around 60%, and therefore, X_L, X_C, and α for the TCSC are chosen according to the reactance of the selected lines. The initial value of the firing angle (α) of the TCSC is kept within capacitive compensating zone. The size of the TCSC module for branch 16 is specified in Section B.5.1 of Appendix B.

7.8.2 Impact of TCSC in the face of power system disturbances

In order to analyze the impact of TCSC in the face of power system disturbances, a MATLAB program is developed for the said 203-bus test system, and a simulation is carried out for three independent types of disturbances: (i) load increase at a particular bus, (ii) outage of transmission line, and (iii) reduction of real and reactive power generation. It is worthwhile to mention that PSS is attached here to all the machines in the network and the TCSC is placed at the line 16, i.e., between buses 152 and 154 (Figure 7.15). The PSS and TCSC controller parameters are set as per Section B.5.1 of Appendix B following literatures [7–9,11–13]. The performance with PSS only and the combined effect of PSS and TCSC controllers are investigated with variation of TCSC firing angle from $\alpha_0 = 145°$ to $\alpha_0 = 160°$. It is to be noted that this range of firing angle is set within $\pm(10° - 15°)$ around the resonant point where TCSC sensitivity is high, i.e., small variation of firing angle results in sharp variation in TCSC reactance.

Case I: **Load increase**

In this case, small-signal performance of the proposed system has been investigated when real and reactive loads of bus 154 are increased from its nominal value ($P_L=0.75$ pu and $Q_L=1.85$ pu) to ($P_L=0.90$ pu and $Q_L=2.159$ pu) and ($P_L=1.15$ pu and $Q_L=2.459$ pu). It has been observed that the damping ratio of the critical swing mode decreases with increasing load but recovers reasonably with the installation of PSS. It has been further observed that the application of both TCSC and PSS controllers enhances the damping ratio significantly over that with PSS alone and this effect is again different for different values ($\alpha_0=145$ - $160°$) of the TCSC firing angle.

Case II: **Generation drop**

The effect of a generation drop on small-signal stability of the system has been investigated here by reducing the total generation (15% and 20%) of three machines (generators 2, 3, and 5) of medium capacity and one machine (generator 20) of higher capacity. It has been found that the damping ratio reduces with the generation drop and improves substantially after application of PSS. Further enhancement of this has been achieved with simultaneous application of PSS and the TCSC controllers, but the impact of TCSC controller is different for different values of TCSC firing angle.

Case III: **Transmission line outage**

The study of small-signal stability of the proposed test system has been extended further when the system is subjected to a contingency like outage of transmission lines 121-152 and 145-149 with ratings 220 kV and 400 kV, respectively. The superiority of TCSC and PSS control over only PSS control is also experienced here. An appreciable enhancement of damping has been observed with variations of TCSC firing angle.

The profile of damping ratio of the critical swing mode with load increase, generation drop, and transmission line outage has been, respectively, plotted in Figure 7.16a–c. It is evident from these plots that the TCSC in addition to the PSS is an effective means for damping small-signal oscillations against all the three cases of power system disturbances. It can be further noticed from these plots that the damping effect of the TCSC controller on the critical swing mode is reasonably better near its resonance point ($\alpha_r=150°$).

• **Time domain study**

A comparative study of the time response analysis of the system with PSS only and with PSS and TCSC controllers has been shown here by finding the angular speed variations of machine 20 inducing different types of disturbances. The deviation of the angular speed response with and without control has been plotted in Figure 7.17a–c for a simulation time of 10 s for all three cases of disturbances. It is evident from these figures that the TCSC controller is an effective FACTS device in mitigating the contingency of transmission line outage in addition to load variation and generation drop.

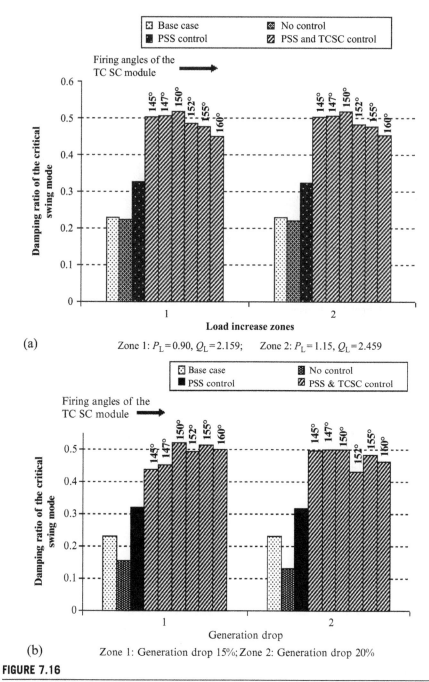

FIGURE 7.16

Profile of damping ratio of the critical swing mode with variation of firing angle with
(a) load increase, (b) generation drop, and

(Continued)

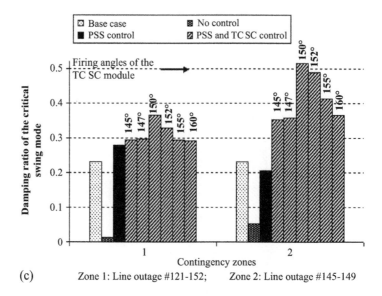

(c)   Zone 1: Line outage #121-152;   Zone 2: Line outage #145-149

FIGURE 7.16, cont'd (c) transmission line outage.

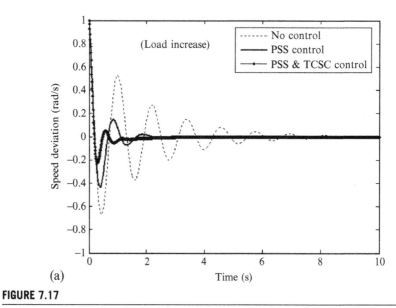

(a)

FIGURE 7.17

The response of angular speed deviation of the machine 20 with (a) load increase,

(Continued)

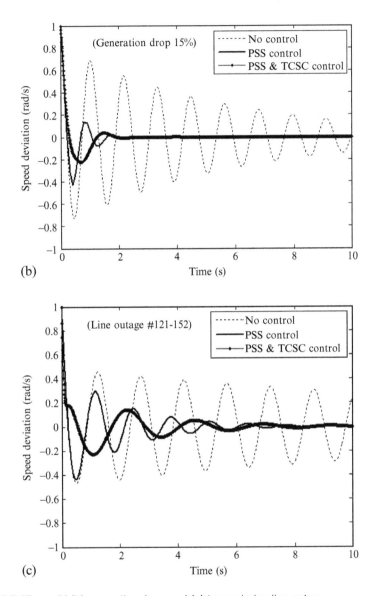

FIGURE 7.17, cont'd (b) generation drop, and (c) transmission line outage.

At this stage, therefore, it may be useful to identify cases of contingency for which the installation of TCSC in addition to PSS is comparatively more effective in mitigating small-signal oscillations. In order to investigate this issue, an indicator being termed as Small Signal Stability Rank (SSSR) has been proposed in [14]. The concept of SSSR is described as follows.

7.8.3 Small-signal stability rank

The SSSR is an index based on the change of the real part of the critical swing mode at a certain contingency with and without control considering the base case as a reference. The TCSC in a power network is a closed-loop controller that considers usually the machine speed or power as its input and introduces a damping so that the critical swing mode moves from a less stable region to a more stable region. The SSSR for a certain contingency is therefore defined by

$$|\text{SSSR}| = \frac{(|\text{Real}(\lambda')| - |\text{Real}(\lambda^\circ)|)}{|\text{Real}(\lambda)|} \tag{7.52}$$

where λ' and λ° are the critical swing modes with and without TCSC control and λ is the critical swing mode for the base case. The magnitude of *SSSR* measures the effect of TCSC on the critical swing mode. A higher value of the *SSSR* implies more effective control to a contingency. The *SSSR* values for different cases of contingencies can be computed following Equation (7.52). The values of SSSR for the said 24-machine, 203-bus system are given in Table 7.10. This result interpreted that PSS and TCSC controllers are more effective in mitigating contingencies like load variation and generation drop compared to that for the outage of transmission line in this multimachine system.

It is also to be noted that the SSSR conceived in this work is based on a change of the real part of the critical swing mode instead of a change of imaginary part, though the imaginary part of the critical swing mode also decides the value of the damping ratio. This is because the system settling time is decided particularly by the real part than the imaginary part. Hence, the structure of SSSR considers the real part of critical swing mode only.

- **Computation of SSSR**

   (i) For $\alpha = 145^\circ$, base case without control $\lambda = -1.0363 \pm j4.380$.
   Load increase ($P_L = 0.90$; $Q_L = 2.15$):

   Without TCSC, $\lambda^\circ = -1.0145 \pm j4.402$; with TCSC, $\lambda' = -2.954 \pm j5.0804$.

Table 7.10 SSSR Values with PSS and TCSC Controllers

TCSC Firing Angle	Magnitudes of SSSR		
	Load Increase ($P_L = 0.90$, $Q_L = 2.15$)	Generation Drop (15%)	Line Outage (121-152)
145°	1.8715	1.4773	0.9212
147°	1.8856	1.4561	0.9039
150°	1.9620	2.1117	1.2124
152°	1.8362	1.9995	1.0447
155°	2.0434	2.0816	0.9394
160°	1.7698	2.0208	0.9220

Following Equation (7.52),

$$|SSSR| = \frac{(2.954 - 1.0145)}{1.0363} = 1.8715$$

Generation drop (15%):

Without TCSC, $\lambda° = -0.7175 \pm j4.5671$; with TCSC, $\lambda' = -2.2485 \pm j4.6177$.

$$|SSSR| = \frac{(2.2485 - 0.7175)}{1.0363} = 1.4773$$

Line outage (121-152):

Without TCSC, $\lambda° = -0.0554 \pm j4.2337$; with TCSC, $\lambda' = -1.0102 \pm j3.2614$.

$$|SSSR| = \frac{(1.0102 - 0.05549)}{1.0363} = 0.9212$$

(ii) For $\alpha = 147°$, base case without control $\lambda = -1.0363 \pm j4.380$.
Load increase ($P_L = 0.90$; $Q_L = 2.15$):

Without TCSC, $\lambda° = -1.0145 \pm j4.402$; with TCSC, $\lambda' = -2.9686 \pm j5.0778$.
Following Equation (7.52),

$$|SSSR| = \frac{(2.9686 - 1.0145)}{1.0363} = 1.8856$$

Generation drop (15%):

Without TCSC, $\lambda° = -0.7175 \pm j4.5671$; with TCSC, $\lambda' = -2.2265 \pm j4.3810$.

$$|SSSR| = \frac{(2.2265 - 0.7175)}{1.0363} = 1.4561$$

Line outage (121-152):

Without TCSC, $\lambda° = -0.0554 \pm j4.2337$; with TCSC, $\lambda' = -0.99229 \pm j3.1765$.

$$|SSSR| = \frac{(0.99229 - 0.05549)}{1.0363} = 0.90398$$

EXERCISE

7.1. Classify types of different FACTS devices. Explain how FACTS devices can modulate power flow in a lossless transmission line through series and shunt compensation.

7.2. Derive the Heffron-Philips state-space model of an SMIB power system with SVC and TCSC controllers. Obtain the system matrix and eigenvalues in each case. The machine and exciter parameters are given as follows:

$$R_e=0, \quad X_e=0.7\,\text{pu}, \quad V_t\angle\theta°=1\angle15°\,\text{pu}, \quad \text{and} \quad V_\infty\angle\theta°=1.05\angle0°\,\text{pu}.$$
$$H=3.7\,\text{s}, \quad T'_{do}=8.5\,\text{s}, \quad K_A=350, \quad T_A=0.3\,\text{s}, \quad R_s=0.0\,\text{pu}, \quad X_q=2.5\,\text{pu},$$
$$X_d=2.8\,\text{pu}, \quad X'_d=0.40\,\text{pu}, \quad D=0, \quad \text{and} \quad \omega_s=314\,\text{rad./s}.$$

Where SVC and TCSC parameters are considered as

	X_L (pu)	X_C (pu)	α (°)	Internal delay (ms)
SVC	0.275	0.4708	155	$T_{SVC}=17$
TCSC	0.0069	0.0289	148	$T_{TCSC}=20$

Comment upon the results.

7.3. Formulate the state-space equations of an SMIB power system with a STATCOM damping controller. Obtain the system matrix and eigenvalues. The parameters of the STATCOM controllers are given by $K_P=2.25$, $K_I=50$, and $K_\omega=6$. Assume time constants of the lead-lag block are $T_1=1.2$ s and $T_2=0.12$ s, respectively.

7.4. The data of an IEEE type 14 bus system are given as in Section B.4 of Appendix B. Assume that a single-stage PSS is equipped with generator 1 and generator 2. Install a TCSC in series with the transmission line between bus 6 and bus 7. Study small-signal stability of the system for the following cases:
  (i) without the application of any controller,
  (ii) with the application of PSS only, and
  (iii) with the installation of PSS and a TCSC controller together.
  For PSS, $K_{PSS}=15.48$, $T_1=2.0$ s, and $T_2=0.0342$ s. For TCSC, $K_{TCSC}=5.57$, $T_1=2.0$ s, and $T_2=0.01$ s.

7.5. For the system given in problem 7.4, install an SVC in bus 10. Investigate the small-signal performance of the system for two independent disturbances:
  (i) real and reactive load increased by 15% of nominal value at bus 9 and
  (ii) the outage of a transmission line (4-13).
  Assume SVC controller parameters are $K_{svc}=20.0$, $T_1=1.0$ s, and $T_2=0.15$ s.

References

[1] N.G. Hingorani, L. Gyugyi, Understanding FACTS: Concepts and Technology of Flexible AC Transmission System, IEEE Press, New York, 2000.
[2] M.A. Abido, Power system stability enhancement using FACTS controllers: a review, Arabian J. Sci. Eng. 34 (1B) (2009) 153–172.

[3] M. L. Kothari and N. Tambey, "Design of UPFC controllers for a multimachine system," *IEEE PES Power System Conference and Exposition*, New York, vol. 3, pp. 1483-1488, 2004.

[4] A.E. Hammad, Analysis of power system stability enhancement by Static Var Compensators, IEEE Trans. Power Syst. 1 (4) (1986) 222–227.

[5] A.B. Khormizi, A.S. Nia, Damping of power system oscillations in multimachine power systems using coordinate design of PSS and TCSC, Int. Conf. Environ. Electr. Eng. EEEIC (2011) 1–4.

[6] M.H. Rashid, Power Electronics-Circuits, Devices, and Applications, Third ed., PHI Pvt. Ltd., New Delhi, 2004.

[7] M.A. Pai, D.P. Sengupta, K.R. Padiyar, Small Signal Analysis of Power Systems, Narosa Publishing House, India, 2004.

[8] C.R. Fuerte-Esquivel, E. Acha, H. Ambriz-Pe'rez, A thyristor controlled series compensator model for the power flow solution of practical power networks, IEEE Trans. Power Syst. 15 (1) (2000) 58–64.

[9] C. A. Ca͂nizares, "Power Flow and Transient Stability Models of FACTS Controllers for Voltage and Angle Stability Studies,"*Proc. of the 2000 IEEE/PES Winter Meeting*, Singapore, 8 pages, Jan. 2000.

[10] L. Gu, J. Wang, Nonlinear coordinated control design of excitation and STATCOM of power systems, Electr. Power Syst. Res. 77 (2007) 788–796.

[11] E.Z. Zhou, O.P. Malik, G.S. Hope, A reduced-order iterative method for swing mode computation, IEEE Trans. Power Syst. 6 (3) (1991) 1224–1230.

[12] P.W. Sauer, M.A. Pai, Power System Dynamics and Stability, Pearson Education Pte. Ltd., Singapore, 1998.

[13] S. K. M. Kodsi and C. A. Canizares, "Modeling and simulation of IEEE 14 bus systems with FACTS controllers," Technical Report, University of Waterloo, (1-46)-3. 2003.

[14] D. Mondal, A. Chakrabarti, A. Sengupta, Investigation of small signal stability performance of a multimachine power system employing PSO based TCSC controller, J. Electr. Syst. 8 (1) (2012) 23–34.

Optimal and Robust Control

8

8.1 INTRODUCTION

It has been explored in the previous chapters that controllers such as PSS and FACTS are effective means for mitigating small-signal oscillations in single machine and in multimachine power systems. However, the performances of these controllers highly depend upon the parameters and their suitable placement in any power network. Several methods are reported in literatures to find these parameters and locations in a power network. This chapter employs heuristic optimization methods to select the optimal location and setting optimal parameters of the FACTS controllers. The conventional optimization techniques are time-consuming, require heavy computational burden, and have slow convergence rates. Many heuristic search methods such as artificial neural network, simulated annealing, fuzzy logic, and particle swarm optimization (PSO) [1–4] have gradually been used for handling power system optimization problems. Each one has its own advantages and drawbacks. They need less computational efforts and have faster convergence characteristics and good accuracy. This chapter gives an overview of the GA and the PSO; they have been used then for parameter optimization and finding optimal location of a TCSC controller for application in a multimachine power system.

It should be pointed out that in reality, a control system without any robustness cannot perform normally because the outputs will be out of their permissible region with the effect of unavoidable disturbances. If the output of a stable closed-loop system is not sensitive to disturbance inputs of the system, in other words, if the influence of disturbance inputs to the outputs of a system is small enough, then we say that this system has enough robustness to disturbances. If under the action of a control, a stable closed-loop system can sufficiently reduce the influence of disturbance inputs to outputs of the system, then we say this control is a robust control. This chapter addresses this issue and a multi-input, single-output (MISO) (four-input, single-output) mixed-sensitivity-based H_∞ robust controller based on LMI (linear matrix inequality) approach with pole placement constraint has been designed for a TCSC in order to achieve robust damping of interarea oscillations in a multimachine power system. In our arrangement, first, the background and general concept of H_∞ control theory have been described, and then, it is employed to design a FACTS (TCSC) controller. Finally, the performance of this controller has been examined for different operating scenarios of a multimachine power system.

8.2 GENETIC ALGORITHM-BASED OPTIMIZATION

Genetic algorithms (GAs) [5] are essentially global search algorithms based on the mechanics of nature (e.g., natural selection, survival of the fittest) and natural genetics. GAs have been used for the optimization of the parameters of control system that are complex and difficult to solve by conventional optimization methods. Particularly, GAs are practical algorithm and easy to implement in the power system analysis. GAs are considered to be robust method because no restrictions on the solution space are made during the process. The power of this algorithm comes from its ability to exploit historical information structures from previous solution and attempt to increase performance of future solution structures. GA maintains a population of individuals that represent the candidate solutions. Each individual is evaluated to give some measure of its fitness to the problem from the objective function. In each generation, a new population is formed by selecting the more fit individuals based on particular selection strategy. Two commonly used genetic operators are crossover and mutation. Crossover is a mixing operator that combines genetic material from selected parents. Mutation acts as a background operator and is used to search the unexplored search space by randomly changing the values at one or more positions of the selected chromosome. Following sections describe an overview and applications of GA based optimization method.

8.2.1 Overview of GA

GA starts with a random generation of initial population, and then, the "selection", "crossover", and "mutation" are preceded until the maximum generation is reached. Important steps of GA are described as follows.

- **Selection**

   The selection of parents to produce successive generations plays an important role in GA. The goal allows the fittest individuals to be more often selected to reproduce. A group of selection methods are available in the literature [6]: "stochastic universal sampling", "uniform", "ranking" and "tournament" etc. *"Stochastic universal sampling"* selection is employed in this book from "Genetic Algorithm and Direct Search Toolbox" in MATLAB. In this selection, parents are created using "roulette wheel" and "uniform sampling", based on expectation and number of parents.

- **Crossover**

   Crossover is an important operator of the GA. It is responsible for the structure recombination (information exchange between mating chromosomes) and the convergence speed of the GA, and it is usually applied with high probability (0.6-0.9). After selection operation, simple crossover proceeds. The main objective of crossover is to reorganize the information of two different individuals and produce a new one. The function *"crossover scattered"* is used in this chapter from "Genetic Algorithm and Direct Search Toolbox" in MATLAB. It is a position-independent crossover function that creates crossover children of the given population.

- **Mutation**

Mutation is a background operator, which produces spontaneous changes in various chromosomes. In artificial genetic systems, the mutation operator protects against some irrecoverable loss. It is an occasional random alteration of the value in the string position. Mutation is needed because even though reproduction and crossover effectively search and recombine extent notions, occasionally, they may lose some potentially useful genetic material. In this book uniform multipoint mutation function, "*mutation uniform*" is employed in MATLAB toolbox. Mutated genes are uniformly distributed over the range of the gene.

8.2.2 Parameter optimization applying GA

- **Optimization problem**

The problem here is on finding the optimal location and setting the optimal parameter of the SVC and the TCSC controller using GA. This results in the minimization of the critical damping index (CDI), which is defined by

$$\text{CDI} = J = (1 - \zeta_i) \tag{8.1}$$

Here, $\zeta_i = -\sigma_i / \sqrt{\sigma_i^2 + \omega_i^2}$ is the damping ratio of the ith critical swing mode. The objective of the optimization is to maximize the damping ratio (ζ) as much as possible. There are four tuning parameters of the SVC and the TCSC controllers, namely, the controllers gain (K), lead and lag time constants (T_1 and T_2), and the location number (N_{loc}). These parameters are to be optimized by minimizing the desired objective function "J" given in Equation (8.1). With a change of parameters of PSS and TCSC controllers, the damping ratio (ζ) as well as J varies. The problem constraints are the bounds on the possible locations and controller parameters. The optimization problem can then be formulated as follows:

$$\text{Minimize } J \, (\text{as in } (8.1))$$

Subject to

$$K^{\min} \le K \le K^{\max} \tag{8.2}$$

$$T_1^{\min} \le T_1 \le T_1^{\max} \tag{8.3}$$

$$T_2^{\min} \le T_2 \le T_2^{\max} \tag{8.4}$$

$$N_{\text{loc}}^{\min} \le N_{\text{loc}} \le N_{\text{loc}}^{\max} \tag{8.5}$$

The objective is to find the optimal locations and parameters for the SVC and TCSC controllers within the inequality constraints given in Equations (8.2)–(8.5). Each "*individual*" in GA is encoded by four parameters: the controllers gain (K), lead and lag time constants (T_1 and T_2), and the location number (N_{loc}). The initial "*population*" is generated randomly for each particle and is kept within a typical range. The minimum and maximum values of the SVC and the TCSC controller parameters along with their location number are depicted in the configurations of the individuals. The entire initial population of size N_{ind} has been calculated by repeating the individuals for N_{ind} times as shown in Figure 8.1a and b corresponding to the SVC and the TCSC controllers, respectively.

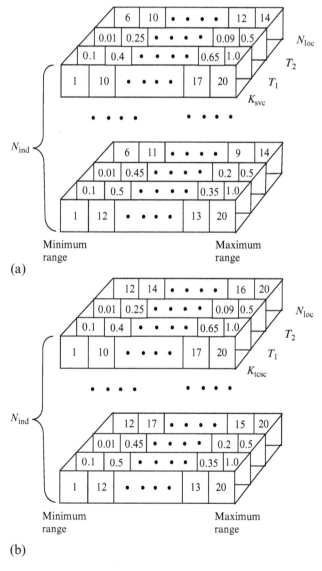

(a)

(b)

FIGURE 8.1

Configurations of individuals and entire population: (a) SVC controller; (b) TCSC controller.

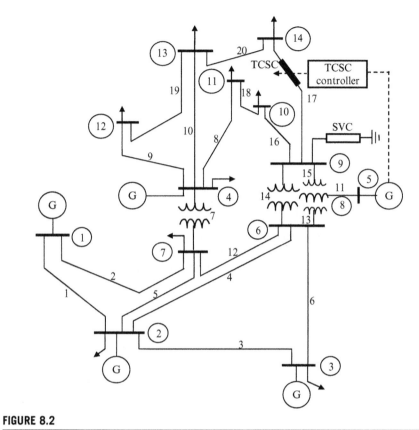

FIGURE 8.2

14-bus test system with application of SVC and TCSC.

The possible locations of the SVC and the TCSC controllers are selected here from the test system given in Figure 8.2. In this system, there are nine load buses (buses 6, 7, 8, 9, 10, 11, 12, 13, and 14) and nine transmission lines or branches (lines 12, 13, 14, 15, 16, 17, 18, 19, and 20) between two load buses. As the SVCs are shunt-connected FACTS devices, the optimal location of the SVC controller is identified among the load buses. Again, TCSCs are series-connected FACTS devices, and therefore, the optimal location of the TCSC controller is identified among the transmission lines between two load buses.

Thus, for the configuration of "individuals" in GA of the SVC controller, all the nine load buses of the test system are proposed to find optimal location of the SVC. Among these nine load buses, the minimum bus number, bus 6, and the maximum bus number, bus 14, are considered as N_{loc}^{min} and N_{loc}^{max}, respectively; as a consequence, other seven buses remain within this minimum range and maximum range (Figure 8.1a).

Similarly, for individual configuration of the TCSC controller, the network branches between two load buses are chosen for locations of the TCSC, and in this case, the minimum line number, line 12, and the maximum line number, line 20, are assigned for N_{loc}^{min} and N_{loc}^{max}, respectively, so that other seven branches remain within this minimum and maximum line number (Figure 8.1b).

In the following section, first, the algorithm of the implemented GA is explained, and then, the application of this method has been shown in the proposed test system.

- **Algorithm of implementation of the GA**

The "Genetic Algorithm and Direct Search Toolbox" consists of a main program associated with a bunch of useful subprograms and routines that are utilized as per requirements. In this work, the main program "*ga.m*" has been implemented for the said optimization problem. In the main function ("ga"), the first argument is the "*Fitnessfcn*" followed by the "*GenomeLength*" and "*Options*". To find the optimal value of the objective function (J), and the optimal parameters of the controllers, this main program uses the user-defined fitness computation program, say, "*gasvc.m*" and "*gatcsc.m*" for SVC and TCSC, respectively, as a subprograms. "*GenomeLength*" sets the dimensions of the design variables that are to be optimized. Several steps of GA starting from "population creation" to final GA "output" can be controlled by the "options" structure created in "*gaoptimset*". In the structure of "gaoptimset", suitable subroutines for the computation of steps such as "selection", "crossover", and "mutation" are declared. A default routine "*PlotFcns.m*" is used by the GA to plot the best value of the fitness function for the specified generation limit. Optimal controller parameters, location, and associated minimized output of the fitness function are evaluated by the following function in "Genetic Algorithm and Direct Search Toolbox" in MATLAB:

function [X, fval, Output, Population, Scores]=*ga* (Fitnessfcn, GenomeLength, Options)	
X, fval	Returns the value of the fitness function "Fitnessfcn" at the solution X
Output	Returns a structure OUTPUT with the following information: <Total generations> <Total function evaluations> <GA termination message>
Population	Returns the final POPULATION at GA termination
Scores	Returns the SCORES of the final POPULATION
Fitnessfcn	String of MATLAB function to be run for parameter optimization (*gasvc. m* and *gatcsc.m*)
GenomeLength	Dimension of inputs to the function (here, number of inputs $=4$; K, T_1, T_2, and N_{loc})
Options	Each of the GA steps (population generation> fitness> scaling> selection> crossover> mutation> scoring> output) is controlled by the options structure configured by 'gaoptimset'

Some important parameter and constant settings in 'gaoptimset' are given in Table 8.1.

Table 8.1 Parameter Settings in 'gaoptimset'

GA Parameters	Value
Population type	Double vector
Maximum generations	200
Number of variables	4
Population size	15
Elite count	2
Population creation	Uniform
Stall generation limit	50

The implemented GA-based algorithm is described here by following steps:

Step 1: Specify parameters for GA: population size, generation limit, number of variables, etc.
Step 2: Generate initial population for the SVC and the TCSC controller parameters: K, T_1, T_2, and N_{loc}.
Step 3: Run small-signal stability analysis program for the proposed test system.
Step 4: Evaluate objective function (J) and hence fitness value for each individual in the current population.
Step 5: Determine and store best individual that minimizes the objective function.
Step 6: Check whether the generation exceeds maximum limit/stall generation limit.
Step 7: If generation < maximum limit, update population for next generation by "*crossover*" and "*mutation*" and repeat from step 3.
Step 8: If generation > maximum limit, stop the program and produce output.

8.2.3 An illustration: GA-based TCSC controller

The validity of the proposed GA-based algorithm has been tested here on the study system given in Figure 8.2. The small-signal models of the multimachine system with SVC and with TCSC controllers described in Chapter 7 (Sections 7.5 and 7.6) are used for eigenvalue computation and small-signal stability analysis. It is to be noted here that the performances of the SVC and the TCSC controller are examined without application of PSS. Therefore, simulation is performed by excluding the state variable of the PSS (ΔV_s) from the said models. The performance of the system has been carried out for two independent disturbances: (i) real and reactive load increased by 15% of nominal value at bus 9 and (ii) outage of a transmission line 4-13. The swing modes of the study system before installation of SVC and TCSC controllers are listed in Table 8.2. It may be observed from Table 8.2 that mode #4 is the critical one as the damping ratio of this mode is smallest compare to other modes. Therefore, parameters of SVC and TCSC controllers and their locations are to be selected in such a way that it can yield maximum damping to this critical swing mode #4.

Table 8.2 Swing Modes Without SVC and TCSC

#	Nominal Load ($P_L = 0.295$ pu, $Q_L = 0.166$ pu)		Load Increased at Bus 9 ($P_L = 0.339$ pu, $Q_L = 0.190$ pu)		Transmission Line (4-13) Outage	
	Swing Modes	Damping Ratio	Swing Modes	Damping Ratio	Swing Modes	Damping Ratio
1	$-1.6071 \pm j7.5211$	0.20896	$-1.5446 \pm j7.5274$	0.2010	$-1.5482 \pm j7.5222$	0.2015
2	$-1.4987 \pm j6.5328$	0.2236	$-1.4244 \pm j6.5313$	0.2130	$-1.4291 \pm j6.5339$	0.2136
3	$-1.2074 \pm j6.1633$	0.19225	$-1.1590 \pm j6.1460$	0.1853	$-1.1501 \pm j6.1659$	0.1833
4	$-0.9461 \pm j5.8552$	0.15953	$-0.8831 \pm j5.8324$	0.1497	$-0.8845 \pm j5.8336$	0.1499

Table 8.3 GA-Based Controller Parameters and Location

GA-Based SVC Parameter	SVC Location	GA-Based TCSC Parameter	TCSC Location
$K_{svc}=11.972$	Bus 9	$K_{tcsc}=9.986$	Branch 17
$T_1=0.8892$		$T_1=0.9967$	
$T_2=0.014$		$T_2=0.1118$	

GA algorithms generate the best set of parameters as well as the best location of the SVC and TCSC controllers (Table 8.3) by minimizing the desired objective function J (8.1). The transmission line compensation (X_{TCSC}/X_{line}) is kept to be around 60% for each of the selected line, and therefore, values of X_L, X_C, and α for the TCSC are chosen according to the line reactances (X_{line}). The initial value of the firing angle (α) of the TCSC is kept within capacitive zone. The maximum iteration number is adopted to be 200 to stop the simulated evolution. The convergence rate of the objective function J toward best solutions with population size 15 and number of generations 200 has been shown in Figure 8.3a and b. The convergence is guaranteed by observing the value of J, which remains unchanged up to eight decimal places. The GA-based SVC and TCSC controllers are installed at their optimal location, and the corresponding values of the damping ratio of the critical swing mode #4 with SVC and TCSC controllers are presented in Table 8.4.

It has been found from Table 8.3 that GA produces optimal location of the SVC controller, which is bus 9, and optimal location of the TCSC controller, which is branch 17. It has been further found from Table 8.4 that installation of GA-based TCSC controllers at their optimal location introduces substantially more damping to the critical swing mode #4 compared to the installation of GA-based SVC controllers.

8.3 PARTICLE SWARM OPTIMIZATION

Particle swarm optimization (PSO) was developed in 1995 by Eberhart and Kennedy [7] rooted on the notion of swarm intelligence of insects, birds, etc. PSO begins with a random population of individuals, here termed as "swarm of particles". Each particle in the swarm is a different possible set of unknown parameters that are to be optimized. The parameters that characterize each particle can be real-valued or may be encoded depending on the particular circumstances. The objective is to efficiently search the solution space by swarming the particles toward the "best-fit solution" with the intention of encountering better solutions through the course of the iteration process and eventually converging on a single best-fit solution. Following sections describe an overview of PSO and its application in power system small-signal stability problem.

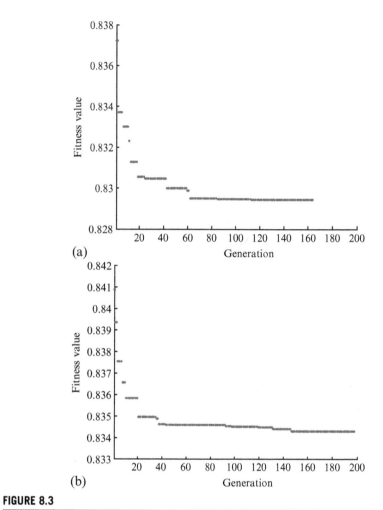

FIGURE 8.3

Convergence rate of the objective function employing GA (a) with SVC; (b) with TCSC.

Table 8.4 Application of GA-Based Controller

Applied Disturbances	With GA-Based SVC		With GA-Based TCSC	
	Critical Swing Mode #4	Damping Ratio	Critical Swing Mode #4	Damping Ratio
Load increased (15%)	$-0.88107 \pm j5.6195$	0.1549	$-0.9764 \pm j5.7114$	0.1685
Line outage (4-13)	$-0.88313 \pm j5.6140$	0.1554	$-0.9247 \pm j5.7731$	0.1581

8.3.1 **Overview of PSO**

The PSO algorithm begins by initializing a random swarm of "M" particles, each having "R" unknown parameters to be optimized. At each iteration, the fitness of each particle is evaluated according to a selected "fitness function". The algorithm stores and progressively replaces the most-fit parameters of each particle ("$pbest_i$", $i = 1, 2, 3, \ldots, M$) as well as a single most-fit particle ($gbest$) as better-fit parameters are encountered. The parameters of each particle (p_i) in the swarm are updated in each iteration (n) according to the following equations [8]:

$$\text{vel}_i(n) = w * \text{vel}_i(n-1) + acc_1 * \text{rand}_1 * (\text{gbest} - p_i(n-1))$$
$$+ acc_2 * \text{rand}_2 * (\text{pbest}_i - p_i(n-1)) \tag{8.6}$$

$$p_i(n) = p_i(n-1) + \text{vel}_i(n) \tag{8.7}$$

where $\text{vel}_i(n)$ is the velocity vector of particle i, normally set to 10-20% of the dynamic range of the variables on each dimension. Velocity changes in Equation (8.6) comprise three parts, that is, the *momentum* part, the *cognitive* part, and the *social* part. This combination provides a velocity getting closer to *pbest* and *gbest*. Every particle's current position is then evolved according to (8.7), which produces a better position in the solution space.

acc_1 and acc_2 are acceleration coefficients that pull each particle toward *gbest* and *pbest_i* positions, respectively, and are often set in the range $\in (0,2)$. Low values of these constants allow particle to roam far from the target regions before being tugged back. On the other hand, high values result in abrupt movement toward or past the target region. $rand_1$ and $rand_2$ are two uniformly distributed random numbers in the ranges $\in (0,1)$. w is the inertia weight of values $\in (0,1)$. Suitable selection of the inertia weight provides a balance between global and local explorations, thus requiring less iteration on an average to find a sufficiently optimal solution. As originally developed, w is often decreased linearly, the purpose being to improve the convergence of the swarm by reducing the inertia weight from an initial value of 0.9 to 0.1 during run. In general inertia, weight w is set according to the equation [9]:

$$w = w_{\text{max}} - \frac{w_{\text{max}} - w_{\text{min}}}{\text{Iter}_{\text{max}}} \times \text{Iter} \tag{8.8}$$

where w_{max} is the final weight and w_{min} is the initial weight. Iter_{max} is maximum iteration number (generations), and Iter is the current iteration number.

- **PSO parameters**

The performance of the PSO is affected by the selection of its parameters. Therefore, a way to find a suitable set of parameters has to be chosen. The selection of the PSO parameters follows the strategy of considering different values for each particular parameter and evaluating its effect on the PSO performance. In this work, different values for the PSO parameters are selected from "PSO toolbox" in MATLAB [10] during implementation of PSO algorithm.

- **Particle**

The "particle" is defined as a vector that contains the desired variables to be optimized. There is a trade-off between the number of particles and the number of iterations of the swarm, and the fitness value of each particle has to be evaluated using a user-defined function at each iteration. Thus, the number of particles should not be large because computational effort could increase dramatically. Swarms of 5-20 particles are normally chosen as appropriate population sizes.

- **Fitness function**

The PSO fitness function used to evaluate the performance of each particle corresponds to a user-defined objective function.

8.3.2 Optimal placement and parameter setting of SVC and TCSC using PSO

This section describes the method of selection of optimal location and parameters of SVC and TCSC controllers using PSO. The small-signal performance of the PSO-based SVC and the TCSC controllers is studied by applying two commonly occurring contingencies, for example, load increase and transmission line outage in a multimachine power system. The behavior of the critical swing mode as well as transient response reveals that the PSO-based TCSC and the SVC controllers are more effective in mitigating small-signal stability problem than their GA-based design even during higher loading.

- **Optimization problem**

The optimization problem presented in Section 8.2.2 has been reconsidered here with the same objective function [CDI$=J=(1-\zeta_i)$] and with finding four tuning parameters of the SVC and TCSC controllers (K, T_1, T_2, and N_{loc}). These parameters are to be optimized by minimizing the objective function J through PSO. Any change of location and parameters of the controllers changes the damping ratio (ζ) and hence J. The constraints of optimization and the optimization problem have been formulated as follows:

$$\text{Minimize } J(\text{as in } (8.1)) \tag{8.9}$$

Subject to

$$K^{\min} \leq K \leq K^{\max} \tag{8.10}$$

$$T_1^{\min} \leq T_1 \leq T_1^{\max} \tag{8.11}$$

$$T_2^{\min} \leq T_2 \leq T_2^{\max} \tag{8.12}$$

$$N_{\text{loc}}^{\min} \leq N_{\text{loc}} \leq N_{\text{loc}}^{\max} \tag{8.13}$$

- **Particle configuration**

The "particle" defined by the vector in Equation (8.14) contains the SVC and the TCSC controller parameters and is given in

$$\text{Particle} : [K \; T_1 \; T_2 \; N_{\text{loc}}] \tag{8.14}$$

Here, K stands for the respective gains of the SVC and the TCSC controllers and is termed K_{svc} and K_{tcsc}, respectively. The initial population is generated randomly for each particle and is kept within a typical range. The minimum and maximum values of the SVC and the TCSC controller parameters along with their location number are given in the particle configurations. The particle configuration corresponding to the SVC and TCSC controllers is presented in Figure 8.4a and b, respectively.

The possible locations of the SVC and the TCSC controllers are selected from the test system given in Figure 8.2. The minimum range ($N_{\text{loc}}^{\text{min}}$) and maximum range ($N_{\text{loc}}^{\text{max}}$) are as described in Section 8.2.2 for individual configuration in GA. Thus, for particle configuration of the SVC controller, bus 6 and bus 14 are considered as $N_{\text{loc}}^{\text{min}}$ and $N_{\text{loc}}^{\text{max}}$, respectively, in Figure 8.4a. Similarly, for particle configuration of the TCSC controller, line 12 and line 20 are assigned for $N_{\text{loc}}^{\text{min}}$ and $N_{\text{loc}}^{\text{max}}$, respectively, in Figure 8.4b.

- **Implementation of PSO algorithm**

The implementation of the PSO algorithm has been described here along with its flowchart in Figure 8.5. To optimize Equation (8.9), MATLAB routines from PSO

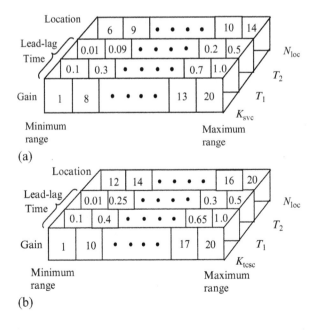

(a)

(b)

FIGURE 8.4

Particle configurations: (a) SVC controller, (b) TCSC controller.

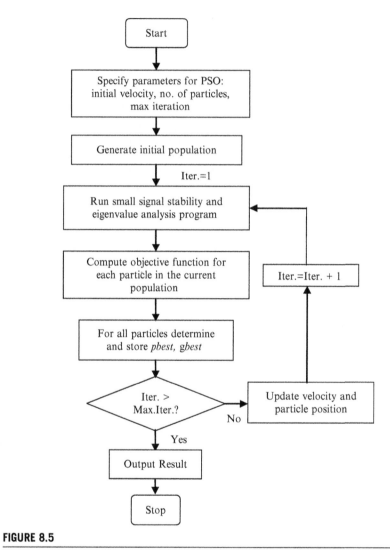

FIGURE 8.5

Flowchart of the implemented PSO algorithm.

toolbox [10] are used. The "PSO toolbox" consists of a main program that is associated with some subprograms and routines that are utilized as per requirements. In this work, the main program "*pso_Trelea_vectorized.m*" has been implemented for "*Common*"-type PSO as a generic particle swarm optimizer. To find the optimal value of the objective function (J), this main program uses the user-defined eigenvalue computation program "*psosvc.m*" or "*psotcsc.m*" as a subprogram for SVC and TCSC, respectively. A default plotting routine "*goplotpso.m*" is used by the PSO algorithm to plot the best value of the objective function *gbest* for the specified generation (epochs) limit. Optimal inputs and associated minimized output of the objective function are evaluated by the following function in PSO toolbox environments:

[OUT, tr, te]=*pso_Trelea_vectorized* (functname, D, mv, VarRange, minmax, PSOparams, plotfcn)

where

OUT	Output of the particle swarm optimizer containing optimal TCSC controller parameters and the best value of the objective function
tr, te	Optional outputs, *gbest* at every iteration and epochs to train, returned as a vector
functname	String of user-defined MATLAB function (say *psosvc.m* or *psotcsc.m*)
D	Dimension of inputs to the function (number of inputs=4; K, T_1, T_2, and N_{loc})
mv	Maximum particle velocity (default=4)
VarRange	Matrix of ranges for each input variable
minmax	0, function-minimized (default set=0)
PSOparams	PSO parameters to select (Table 8.5)
plotfcn	Optional name of plotting function, default "*goplotpso.m*"

The PSO parameters required to be specified in the PSO algorithm are given in Table 8.5. Choice of these parameters affects the performance and the speed of convergence of the algorithm. The PSO algorithm generates the best set of parameters as well as the best location (Table 8.6) corresponding to both the SVC and the TCSC controllers by minimizing the objective function "*J*". The convergence rate of the objective function "*J*" toward *gbest* with the number of particles 15 and generations 200 has been shown in Figure 8.6a and b. The effectiveness of the design is demonstrated through the simulation of the problem carried out in the next section.

Table 8.5 Parameters Used for PSO Algorithm

PSO Parameters	Value	PSO Parameters	Value
Swarm size	15	Epochs before error gradient criterion terminates run	100
Dimension of inputs	4	acc_1, acc_2	2, 2
Maximum generation (epoch)	200	w_{start}, w_{end}	0.9, 0.4
Number of particles	5	$rand_1$, $rand_2$	(0, 1)
Minimum error gradient terminates run	$1 \times e^{-8}$	PSO type	Common "0"

Table 8.6 PSO-Based Controller Parameters and Location

PSO-Based SVC Parameter	SVC Location	PSO-Based TCSC Parameter	TCSC Location
$K_{svc}=20.0$ $T_1=1.0$ $T_2=0.15$	Bus 10	$K_{tcsc}=16.809$ $T_1=1.0$ $T_2=0.2264$	Branch 16

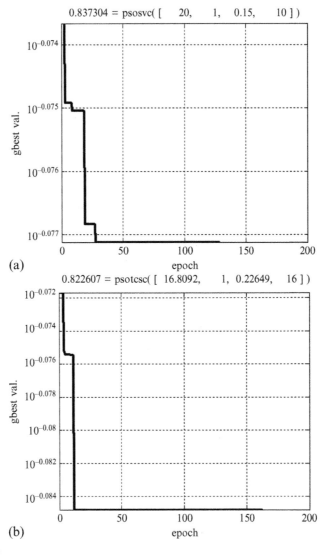

(a)

(b)

FIGURE 8.6

Convergence rate of the objective function employing PSO (a) with SVC; (b) with TCSC.

Table 8.7 Application of PSO-Based Controllers

Applied Disturbances	With PSO-Based SVC		With PSO-Based TCSC	
	Critical Swing Mode #4	Damping Ratio	Critical Swing Mode #4	Damping Ratio
Load increased (15%)	$-0.98121 \pm j6.0070$	0.16121	$-1.0611 \pm j5.7341$	0.1819
Line outage (4-13)	$-0.98224 \pm j6.0568$	0.16008	$-1.0602 \pm j5.7519$	0.1812

8.3.3 Performance Study of PSO-based SVC and TCSC

The validity of the proposed PSO algorithms has been tested on the study system (Figure 8.2). The performance of this system has been investigated further, applying PSO-based SVC and TCSC for two said disturbances: (i) real and reactive load increased by 15% of nominal value at bus 9 and (ii) outage of a transmission line (4-13). The damping ratio of the critical swing mode with the application of PSO-based SVC and the PSO-based TCSC controllers has been presented in Table 8.7. Comparing this result with the results (without controller) given in Table 8.2, it is found that both the controllers substantially improve damping of the critical swing mode for both types of disturbances. It has been further observed that the TCSC controller adds more damping in the system with respect to the SVC controller.

8.4 IMPLICATION OF SVC AND TCSC CONTROLLERS ON CRITICAL LOADING

In order to study the effect of critical loading on system stability, the real power load (P_L) at bus 9 is increased from its nominal value $P_L = 0.295$ pu and $Q_L = 0.166$ pu in steps up to the point of critical loading, keeping reactive power load (Q_L) constant. In each case, the eigenvalues of the system matrix are checked for stability. It has been found that without any controller, at load $P_L = 2.60$ pu and $Q_L = 0.166$ pu, Hopf bifurcation [11] takes place for the critical swing mode #4 that moves toward right half of the s-plane and thus leads to low-frequency oscillatory instability of the system. When the SVC and the TCSC controllers are installed individually at their optimal locations, it has been observed that there is no Hopf bifurcation and stable operating condition is restored in the system. The swing modes of the system without and with SVC and TCSC controllers are presented in Tables 8.8 and 8.9, respectively. This implies that inclusion of SVC and TCSC can put off the Hopf bifurcation until further increase of load levels.

Table 8.8 Application of PSO- and GA-Based SVC with Hopf Bifurcation Load

	Hopf Bifurcation Load ($P_L = 2.60$ pu, $Q_L = 0.166$ pu)					
	Without Control		With PSO-Based SVC		With GA-Based SVC	
#	Swing Modes	Damping Ratio	Swing Modes	Damping Ratio	Swing Modes	Damping Ratio
1	$-1.1190 \pm j7.7098$	0.14363	$-1.2337 \pm j7.5753$	0.16074	$-1.1871 \pm j7.7211$	0.15196
2	$-1.6357 \pm j5.9069$	0.26687	$-1.4921 \pm j6.0160$	0.24072	$-1.5778 \pm j6.059$	0.25201
3	$-0.9230 \pm j2.5144$	0.34461	$-1.776 \pm j2.7131$	0.54768	$-0.77007 \pm j2.6005$	0.28393
4	$0.0072 \pm j4.6175$	-0.00156	$-0.9712 \pm j3.4139$	0.27363	$-0.20113 \pm j4.7623$	0.04219

Table 8.9 Application of PSO- and GA-Based TCSC with Hopf Bifurcation Load

	Hopf Bifurcation Load ($P_L = 2.60$ pu, $Q_L = 0.166$ pu)					
	Without Control		With PSO-Based TCSC		With GA-Based TCSC	
#	Swing Modes	Damping Ratio	Swing Modes	Damping Ratio	Swing Modes	Damping Ratio
1	$-1.1190 \pm j7.7098$	0.14363	$-1.1172 \pm j7.6664$	0.1442	$-1.0970 \pm j7.6775$	0.1414
2	$-1.6357 \pm j5.9069$	0.26687	$-1.5937 \pm j5.7222$	0.2683	$-1.5572 \pm j5.6456$	0.2658
3	$-0.9230 \pm j2.5144$	0.34461	$-3.194 \pm j2.4392$	0.7947	$-2.6778 \pm j2.5202$	0.7282
4	$0.0072 \pm j4.6175$	-0.00156	$-1.1218 \pm j3.6820$	0.2914	$-1.0959 \pm j3.7960$	0.2773

8.5 COMPARISON BETWEEN PSO- AND GA-BASED DESIGNS

The performance comparisons between PSO- and GA-based designs have been illustrated here on the basis of the results obtained in Tables 8.4 and 8.7. It is evident that both PSO and GA can efficiently handle the proposed optimization problem and generate satisfactory results. But the PSO-based SVC and TCSC controllers introduce more damping to the critical swing mode #4 compared to the GA-based SVC and TCSC controllers even during critical loading (Tables 8.10 and 8.11). This implies that PSO-based SVC and TCSC controllers can mitigate small-signal oscillations more efficiently than the corresponding GA-based controllers. Again, the plots of convergence rate of the objective function (Figures 8.3 and 8.6) indicate that PSO method has more fast and stable convergence characteristics than GA. The time response plots (Figure 8.7a–c) of rotor speed deviation of the machine #1 also interpret that the PSO-based controllers introduce reasonably more damping compared to the GA-based controllers in case of both types of disturbances. Summarizing the results mentioned earlier, it is possible to conclude that the PSO-based optimization method is superior to the GA-based one.

8.6 H_∞ OPTIMAL CONTROL

The linear quadratic regulator, Kalman filter, and linear quadratic Gaussian problems can all be posed as 2-norm optimization problems [12]. These optimization problems can be alternatively posed using the system H_∞ norm as a cost function. The H_∞ norm

Table 8.10 Comparison Between PSO- and GA-Based SVCs

Controllers	Damping Ratio of Critical Swing Mode #4		
	Load at Bus #9 (15% More than Nominal)	Line Outage (#4-13)	Critical Load at Bus #9
PSO-based SVC	0.16121	0.16008	0.27363
GA-based SVC	0.15490	0.15540	0.04219

Table 8.11 Comparison Between PSO- and GA-Based TCSCs

Controllers	Damping Ratio of Critical Swing Mode #4		
	Load at Bus 9 (15% More than Nominal)	Line Outage (4-13)	Critical Load at Bus 9
PSO-based TCSC	0.18196	0.1812	0.2914
GA-based TCSC	0.16852	0.1581	0.2773

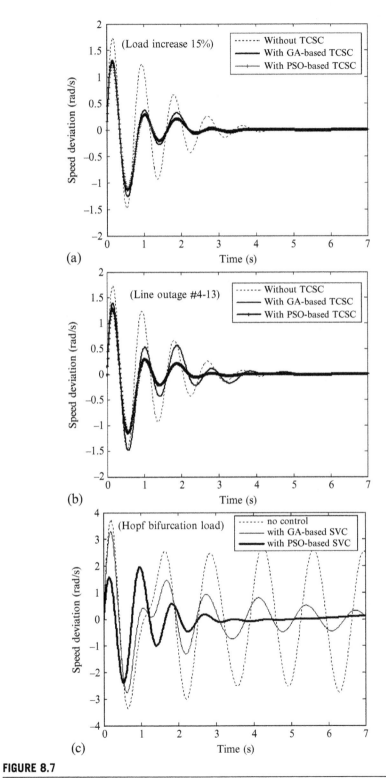

FIGURE 8.7

Rotor speed deviation response of machine #1. (a) Load increase at bus 9 (15% more than nominal); (b) line outage 4-13; (c) Hopf bifurcation load at bus 9.

is the worst-case gain of the system and therefore provides a good match to engineering specifications, which are typically given in terms of bounds on errors and controls.

8.6.1 Background

The terms H_∞ norm and H_∞ control are not terms that convey a lot of engineering significance. When we talk about H_∞, we are talking about a design method that aims to minimize the peak value of one or more selected transfer functions. The H_∞ norm of a stable scalar transfer function $F(s)$ is the peak value of $|F(j\omega)|$ as a function of frequency (ω), that is,

$$\|F(s)\|_\infty \overset{\Delta}{=} \max_\omega \ |F(j\omega)| \tag{8.15}$$

Strictly speaking, "max" (the maximum value) should be replaced by "sup" (supremum, the least upper bound) because the maximum may only be approached as $\omega \to \infty$ and may therefore not actually be achieved.

The symbol ∞ comes from the fact that the maximum magnitude over frequency may be written as

$$\max_\omega \ |F(j\omega)| = \lim_{n\to\infty} \left(\int_{-\infty}^{\infty} |F(j\omega)|^n d\omega \right)^{1/n} \tag{8.16}$$

Essentially, by rising $|F(j\omega)|$ to an infinite power, we pick out its peak value. H_∞ is the set of transfer functions with bounded ∞ norm, which is the set of stable and proper transfer functions.

8.6.2 Algorithms for H_∞ control theory

Given a proper continuous time, linear time-invariant (LTI) plant $P(s)$ maps exogenous inputs "d" and control inputs "u" to controlled outputs "z" and measured outputs "y" (Figure 8.8). That is,

$$\begin{pmatrix} z(s) \\ y(s) \end{pmatrix} = P(s) \begin{pmatrix} d(s) \\ u(s) \end{pmatrix} \tag{8.17}$$

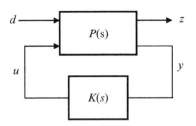

FIGURE 8.8

General LTI design setup.

and given some dynamic output feedback control law such that

$$u = K(s)y = \begin{bmatrix} A_k & B_k \\ C_k & D_k \end{bmatrix} y \qquad (8.18)$$

specified in the state space through the parameter matrices (A_k, B_k, C_k, and D_k) of the controller $K(s)$ and with the partitioning

$$P(s) = \begin{pmatrix} P_{11}(s) & P_{12}(s) \\ P_{21}(s) & P_{22}(s) \end{pmatrix} \qquad (8.19)$$

then, the closed-loop transfer function from disturbance d to controlled output z is

$$S(P, K) = P_{11} + P_{12}K(I - P_{22}K)^{-1}P_{21} \qquad (8.20)$$

The overall control objective is to minimize the H_∞ norm of the transfer function from d to z. This is done by finding a controller K that, based on the information in y, generates a control signal u that counteracts the influence of d on z, thereby minimizing the closed-loop norm of the transfer function from d to z.

Thus, intension is to minimize

$$\|S(P, K)\|_\infty = \left\| \begin{bmatrix} A & B \\ C & D \end{bmatrix} \right\|_\infty \quad \text{over all } K \text{ that stabilizes } P, \text{ which renders } A \text{ stable.}$$

In practice, we calculate the suboptimal solution rather than optimal solution. The suboptimal H_∞ control problem of parameter γ consists of finding a controller $K(s)$ such that [13]

- the closed-loop system is internally stable;
- the H_∞ norm of $S(P, K)$ (the maximum gain from d to z) is strictly less than γ, where $\gamma > 0$ is some prescribed performance level.

Therefore, the problem can be reformulated as follows: try to find a controller $K(s)$ such that

$$K \text{ stabilizes } P \text{ and achieves } \|S(P, K)\|_\infty < 1$$

This condition reads in state space as

$$\|C(sI - A)^{-1}B + D\|_\infty < 1 \qquad (8.21)$$

We shall now introduce some minimal realization of the plant P as is usual in state-space approaches to H_∞ control:

$$P(s) = \begin{pmatrix} D_{11} & D_{12} \\ D_{21} & D_{22} \end{pmatrix} + \begin{pmatrix} C_1 \\ C_2 \end{pmatrix} (sI - A)^{-1} (B_1 \quad B_2) \qquad (8.22)$$

This realization corresponds to the state-space equations:

$$\dot{x} = Ax + B_1 d + B_2 u \qquad (8.23)$$

$$z = C_1 x + D_{11} d + D_{12} u \qquad (8.24)$$

$$y = C_2 x + D_{21} d + D_{22} u \qquad (8.25)$$

The problem dimensions are summarized by

$$A \in R^{n \times n}; D_{11} \in R^{l_1 \times m_1}; D_{22} \in R^{l_2 \times m_2}$$

A set of standard well-posed constraints is imposed on the setup:

- For output feedback stabilizability, the pair (A, B_2) and (C_2, A) must be, respectively, stabilizable and detectable.
- For nonsingularity, D_{21} must be right invertible (full measurement noise), D_{12} must be left invertible (full control penalty), and matrices

$$\begin{bmatrix} A - sI & B_2 \\ C_1 & D_{12} \end{bmatrix}, \begin{bmatrix} A - sI & B_1 \\ C_2 & D_{21} \end{bmatrix}$$

  must be, respectively, left and right invertible for all s.
- $D_{22} = 0$. Note that the assumption that $D_{22} = 0$ is a temporary assumption leading to a simplified form of solution that can always be reversed.

However, in contrast with the case of H_2 optimization, basic H-infinity algorithms solve a *suboptimal* controller design problem, formulated as that of finding whether, for a given $\gamma > 0$, a controller achieving the closed-loop L_2 gain $\|T_{zd}\|_\infty < \gamma$ exists and, in case the answer is affirmative, calculating one such controller.

It might be noticed here that the term "suboptimal" is used rather than "optimal". The reason for that is that it is often not necessary and sometimes even undesirable to design an optimal controller. One of the most compelling reasons is that the optimal closed-loop transfer matrix T_{zd} can be shown to have a *constant* largest singular number over the complete frequency range. In particular, this means that the optimal controller is not strictly proper, and the optimal frequency response to the cost output does not roll off at high frequencies. A suboptimal controller may also have nice properties (e.g., lower bandwidth) over the optimal ones. However, knowing the achievable optimal (minimum) H_∞ norm may be useful theoretically since it sets a limit on what can be achieved.

8.6.3 Mixed-sensitivity-based H_∞ controller: An LMI approach

The principal aim of this robust controller design is to minimize interarea oscillations in power systems that may occur due to various reasons, for example, variation of load demand, generation drop, and transmission line outage. The design objective is to find an internally stabilizing damping controller that satisfies an infinity norm constraint while ensuring that the closed-loop poles lie in specific locations in the complex plane.

The configuration of the closed-loop system together with the H_∞ controller is proposed in Figure 8.9. Here, $G(s)$ is the open-loop plant, $K(s)$ is the controller

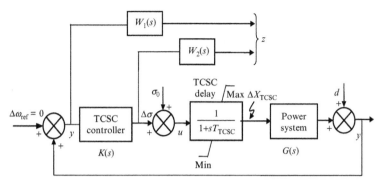

FIGURE 8.9

The closed-loop system along with the H_∞ controller $K(s)$.

to be designed, and $W_1(s)$ and $W_2(s)$ are frequency-dependent weights for shaping the characteristics of the closed-loop plant. The input to the controller is the normalized speed deviation (Δv), and the output signal is the deviation in thyristor conduction angle ($\Delta \sigma$). The problem is to minimize the weighted sensitivity transfer function $S(s)[=(I-G(s)K(s))^{-1}]$, which ensures disturbance rejection and complementary sensitivity transfer function $K(s)S(s)[=K(s)(I-G(s)K(s))^{-1}]$ that ensures robustness in design and minimizes the control effort.

The state-space description of the augmented plant is given in [14]

$$\dot{x}_\text{p} = A_\text{p}x_\text{p} + B_\text{p1}d + B_\text{p2}u \tag{8.26}$$

$$z = C_\text{p1}x_\text{p} + D_\text{p11}d + D_\text{p12}u \tag{8.27}$$

$$y = C_\text{p2}x_\text{p} + D_\text{p21}d + D_\text{p22}u \tag{8.28}$$

where x_p is the state vector of the augmented plant, u is the plant input, y is the measured signal modulated by the disturbance input d, and z is the controlled output.

The controller $K(s)$ can be realized by the following state-space equations:

$$\dot{\hat{x}} = A_k\hat{x} + B_k y \tag{8.29}$$

$$u = C_k\hat{x} + D_k y \tag{8.30}$$

The state-space representation of the closed-loop plant is then given in

$$\dot{\chi} = A_\text{cl}\chi + B_\text{cl}d \tag{8.31}$$

$$z = C_\text{cl}\chi + D_\text{cl}d \tag{8.32}$$

where $\dot{\chi} = \begin{bmatrix} \dot{x}_p \\ \dot{\hat{x}} \end{bmatrix}$, $A_{cl} = \begin{bmatrix} A_p + B_{p2}D_kC_{p2} & B_{p2}C_k \\ B_kC_{p2} & A_k \end{bmatrix}$, $B_{cl} = \begin{bmatrix} B_{p1} + B_{p2}D_kD_{p21} \\ B_kD_{p21} \end{bmatrix}$,

$C_{cl} = [C_{p1} + D_{p12}D_kC_{p2} \quad D_{p12}C_k]$, and $D_{cl} = D_{p11} + D_{p12}D_kD_{p21}$

Without loss of generality, D_{p22} can be set to zero to make the derivation simpler and the plant becomes strictly proper. The transfer function from "d" to "z" can be found as

$$T_{zd} = \begin{bmatrix} W_1(s)S(s) \\ W_2(s)K(s)S(s) \end{bmatrix} = C_{cl}(sI - A_{cl})^{-1}B_{cl} + D_{cl} \qquad (8.33)$$

The objective of the mixed-sensitivity problem is to find an internally stabilizing controller $K(s)$ that minimizes the transfer function from "d" to "z" and meets the following requirement [15]:

$$\|T_{zd}\|_\infty < \gamma \qquad (8.34)$$

where γ is a designable parameter and $S(s)$ is the sensitivity transfer function. In an LMI formulation, the objective (8.34) can be achieved in a suboptimal sense if there exists a solution $X_{cl} = X_{cl}^T > 0$ such that the *bounded real lemma* [16] given in

$$\begin{bmatrix} A_{cl}^T X_{cl} + X_{cl}A_{cl} & B_{cl} & X_{cl}C_{cl}^T \\ B_{cl}^T & -I & D_{cl}^T \\ C_{cl}X_{cl} & D_{cl} & -\gamma^2 I \end{bmatrix} < 0 \qquad (8.35)$$

is satisfied and then the resulting design problem reduces to an LMI problem.

It is well known that satisfactory transient response of a system can be achieved by placing all the closed-loop poles of the system in a specific region of the s-plane. In an LMI framework, such a class of convex region of the complex plane is called an "LMI region" and can be assigned by clustering all the closed-loop poles inside a conic sector (Figure 8.10), which ensures that the damping ratio of poles lying in this sector is at least $\zeta = \cos\frac{\theta}{2}$. The problem therefore reduces to

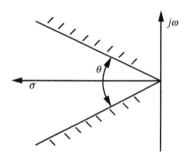

FIGURE 8.10

Conic sector LMI region of closed-loop poles.

minimization of γ (8.34) under LMI-based H_∞ control with pole placement constraints.

Pole clustering in LMI regions can be formulated as an LMI optimization problem. It is shown in [17] that the system matrix, A_{cl} of the closed-loop system (8.31)–(8.32), has all its poles inside the conical sector if and only if there exists a solution $X_c = X_c^T > 0$ such that

$$
\begin{bmatrix}
\sin\dfrac{\theta}{2}\left(A_{cl}X_c + X_c A_{cl}^T\right) & \cos\dfrac{\theta}{2}\left(A_{cl}X_c - X_c A_{cl}^T\right) \\
\cos\dfrac{\theta}{2}\left(X_c A_{cl}^T - A_{cl}X_c\right) & \sin\dfrac{\theta}{2}\left(A_{cl}X_c + X_c A_{cl}^T\right)
\end{bmatrix} < 0
\qquad (8.36)
$$

or equivalently, this can be expressed in Kronecker product form given in

$$
\left[\eta \otimes A_{cl}X_c + \eta^T \otimes X_c A_{cl}^T\right] < 0
\qquad (8.37)
$$

where $\eta = \begin{bmatrix} \sin\dfrac{\theta}{2} & \cos\dfrac{\theta}{2} \\ -\cos\dfrac{\theta}{2} & \sin\dfrac{\theta}{2} \end{bmatrix}$

The inequalities in Equations (8.35) and (8.37) are not jointly convex as the respective solutions of these equations $X_{cl} \neq X_c$. The convexity can be accomplished by seeking a common solution, $X_{cl} = X_c = X_d$. It is to be noted that the inequalities in (8.35) and (8.37) contain nonlinear terms $A_{cl}X_d$, and $C_{cl}X_d$ where A_{cl} and C_{cl} contain unknown matrices of the controller given in Equations (8.29) and (8.30) and the resulting problem therefore cannot be handled by LMI optimization directly. To convert the problem into a linear one, a change of controller variables is necessary and the transformation [17,18] gives the following simplified LMIs in terms of new controller variables:

$$
\begin{bmatrix} Q & I \\ I & S \end{bmatrix} > 0
\qquad (8.38)
$$

$$
\begin{bmatrix} \Pi_{11} & \Pi_{21}^T \\ \Pi_{21} & \Pi_{22} \end{bmatrix} < 0
\qquad (8.39)
$$

$$
\left[\eta \otimes \Psi + \eta^T \otimes \Psi^T\right] < 0
\qquad (8.40)
$$

The matrix variables Π_{11}, Π_{21}, Π_{22}, and ψ are given in

$$
\Pi_{11} =
\begin{bmatrix}
A_p Q + Q A_p^T + B_{p2}\hat{C} + \hat{C}^T B_{p2}^T & B_{p1} + B_{p2}\hat{D}D_{p21} \\
\left(B_{p1} + B_{p2}\hat{D}D_{p21}\right)^T & -\gamma I
\end{bmatrix},
$$

$$
\Pi_{21} =
\begin{bmatrix}
\hat{A} + \left(A_p + B_{p2}\hat{D}C_{p2}\right)^T & SB_{p1} + \hat{B}D_{p21} \\
C_{p1}Q + D_{p12}\hat{C} & D_{p11} + D_{p12}\hat{D}D_{p21}
\end{bmatrix},
$$

$$\Pi_{22} = \begin{bmatrix} A_p^T S + SA_p + \hat{B}C_{p2} + C_{p2}^T \hat{B}^T & (C_{p1} + D_{p21}\hat{D}C_{p2})^T \\ C_{p1} + D_{p12}\hat{D}C_{p2} & -\gamma I \end{bmatrix},$$

$$\Psi = \begin{bmatrix} A_p Q + B_{p2}\hat{C} & A_p + B_{p2}\hat{D}C_{p2} \\ \hat{A} & SA_p + \hat{B}C_{p2} \end{bmatrix}$$

The new controller variables are then defined as

$$\hat{A} = NA_k M^T + NB_k C_{p2} Q + SB_{p2} C_k M^T + S\left(A_p + ^P B_{p2} D_k C_{p2}\right) Q \tag{8.41}$$

$$\hat{B} = NB_k + SB_{p2}D_k \tag{8.42}$$

$$\hat{C} = C_k M^T + D_k C_2 Q \tag{8.43}$$

$$\hat{D} = D_k \tag{8.44}$$

where Q, S, M, and N are submatrices of X_d. The LMIs in Equations (8.38)–(8.40) are solved for $\hat{A}$, $\hat{B}$, $\hat{C}$, and $\hat{D}$ employing interior-point optimization methods [19]. Once $\hat{A}, \hat{B}, \hat{C}$, and $\hat{D}$ are obtained, the controller variables A_k, B_k, C_k, and D_k are recovered by solving (8.41)–(8.44). Moreover, if M and N are square and invertible matrices, then A_k, B_k, C_k, and D_k are unique. For full-order controller design, one can always assume that M and N have full row rank. Hence, the controller variables A_k, B_k, C_k, and D_k can always be replaced with $\hat{A}, \hat{B}, \hat{C}$, and $\hat{D}$, without loss of generality. The derivation of LMIs (8.38)–(8.40) can be obtained as follows.

Considering inequalities (8.35) and (8.37), let common solutions X_d and X_d^{-1} partitioned as

$$X_d = \begin{pmatrix} Q & M \\ M^T & U \end{pmatrix} \quad \text{and} \quad X_d^{-1} = \begin{pmatrix} S & N \\ N^T & V \end{pmatrix} \tag{8.45}$$

where Q and S are $n \times n$ symmetric matrices. The matrices M and N have full row rank with dimension $(n \times k)$. n and k are the order of the system (A_p) and the order of the controller $K(s)$, respectively. Here, expressions U and V are not relevant for the present purpose.

The identity $X_d X_d^{-1} = I$ together with (8.38) gives $MN^T = I - QS$. Thus, M and N have full row rank and square invertible when $I - QS$ is invertible. Let $X_d > 0$ and $K(s)$ be the solutions of inequalities (8.35) and (8.37) and partition X_d as in Equation (8.45). It is readily verified that X_d satisfies the identity $X_d \Omega_2 = \Omega_1$ with

$$\Omega_1 := \begin{pmatrix} Q & I \\ M^T & 0 \end{pmatrix} \quad \text{and} \quad \Omega_2 := \begin{pmatrix} I & S \\ 0 & N^T \end{pmatrix} \tag{8.46}$$

As M and N matrices have full row rank that makes Ω_2 a full column rank matrix with pre- and postmultiplying the inequality $X_d > 0$ by Ω_2^T and Ω_2, respectively. Using Equation (8.46) and Ω_2 full column rank, this yields $\Omega_2^T \Omega = \begin{pmatrix} Q & I \\ I & S \end{pmatrix} > 0$ that gives the LMI condition (8.38).

Next, pre- and postmultiplying the inequality (8.35) by the diag (Ω_2^T, I, I) and diag (Ω_2, I, I), respectively; carrying out the matrix product; and performing the change of controller variables (8.41)–(8.44) evaluate the LMI condition (8.39).

Similarly, the LMI condition (8.40) is derived from Equation (8.37) by pre- and postmultiplying the block diagonal matrices diag $(\Omega_2^T, \ldots, \Omega_2^T)$ and diag $(\Omega_2, \ldots, \Omega_2)$, respectively, and carrying out the matrix product with change of controller variable Equations (8.41)–(8.44).

8.6.4 Design of an H_∞ TCSC controller

The LMI formulations described in the foregoing section are now applied here to design a robust TCSC damping controller for application in a multibus test system. The block diagram of a MISO, TCSC controller model is shown in Figure 8.11. This controller is assumed to be designed based on H_∞ control theory in an LMI framework. The input signal is the speed deviation ($\Delta\omega$), and the output signal is the deviation in thyristor conduction angle ($\Delta\sigma$). This model utilizes the concept of a variable series reactance (ΔX_{TCSC}), which can be modulated through appropriate variation of the firing angle (α).

Following the standard guidelines of mixed-sensitivity design, weights $W_1(s)$ and $W_2(s)$ are chosen as low- and high-pass filters, respectively. The weights $W_1(s)$ and $W_2(s)$ are worked out to be

$$W_1(s) = \frac{2}{s+1.5}; \quad W_2(s) = \frac{0.5s+1}{0.25s+1}$$

The multiobjective H_∞ synthesis program for disturbance rejection and control effort optimization features of LMI was accessed by suitably chosen arguments of the function *hinfmix* of the *LMI Toolbox* in MATLAB [20]. The pole placement

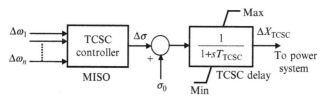

FIGURE 8.11

TCSC controller model.

objective in the LMI (8.37) has been achieved by defining the conical sector with $\frac{\theta}{2} = 67.5°$, which provides a desired minimum damping $\zeta = 0.39$ for all the closed-loop poles. The order of the controller obtained from the LMI solution was equal to the reduced plant order plus the order of the weights, which was quite high (18th order), posing difficulty in practical implementation. Therefore, the controller was reduced to a seventh-order one by the "balanced truncation" without significantly affecting the frequency response. This reduced-order controller has been tested on the full-order system against varying generation, load power change, and transmission line outage.

- **Controller design steps**

    Step 1: Derive full-order system matrix ($[A_{sys}]_{197 \times 197}$) of the 24-machine, 203-bus study system in MATLAB.
    Step 2: Obtain reduced 10th-order model $[A_{sys}]_{10 \times 10}$ of the full-order system applying "balanced truncation" method in MATLAB (using function "*balmr*" setting desired arguments).
    Step 3: Pack system matrices into "TREE" variable using function "*mksys*" in Robust Control Toolbox.
    Step 4: Work out frequency-dependent weighting functions $W_1(s)$ and $W_2(s)$.
    Step 5: Create augmented plant using MATLAB function "*augtf*" with reduced-order state-space system and weighting functions $W_1(s)$ and $W_2(s)$ for H_∞ control system design.
    Step 6: Obtain standard LTI state-space realization $P(s)$ of the augmented plant by the function "*ltisys*".
    Step 7: Implement multiobjective H_∞ synthesis problem with pole placement design in LMI Control Toolbox following function $[gopt, h2opt, K, Q, S] = hinfmix\ (P, r, obj, region)$
    where
- P is the system matrix of the LTI plant $P(s)$;
- r is a three-entry vector listing the lengths of z, y, and u;
- $obj = [\gamma_0, v_0, \alpha, \beta]$ is a four-entry vector specifying the H_∞ constraints and criterion. $obj = [0\ 0\ 1\ 0]$ has been set for H_∞ control design;
- $region$ specifies the LMI region for pole placement. "*lmireg*" has been used to interactively generate the matrix region (conical sector with $\frac{\theta}{2} = 67.5°$).

    The outputs *gopt* and *h2opt* are the guaranteed H_∞ performances, K is the controller system matrix, and Q and S are the optimal values of the variables Q and S equation (8.38).

    Step 8: Extract controller system matrices using function "*ltiss*".
    Step 9: Obtain reduced seventh-order model of the controller by "balanced truncation" method using function "*balmr*" setting desired arguments.

    The state variable representation of the four-input, single-output controller for the TCSC is obtained as

$$A_{K_{\text{tcsc}}} = \begin{bmatrix} -0.24504 & -1854.6 & -0.02038 & 0.01947 & -0.29701 & 0.10962 & 0.04966 \\ 1836.20 & -2429.8 & -2.2824 & 1.79380 & -29.793 & 11.0130 & 4.98410 \\ -0.01863 & 1.1139 & -0.00620 & 0.03802 & -0.16872 & 0.06247 & 0.02840 \\ -0.01692 & 1.0728 & -0.03760 & -0.00604 & 0.19162 & -0.07132 & -0.03225 \\ 0.08566 & 1.24420 & 0.00195 & 0.03433 & -24.094 & 18.1920 & 8.44730 \\ -0.04555 & -8.1658 & -0.00913 & 0.00526 & -0.8256 & -6.7779 & -5.9537 \\ -0.04074 & 0.22424 & -0.02213 & -0.01573 & -2.1844 & -0.05223 & -3.8335 \end{bmatrix}$$

$$C_{K_{\text{tcsc}}} = \begin{bmatrix} -16.897 & -1668.7 & -0.7532 & 0.6376 & -10.229 & 3.7765 & 1.7115 \end{bmatrix}$$

$$D_{K_{\text{tcsc}}} = \begin{bmatrix} 9.6252 & 7.0393 & 14.664 & 13.799 \end{bmatrix}$$

$$B_{K_{\text{tcsc}}} = \begin{bmatrix} 11.886 & 9.1837 & 7.5342 & 1.7663 \\ 249.64 & -201.2 & -734.31 & -1463.8 \\ 0.4711 & 0.2125 & 0.5264 & 0.1518 \\ 0.1889 & 0.3467 & 0.4495 & 0.2202 \\ 1.160 & -9.6148 & 3.2768 & 0.3184 \\ 2.0539 & 1.5970 & -1.0910 & -2.5106 \\ 1.6676 & 0.1484 & 0.2874 & 0.2088 \end{bmatrix}$$

It has been computed in Section 7.8.1 (Chapter 7) that the full-order system without controller has a total of 23 electromechanical swing modes at base case including 11 numbers having a frequency range 0.2-1.0 Hz, which are identified as interarea modes of the system and are listed here separately in Table 8.12. It is evident that the damping ratio of the interarea mode #4 is smallest among these 11 interarea modes and is referred to as the *critical interarea mode*. It is to be noted that the critical swing mode #16 in table 7.9 (Chapter 7) is redefined here as critical interarea mode #4 in Table 8.12. The right eigenvector and participation factor analysis confirms that this mode is involved primarily with machines 4, 13, 24, and 20; in particular, it has strong association with machine 20. These machines are belonging to four different areas as illustrated in Figure 8.12. Therefore, the measuring zone of control input signals is chosen from these four areas. The TCSC module has been placed in branches 152-154. The values of X_L and X_C considered for the design of TCSC are given in Section B.5.1 of Appendix B. The initial value of the firing angle (α) of the TCSC is kept within the capacitive zone with compensation of the TCSC being 56%.

Table 8.12 Interarea Modes with Frequency (0.2-1.0 Hz) at Nominal Operating Condition

#	Swing Modes	Frequency (f)	Damping Ratio (ζ)
1	$-3.0431 \pm j6.0689$	0.9659	0.4482
2	$-3.3906 \pm j5.8905$	0.9375	0.4988
3	$-3.3215 \pm j5.7429$	0.9140	0.5006
4	$-1.0363 \pm j4.3800$	0.6971	0.2302
5	$-3.4131 \pm j5.0370$	0.8016	0.5609
6	$-3.2044 \pm j4.9464$	0.7872	0.5437
7	$-3.8535 \pm j4.0865$	0.6503	0.6860
8	$-3.2680 \pm j4.5548$	0.7249	0.5829
9	$-3.3024 \pm j4.3813$	0.6973	0.6019
10	$-3.3031 \pm j4.4201$	0.7034	0.5986
11	$-2.9858 \pm j2.5168$	0.4005	0.7646

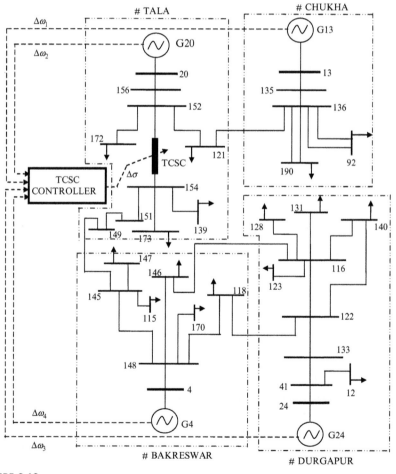

FIGURE 8.12

Part of the 14-area, 24-machine, and 203-bus study system with TCSC controller.

In the following section, the behavior of this interarea mode #4 and the small-signal performance of the proposed closed-loop system have been investigated applying the reduced-order LMI-based TCSC controller in the full-order plant.

8.6.5 Performance of the closed-loop H_∞ control

To examine the performance of the robust TCSC controller, the dynamic behavior of the system has been investigated inducing interarea oscillations for three commonly occurring power system disturbances, that is, load increase, generation drop, and transmission line outage for their small and wide variations. In each case, the eigenvalue as well as time domain analysis of the system has been carried out. The results of simulations are illustrated as follows:

- **Small disturbance performance**

At first, the real and reactive loads of bus 154 are increased in small steps from its nominal value ($P_L=0.75$ pu and $Q_L=1.85$ pu). With increase in load, there is a deterioration of the damping ratio of the critical swing mode that has been observed. However, substantial improvement is found with the installation of the controller. Second, the effect of generation drop on interarea oscillations has been investigated, and it is found that the stability of the system improves adequately in the presence of the controller. The performance of the controller is further verified using a contingency like the outage of tie lines 145-149 and 118-122 separately. The outage of the tie line 145-149 shifted the interarea mode toward a more critical position, but the incorporation of the robust TCSC controller shows noticeable enhancement of damping. The damping action of the controller is also found to be satisfactory with respect to the outage of the line 118-122. Table 8.13 contains the results of both without and with control conditions.

The performance robustness of the controller is now demonstrated by computing the angular speed response of machine 20. The deviation of angular speed response of machine 20 with and without control has been plotted in Figure 8.13a–c for a simulation time of 10 s. It appears that the controller exhibits superior damping characteristics for the case of generation drop and transmission line outage compared to the case of increase of load power demand.

- **Large disturbance performance**

The robust performance of the closed-loop system is evaluated again in the face of large variations of system disturbances that include real and reactive load increase (40% more than nominal) in selected buses, drop in real power generations (total 40%) in some designated generators buses, and simultaneous tripping of three transmission lines (42-53, 118-122, and 145-149). The TCSC controller provides very good damping characteristics in all these contingencies. The damping effect on the critical interarea mode without and with control action has been presented in Table 8.14. It has been observed that simultaneous occurrence of three-tie-line outage pushes the critical mode to the right half of the s-plane, resulting in the instability

Table 8.13 Critical Interarea Mode Without and with Robust TCSC Controller for Small Disturbance

Power System Disturbances		Without Control		With Robust TCSC Controller	
		Critical Interarea Mode	Damping Ratio	Critical Interarea Mode	Damping Ratio
Load increase	$P_L = 0.90$, $Q_L = 2.15$	$-1.0145 \pm j4.4020$	0.22458	$-2.8832 \pm j6.2856$	0.41693
	$P_L = 1.15$, $Q_L = 2.45$	$-0.9983 \pm j4.4352$	0.21961	$-2.8577 \pm j6.3978$	0.40784
Generation drop	Total 15%	$-0.7175 \pm j4.5671$	0.15520	$-2.8392 \pm j6.3784$	0.40666
	Total 20%	$-0.6118 \pm j4.6395$	0.13076	$-2.8404 \pm j6.3768$	0.40689
Line outage	145-149	$-0.2040 \pm j3.9457$	0.05164	$-2.9385 \pm j6.3567$	0.41961
	118-122	$-1.0140 \pm j4.2986$	0.22960	$-3.1808 \pm j6.7737$	0.42505

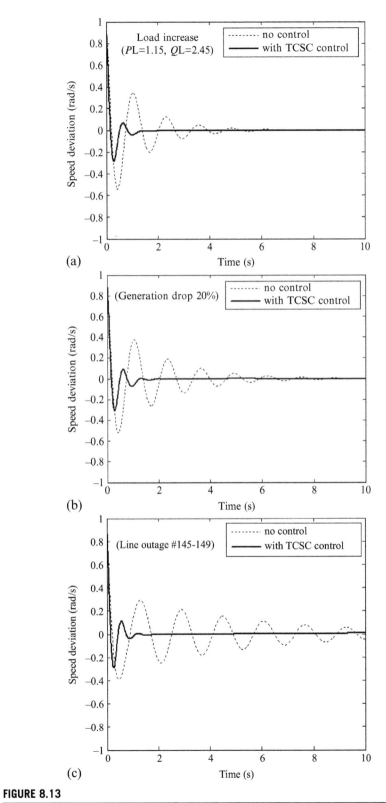

FIGURE 8.13

Small disturbance dynamic response. (a) Load increase; (b) generation drop; (c) line outage.

Table 8.14 Critical Interarea Mode Without and with Robust TCSC Controller for Large Disturbance

Power System Disturbances		Without Control		With Robust TCSC Control	
		Critical Interarea Mode	Damping Ratio	Critical Interarea Mode	Damping Ratio
Load increase	40% more than nominal value	$-0.75328 \pm j4.9973$	0.14905	$-2.7942 \pm j6.6364$	0.38805
Generation drop	Total 40% (gen 2, 3, 5, and 20)	$-0.16153 \pm j4.9112$	0.03287	$-2.9344 \pm j8.2818$	0.33398
Line outage	42-53, 118-122, 145-149	$0.00605 \pm j3.9016$	-0.00155	$-3.0273 \pm j7.1389$	0.39041

of the system. In this situation, installation of the TCSC controller shows significant improvement of damping and brings back the system under stable operating condition, establishing the need for the robust controller.

The dynamic behavior of the system with respect to generator speed has also been investigated exciting the interarea oscillations following these large disturbances. It is evident that the behavior of the system is sufficiently oscillatory for all three cases. The angular speed responses of generator 20 with different power system disturbances have been plotted in Figure 8.14a–c. It is visible that LMI-based TCSC controller introduces adequate improvement on system oscillations and provides a reasonable settling time (2-3 s). A view of Figure 8.14c shows that the controller achieves a higher level of damping for the tie-line outage oscillations compared to the case of load increase and generation drop.

The results mentioned earlier reported one important conclusion that the interarea oscillations can be damped effectively when TCSC has been placed in the branch associated with a particular machine participating strongly to the interarea mode. The need of robust TCSC controller for control of interarea oscillations in the face of large variations of power system disturbances has also been established.

8.7 MULTIAREA CLOSED-LOOP CONTROL

Once the design and simulations have been performed, the next credible step would be to implement the closed-loop control, and for this requirement, a simple feedback control scheme has been proposed here. The proposed possible schematic diagram has been depicted in a block diagram in Figure 8.15. The rotor speed is detected by the digital proximity pickup (Figure 8.16), which can usually measure speed up to

15000 rpm. The frequency of the pulses delivered through the proximity sensor will depend upon the number of teeth of the rotor and its speed of rotation.

Here, the objective is to realize a closed-loop control system using TCSC controller in a multiarea power system. To accomplish this task, the power system needs to install required numbers of measuring equipments at remote nodes for collections of auxiliary control input (rotor speed, $\Delta\omega$); data corresponding to the generators have high participation in interarea oscillations. These machines may be located

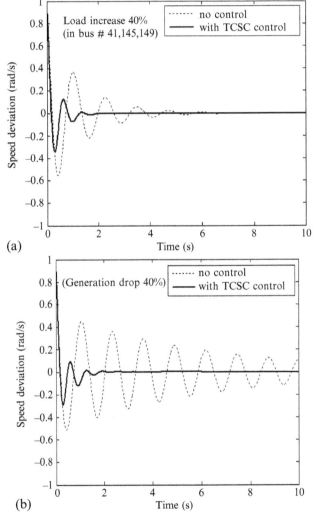

FIGURE 8.14

Large disturbance dynamic response. (a) Load increase; (b) generation drop;

(Continued)

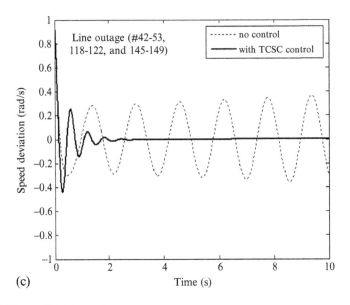

FIGURE 8.14, cont'd (c) line outage.

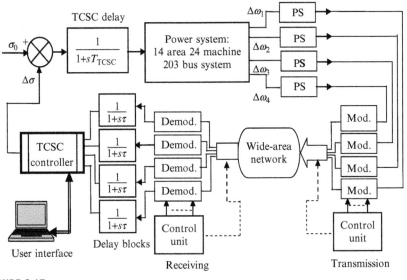

FIGURE 8.15

Configuration of remote feedback control scheme.

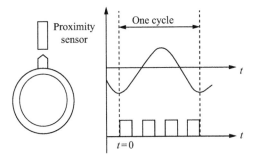

FIGURE 8.16

Configuration of proximity sensor.

in different areas or substations. Generator speed signals measured by the proximity sensor (PS) are modulated first for transmission through the wide-area communication network. The Ethernet or high-speed fiber-optic link may be suitable for this purpose. In the receiving end of the centralize control station, transmitted signals can be demodulated and filtered out to receive input signals to the controller.

One of the important concerns in multiarea measurement-based control is the signal transmission delay. The value of delay is generally considered as 0.05-1.0 s [21], which can vary depending on the distance of the controller site from the sensor location and the mechanism of data acquisition. The delays for the remote signals can be modeled by a first-order filter [22] in the feedback path where the equivalent time constant τ is representing the delay. Denoting delayed speed signal by $\Delta\omega_\tau$ corresponding to the original signal $\Delta\omega$, the state-space formulations of the transportation delay blocks can be obtained as

$$\Delta\dot{\omega}_\tau = A_\tau \Delta\omega_\tau + B_\tau \Delta\omega \tag{8.47}$$

$$\Delta y_\tau = C_\tau \Delta\omega_\tau \tag{8.48}$$

where $\Delta\omega_\tau = [\Delta\omega_{1\tau} \ \Delta\omega_{2\tau} \ \cdots \ \Delta\omega_{\mu\tau}]^T$, $A_\tau = \text{diag}(-2/\tau_1, -2/\tau_2, \ldots, -2/\tau_\mu)$, $B_\tau = \text{diag}(2/\tau_1, 2/\tau_2, \ldots, 2/\tau_\mu)$, and C_τ is the identity matrix. μ denotes the numbers of measured signals. For the given study system, $\mu = 4$ for four numbers of selected input signals of the controller. The speed input signals measured by the PS are transmitted via wide-area communication network to the controller. The controller produces output control action signal and thyristor conduction angle ($\Delta\sigma$), which introduces additional damping to the interarea mode executing TCSC reactance in phase with the speed deviations of the generators.

EXERCISE

8.1. Explain the different steps of particle swarm optimization (PSO) and GA for application in parameter optimization of a power system damping controllers. What are the advantages of PSO over GA?

8.2. Explain the terms, H_∞ norm and H_∞ control. Discuss the mixed-sensitivity H_∞ control theory based on LMI approach. What is the significance of the LMI region?

8.3. In a single-machine infinite bus system, the machine and exciter parameters are given as follows:

$R_e=0$, $X_e=0.5$ pu, $V_t \angle \theta° = 1 \angle 15°$ pu, and $V_\infty \angle \theta° = 1.05 \angle 0°$ pu. $H=3.2$ s, $T'_{do}=9.6$ s, $K_A=400$, $T_A=0.2$ s, $R_s=0.0$ pu, $X_q=2.1$ pu, $X_d=2.5$ pu, $X'_d=0.39$ pu, $D=0$, and $\omega_s=314$ rad/s

(i) Design an LMI-based H_∞ TCSC controller for this system and ensure that all the closed-loop poles of the system are lying in a feasible LMI region.
(ii) Repeat the problem (i) for the design of an LMI-based H_∞ SVC controller.

8.4. Consider a WSCC-type 3-machine, 9-bus system. Find the optimal location and parameters of the TCSC controller applying PSO- and GA-based optimization method separately. Investigate the characteristics of the critical swing mode of the system for both the optimization methods.

8.5. What is Hopf bifurcation? For the system given in problem 8.4, investigate the point of Hopf bifurcation load. It is assumed that real power load (P_L) is increased in steps at load bus. 5.
 (i) Show that installation of an SVC at load bus 8 can put off Hopf bifurcation.
 (ii) Repeat the problem (i) when a TCSC is installed in line 5-7.

References

[1] D.P. Kothari, Application of neural networks to power systems, Proc. IEEE Int. Conf. Ind. Technol. 1 (2000) 621–626.
[2] S. Kirkpatrick, C.D. Gellat, M.P. Vecchi, Optimization by simulated annealing, Science 220 (1983) 671–680.
[3] M.L. Kothari, T.J. Kumar, A new approach for designing fuzzy logic power system stabilizer, IEEE International Power Engineering Conference (IPEC-2007), 2007, pp. 419-424.
[4] D. Mondal, A. Chakrabarti, A. Sengupta, Optimal placement and parameter setting of SVC and TCSC using PSO to mitigate small signal stability problem, Int. J. Electric Power Energy Syst. 42 (1) (2012) 334–340. www.elsevier.com.
[5] D.E. Goldberg, Genetic Algorithms in Search Optimization and Machine Learning, Addison-Wesley Publishing Company, Inc., New York, 1989.
[6] X.P. Wang, L.P. Cao, Genetic Algorithms—Theory, Application and Software Realization, Xi'an Jiaotong University, Xi'an, China, 1998.
[7] J. Kennedy, R. Eberhart, Particle swarm optimization, IEEE Int. Conf. Neural Netw. 4 (1995) 1942–1948.
[8] D.J. Krusienski, W.K. Jenkins, Design and performance of adaptive systems based on structured stochastic optimization strategies, IEEE Circuits and Systems Mag. (First quarter), (2005) 8–20.

[9] G.I. Rashed, H.I. Shaheen, S.J. Cheng, Optimum location and parameter setting of TCSC by both genetic algorithm and particle swarm optimization, IEEE 2nd International Conference on Industrial Electronics and Applications (ICIEA-2007), 2007, pp. 1141-1147.

[10] B. Birge, Particle Swarm Optimization Toolbox, Available: www.mathworks.com.

[11] M.A. Pai, D.P. Sengupta, K.R. Padiyar, Small Signal Analysis of Power Systems, Narosa Publishing House, India, 2004.

[12] K. Zhou, J.C. Doyle, Essentials of robust control, Prentice-Hall, New Jersey, 1998.

[13] S.Z. Sayed Hassen, Robust and gain-scheduled control using Linear Matrix Inequalities, M. Eng. Science Thesis, Monash University, Australia, 23rd April 2001.

[14] S. Skogestad, I. Postlethwaite, Multivariable Feedback Control Analysis and Design, John Wiley and Sons, New York, 1996.

[15] K. Zhou, J.C. Doyle, K. Glover, Robust and Optimal Control, Prentice Hall, New Jersey, 1995.

[16] P. Gahinet, P. Apkarian, A linear matrix inequality approach to H_∞ control, Int. J. Robust Nonlinear Control 4 (4) (1994) 421–448.

[17] M. Chilali, P. Gahinet, H_∞ design with pole placement constraints: An LMI approach, IEEE Trans. Automatic Control 41 (3) (1996) 358–367.

[18] C. Scherer, P. Gahinet, M. Chilali, Multiobjective output-feedback control via LMI optimization, IEEE Trans. on Automatic Control 42 (7) (1997) 896–911.

[19] Y. Nesterov, A. Nemirovski, Interior Point Polynomial Methods in Convex Programming; Theory and Applications, SIAM, Philadelphia, PA, 1994.

[20] Matlab Users Guide, The Math Works Inc., USA, 1998. www.mathworks.com.

[21] B. Chaudhuri, R. Majumder, B.C. Pal, Wide-area measurement-based stabilizing control of power system considering signal transmission delay, IEEE Trans. Power Syst. 19 (4) (2004) 1971–1979.

[22] B. Chaudhuri, B.C. Pal, Robust damping of multiple swing modes employing global stabilizing signals with a TCSC, IEEE Trans. Power Syst. 19 (1) (2004) 499–506.

Nomenclature

B_{svc}	SVC equivalent susceptance at fundamental frequency (pu)
D	damping constant (pu)
E_d'	d-axis component of voltage behind transient reactance (pu)
E_q'	q-axis component of voltage behind transient reactance (pu)
E_{fd}	exciter output (pu)
$f_E(E_{fd})$	exciter saturation function (pu)
$Gex(s)$	exciter transfer function
H	machine inertia constant (s)
H_∞	set of transfer functions with bounded ∞ norm
I_d	d-axis component of current (pu)
I_q	q-axis component of current (pu)
K_A	automatic voltage regulator gain
$K_1 - K_2$	K-constants of Heffron-phillips linearized model generator
K_{svc}	SVC controller gain
K_{TCSC}	TCSC controller gain
K_E	exciter constant (self excited) (pu)
K_F	rate feedback stabilizer gain
K_{PSS}	power system stabilizer gain
N_{loc}	FACTS (SVC, TCSC) location number
N_{ind}	GA population size
P_{acc}	electrical accelerating power
P_G	real power generation (pu)
P_L	real power load (pu)
P_{st}	power flow between node s and t (pu)
Q_G	reactive power generation (pu)
Q_L	reactive power load (pu)
R_e	external equivalent resistance (SMIB system) (pu)
R_s	stator resistance (pu)
R_F	rate feedback signal (pu)
T_A	regulator time constant (s)
T_E	exciter time constant (s)
T_F	feedback stabilizer time constant (s)
T_1	lead-time constant of the PSS (s)
T_2	lag-time constant of the PSS (s)
T_{do}'	d-axis open circuit time constant (s)
T_{qo}'	q-axis open circuit time constant (s)
T_{PSS}	torque introduced by PSS (pu)
T_{svc}	SVC internal delay (ms)
T_{TCSC}	TCSC internal delay (ms)
T_M	mechanical starting time constant (s)
T_w	washout time constant (s)
V_{ref}	ref. input voltage of regulator (pu)
V_R	output of the regulator (pu)
V_s	PSS output (pu)

V_{sc}	STATCOM bus voltage (pu)
V	machine terminal voltage (pu)
V_∞	infinite bus voltage
V_0	output of auxiliary controller in SVC (pu)
X_d	d-axis component of synchronous reactance (pu)
X_q	q-axis component of synchronous reactance (pu)
X_d'	d-axis component of transient reactance (pu)
X_q'	q-axis component of transient reactance (pu)
X_{svc}	SVC equivalent reactance at fundamental frequency (pu)
X_e	external equivalent reactance (SMIB system) (pu)
X_C	capacitive reactance (pu)
X_L	inductive reactance (pu)
X_{TCSC}	TCSC reactance at fundamental frequency (pu)
X_{st}	transmission line reactance between node s and t (pu)
Y_{ik}	transmission line admittance between node i and k (pu)
θ	bus voltage angle (°)
Δ	deviation operator
δ	torque angle (°)
δ_{sc}	STATCOM bus angle (°)
ω	rotor speed (rad./s)
ω_s	synchronous speed (rad./s)
α	firing angle of thyristor (°)
α_r	firing angle of thyristor at resonance point (°)
σ	conduction angle of thyristor (°)

Fundamental Concepts

A.1 GENERALIZED CONCEPT OF STABILITY-BRIEF REVIEW

The stability of a linear system is entirely independent of the input, and the state of a stable system with zero input will always return to the origin of the state space, independent of the finite initial state. In contrast, the stability of a nonlinear system depends on the type and magnitude of input, and the initial state. These factors have to be taken into account in defining the stability of a nonlinear system. In control system theory, the stability of a non linear system is classified into the following categories, depending on the region of state space in which the state vector ranges; Local Stability, Finite Stability and Global Stability.

A.1.1 Local Stability

Consider a nonlinear autonomous system described by the following state equations;

$$\dot{x} = f(x, u) \tag{A.1}$$

$$y = g(x, u) \tag{A.2}$$

where x is the state vector $(n \times 1)$; u is the vector $(r \times 1)$ of inputs to the system and y is the vector $(m \times 1)$ of outputs. This nonlinear system is said to be *locally stable* about an equilibrium point if, when subjected to small perturbation $(\Delta x, \Delta u)$, it remains within a small region surrounding the equilibrium point.

   If, as t increases, the system returns to the original state, it is said to be *asymptotically stable in-the-small* or stability under small disturbances i.e., local stability conditions can be studied by linearizing the nonlinear system equations about the desired equilibrium point.

A.1.2 Finite Stability

If the state of a system after perturbation remains within a finite region R, it is said to be stable within R. If, further the state of the system after perturbation returns to the original equilibrium point from any initial point $x(t_0)$ within R, it is said to be *asymptotically* stable within the finite region R Figure A.1.

271

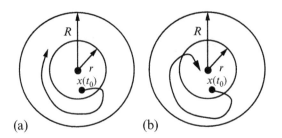

FIGURE A.1

Stability in nonlinear system. (a) Local stability or Finite stability. (b) Asymptotic stability

A.1.3 **Global Stability**

The system is said to be *globally stable* or *asymptotically stable in-the-large* if R includes the entire finite space and the state of the system after perturbation from every initial point regardless of how near or far it is from the equilibrium point within the finite space returns to the original equilibrium point as $t \rightarrow \infty$.

A.2 **ASPECT OF LINEARIZATION**

A.2.1 **Linearization of a Nonlinear Function**

Consider a nonlinear function $y = f(x)$ as shown in Figure A.2. Assume that it is necessary to operate in the vicinity of point a on the curve (the operating point) whose co-ordinates are x_a, y_a.

For the small perturbations Δx and Δy about the operating point a let

$$\Delta x = x \qquad\qquad (A.3)$$

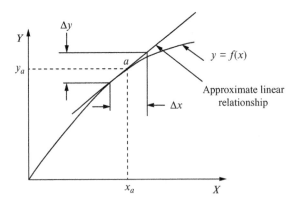

FIGURE A.2

Linearization of a nonlinear function

$$\Delta y = y \qquad \text{(A.4)}$$

If the slope at the operating point is $\dfrac{dy}{dx}\Big|_a$, then the approximate linear relationship becomes

$$\Delta y = \frac{dy}{dx}\Big|_a \Delta x \qquad \text{(A.5)}$$

i.e.,

$$y = \frac{dy}{dx}\Big|_a x. \qquad \text{(A.6)}$$

A.2.2 Linearization of a Dynamic System

The behaviour of a dynamic system, such as power system, may be described in the following form

$$\dot{x} = f(x, u) \qquad \text{(A.7)}$$

$$y = g(x, u) \qquad \text{(A.8)}$$

where x is the state vector ($n \times 1$); u is the vector ($r \times 1$) of inputs to the system and y is the vector ($m \times 1$) of outputs. Here the procedure for linearization of equation (A.7) and (A.8) has been described. Let x_0 be the initial state vector and u_0 the input vector corresponding to the equilibrium point about which the small signal performance is to be investigated. Since x_0 and u_0 satisfy equation (A.7), we have

$$\dot{x}_0 = f(x_0, u_0) = 0 \qquad \text{(A.9)}$$

Let us perturbed the system from the above state, by letting

$$x = x_0 + \Delta x \quad u = u_0 + \Delta u$$

The state must satisfy equation (A.7). Hence,

$$\dot{x} = \dot{x}_0 + \Delta\dot{x} = f[(x_0 + \Delta x), (u_0 + \Delta u)] \qquad \text{(A.10)}$$

As the perturbations are assumed to be small, the nonlinear functions $f(x, u)$ can be expressed in terms of Taylor's series expansion. With terms involving second and higher order of Δx and Δu neglected, we may write

$$\dot{x}_i = \dot{x}_{i0} + \Delta\dot{x}_i = f_i[(x_0 + \Delta x), (u_0 + \Delta u)]$$

$$= f_i(x_0, u_0) + \frac{\partial f_i}{\partial x_1}\Delta x_1 + \cdots + \frac{\partial f_i}{\partial x_n}\Delta x_n + \frac{\partial f_i}{\partial u_1}\Delta u_1 + \cdots + \frac{\partial f_i}{\partial u_r}\Delta u_r$$

Since $\dot{x}_{i0} = f_i(x_0, u_0)$, we obtain

$$\Delta\dot{x}_i = \frac{\partial f_i}{\partial x_1}\Delta x_1 + \cdots + \frac{\partial f_i}{\partial x_n}\Delta x_n + \frac{\partial f_i}{\partial u_1}\Delta u_1 + \cdots + \frac{\partial f_i}{\partial u_r}\Delta u_r$$

with $i = 1, 2, 3, \ldots, n$. In a like manner, from equation (A.8), we have

$$\Delta y_j = \frac{\partial g_j}{\partial x_1}\Delta x_1 + \cdots + \frac{\partial g_j}{\partial x_n}\Delta x_n + \frac{\partial g_j}{\partial u_1}\Delta u_1 + \cdots + \frac{\partial g_j}{\partial u_r}\Delta u_r$$

with $i = 1, 2, 3, \ldots, m$. Therefore, the linearized forms of equations (A.7) and (A.8) are

$$\Delta \dot{x} = A\Delta x + B\Delta u \tag{A.11}$$

$$\Delta y = C\Delta x + D\Delta u \tag{A.12}$$

Where

$$[A]_{n\times n} = \begin{bmatrix} \dfrac{\partial f_1}{\partial x_1} & \cdots & \dfrac{\partial f_1}{\partial x_n} \\ \cdots & \cdots & \cdots \\ \dfrac{\partial f_n}{\partial x_1} & \cdots & \dfrac{\partial f_n}{\partial x_n} \end{bmatrix} \quad [B]_{n\times r} = \begin{bmatrix} \dfrac{\partial f_1}{\partial u_1} & \cdots & \dfrac{\partial f_1}{\partial u_r} \\ \cdots & \cdots & \cdots \\ \dfrac{\partial f_n}{\partial u_1} & \cdots & \dfrac{\partial f_n}{\partial u_r} \end{bmatrix}$$

$$[C]_{m\times n} = \begin{bmatrix} \dfrac{\partial g_1}{\partial x_1} & \cdots & \dfrac{\partial g_1}{\partial x_n} \\ \cdots & \cdots & \cdots \\ \dfrac{\partial g_m}{\partial x_1} & \cdots & \dfrac{\partial g_m}{\partial x_n} \end{bmatrix} \quad [D]_{m\times r} = \begin{bmatrix} \dfrac{\partial g_1}{\partial u_1} & \cdots & \dfrac{\partial g_1}{\partial u_r} \\ \cdots & \cdots & \cdots \\ \dfrac{\partial g_m}{\partial u_1} & \cdots & \dfrac{\partial g_m}{\partial u_r} \end{bmatrix}$$

In equations (A.11) and (A.12), A is the state or plant matrix, B is the control or input matrix, C is the output matrix and D is the feed-forward matrix. These partial derivatives are evaluated at the equilibrium point about which the small perturbation is being analyzed.

A.3 SYSTEM MATRIX AND ITS EIGEN PROPERTIES
A.3.1 Eigenvalues and Eigenvectors

The single machine or multimachine linearized dynamic model of a power system can be written in simple form as

$$\Delta \dot{X}(t) = A\Delta X(t) + E\Delta U(t) \tag{A.13}$$

where ΔX: state vector ($r \times 1$), $r =$ total number of states.

    A: system matrix ($r \times r$).

    E: input matrix.

    ΔU: input vector.

    The *eigenvalues* of the matrix A are given by the values of the scalar parameter λ for which there exist non-trivial solutions (i.e. other than $\phi = 0$) to the equation

$$A\phi = \lambda\phi \tag{A.14}$$

To find the eigenvalues, equation (A.14) may be written in the form

$$(A - \lambda I)\phi = 0 \tag{A.15}$$

For a non-trivial solution

$$\det(A - \lambda I)\phi = 0 \tag{A.16}$$

Expansion of this determinant gives the '*characteristics equation*'. The r solutions of (A.16) $\lambda = \lambda_1, \lambda_2, \ldots, \lambda_r$ are the eigenvalues of A.

For any eigenvalue λ_p the r-column vector ϕ_p which satisfies equation (A.14) is called the **right-eigenvector** of A associated with the eigenvalue λ_p.

Thus, we get

$$A\phi_p = \lambda_p \phi_p \quad p = 1, 2, \ldots, r \tag{A.17}$$

The right-eigenvector has the form

$$\phi_p = \begin{bmatrix} \phi_{1p} & \phi_{2p} & \cdots & \phi_{rp} \end{bmatrix}^{\mathrm{T}}$$

Similarly, the r-row vector ψ_p which satisfies the equation

$$\psi_p A = \lambda_p \psi_p, \quad p = 1, 2, \ldots, r \tag{A.18}$$

is called the **left-eigenvector** associated with the eigenvalue λ_p.

The left-eigenvector has the form

$$\psi_p = \begin{bmatrix} \psi_{1p} & \psi_{2p} & \cdots & \psi_{rp} \end{bmatrix}$$

The left and right-eigenvectors corresponding to different eigenvalues are orthogonal, *i.e.*,

$$\psi_q \phi_p = 0 \tag{A.19}$$

where $\lambda_p \neq \lambda_q$ and

$$\psi_p \phi_p = \alpha_p \tag{A.20}$$

where $\lambda_p = \lambda_q$ and α_p is a non zero constant. It is normal practice to normalized these vectors so that

$$\psi_p \phi_p = 1 \tag{A.21}$$

A.3.2 Effect of Right and Left Eigenvectors on System States

Referring to the sate equation (A.13) for the autonomous system (with zero input) the system equation is given by

$$\Delta \dot{X}(t) = A \Delta X(t) \tag{A.22}$$

In order to avoid cross-coupling between the state variables, consider a new state vector Z related to the original state vector X by the similarity transformation

$$\Delta X = \Phi Z \tag{A.23}$$

Where Φ is the modal matrix of A and is defined by $\Phi = \begin{bmatrix} \phi_1 & \phi_2 & \cdots & \phi_r \end{bmatrix}$ and $\Phi^{-1} = \Psi = \begin{bmatrix} \psi_1^{\mathrm{T}} & \psi_2^{\mathrm{T}} & \cdots & \psi_r^{\mathrm{T}} \end{bmatrix}^T$

The $\Phi^{-1}A\Phi$ will transform the matrix A into a diagonal matrix Λ with the eigenvalues $\lambda_1, \lambda_2, \ldots, \lambda_r$ are the diagonal elements. Therefore, after substitution of equation (A.23) into (A.22) gives

$$\Phi\dot{Z} = A\Phi Z \tag{A.24}$$

The new state equation can be written as

$$\dot{Z} = \Phi^{-1}A\Phi Z \tag{A.25}$$

this becomes

$$\dot{Z} = \Lambda Z \tag{A.26}$$

where Λ is a diagonal matrix consisting of eigenvalues of matrix A. Equation (A.26) represents r nos. uncoupled first order equations

$$\dot{Z}_p = \lambda_p Z_p, \quad p = 1, 2, \ldots, r \tag{A.27}$$

and the solution with respect to time t of this equation is given by

$$Z_p(t) = Z_p(0)e^{\lambda_p t} \tag{A.28}$$

where $Z_p(0)$ is the initial value of the state Z_p.

The response in terms of original state vector is given by

$$\Delta X = \Phi Z = [\phi_1 \quad \phi_2 \quad \cdots \quad \phi_r] \begin{bmatrix} Z_1(t) \\ Z_2(t) \\ \cdot \\ \cdot \\ \cdot \\ Z_r(t) \end{bmatrix} \tag{A.29}$$

Using equation (A.28) in equation (A.29) results in

$$\Delta X(t) = \sum_{p=1}^{r} \phi_p Z_p(0)e^{\lambda_p t} \tag{A.30}$$

Again from Equation (A.29), we get

$$Z(t) = \Phi^{-1}\Delta X(t) \tag{A.31}$$

$$Z(t) = \Psi \Delta X(t) \tag{A.32}$$

This implies that

$$Z_p(t) = \psi_p \Delta X(t) \tag{A.33}$$

with $t = 0$, it follows that

$$Z_p(0) = \psi_p \Delta X(0) \tag{A.34}$$

By using C_p to denote the scalar product $C_p = \psi_p \Delta X(0)$, this represents the magnitude of the excitation of the p th mode resulting from the initial condition. Therefore, equation (A.30) may be written as

$$\Delta X(t) = \sum_{p=1}^{r} \phi_p C_p e^{\lambda_p t} \tag{A.35}$$

In other words, the time response of the p th state variable is given by

$$\Delta X_p(t) = \phi_{1p} C_1 e^{\lambda_1 t} + \phi_{2p} C_2 e^{\lambda_2 t} + \cdots + \phi_{rp} C_r e^{\lambda_r t} \tag{A.36}$$

Equation (A.36) indicates that the right-eigenvector entries $\phi_{\kappa p}$, ($\kappa = 1, 2, \ldots, r$) measures the relative *activity of the state* variables participating in the oscillation of certain mode (λ_p). For example, the degree of activity of the state variable X_p in the pth mode is given by the element $\phi_{\kappa p}$ of the right eigenvector ϕ_p.

Similarly, the effect of *left-eigenvector* on system state variable can be illustrated as follows:

The transformed state vector Z is related to the original state vector X by the equation

$$\begin{aligned} \Delta X(t) &= \Phi Z(t) \\ &= [\phi_1 \ \phi_2 \ \cdots \ \phi_r] Z(t) \end{aligned} \tag{A.37}$$

and by the equation (A.32)

$$\begin{aligned} Z(t) &= \Psi \Delta X(t) \\ &= [\psi_1^T \ \psi_2^T \ \cdots \ \psi_r^T]^T \Delta X(t) \end{aligned} \tag{A.38}$$

Again from equation (A.27) we get

$$\dot{Z}_p = \lambda_p Z_p, \quad p = 1, 2, \ldots, r$$

Thus the variables $\Delta X_1, \Delta X_2, \ldots, \Delta X_r$ are the original state variables represent the dynamic performance of the system. The variables $Z_1, Z_2, \ldots, Z_r$ are the transformed state variables such that each variable is associated with only one eigenvalue *i.e.*, they are directly related to the electromechanical modes. As seen from equation (A.38), the left eigenvector ψ_p identifies which combination of the original state variables displays only the p th mode. Thus the κ th element of the right-eigenvector ϕ_p measures the activity of the variable X_κ in the p th mode, and the κ th element of the left-eigenvector ψ_p weighs the contribution of this activity to the pth mode.

A.4 WHAT ARE SEMI-DEFINITE PROGRAMMING (SDP) PROBLEMS?

A wide variety of problems in systems and control theory can be formulated as a semi-definite programming problem of the form

minimize $C^T x$ subject to $F(x) \geq 0$

where $x \in R^m$ is the variable, $F(x) = F_0 + \sum_{i=1}^{m} x_i F_i$, the matrices $c \in R^m$ and $F_i = F_i^T \in R^{n \times n}$, $i = 0, 1, \ldots, m$ are given. The matrix $F(x)$ is positive semi-definite and the constraint $F(x) \geq 0$ is called a linear matrix inequality (LMI). SDP problems are convex programming problems with a linear objective function and LMI constraints.

Semi-definite programming problems can be recast in the form:

$$Ax \leq b$$

A.4.1 What is a linear matrix inequality?

A linear matrix inequality (LMI) has the form:

$$F(x) = F_0 + \sum_{i=1}^{m} x_i F_i > 0 \qquad \text{(A.39)}$$

where $x \in R^m$ is the variable. $x = (x_1, x_2, \ldots, x_m)$ is a vector of unknown scalars (the decision or optimization variables) and the symmetric matrices $F_i = F_i^T \in R^{n \times n}, i = 0,$ $1, \ldots, m$ are given. '> 0' stands for "positive definite", i.e., the largest eigenvalue of $F(x)$ is positive.

Note that the constraints $F(x) > 0$ and $F(x) < G(x)$ are special case of (A.39), since they can be rewritten as $-F(x) < 0$ and $F(x) - G(x) < 0$, respectively.

The LMI (A.39) is a convex constraint on x since $F(y) > 0$ and $F(z) > 0$ imply that $F\left(\frac{y+z}{2}\right) > 0$. As a result,

- Its solution set, called the feasible set, is a convex subset of R^m
- Finding a solution x to (A.39), if any, is a convex optimization problem.

Convexity has an important consequence, even though (A.39) has no analytical solution in general, it can be solved numerically with guarantees of finding a solution when one exists.

In control systems there are a number of problems which lead to the solution of an LMI. For example,

- Lyapunov equation: $A^T P + PA = -Q$

  Lyapunov theorem: The linear time-invariant dynamical system described by;

$$\dot{x}(t) = Ax(t)$$

where $x \in R^n$ is the variable and the matrix $A \in R^{n \times n}$ is *stable* if and only if given any positive definite symmetric matrix $Q \in R^{n \times n}$ there exists a unique positive definite symmetric matrix P satisfying the Lyapunov's equation:

$$A^T P + PA = -Q < 0 \qquad \text{(A.40)}$$

The Lyapunov equation (A.40) is in the form of an LMI. This LMI could be rewritten in the form of (A.39). Indeed, considering $n = 2$, and defining:

$$P = \begin{bmatrix} x_{11} & x_{12} \\ x_{21} & x_{22} \end{bmatrix}; \quad P_1 = \begin{bmatrix} 1 & 0 \\ 0 & 0 \end{bmatrix}; \quad P_2 = \begin{bmatrix} 0 & 1 \\ 1 & 0 \end{bmatrix}; \quad P_3 = \begin{bmatrix} 0 & 0 \\ 0 & 1 \end{bmatrix}$$

we can write with $x_1 = x_{11}, \ x_2 = x_{12} = x_{21}, \ x_3 = x_{22}$

$$P = x_{11} \begin{bmatrix} 1 & 0 \\ 0 & 0 \end{bmatrix} + x_{12} \begin{bmatrix} 0 & 1 \\ 1 & 0 \end{bmatrix} + x_{22} \begin{bmatrix} 0 & 0 \\ 0 & 1 \end{bmatrix}$$

$$= x_1 P_1 + x_2 P_2 + x_3 P_3$$

Therefore,

$$A^{\mathrm{T}}P + PA$$

$$= x_1\left(A^{\mathrm{T}}P_1 + P_1A\right) + x_2\left(A^{\mathrm{T}}P_2 + P_2A\right) + x_3\left(A^{\mathrm{T}}P_3 + P_3A\right)$$

$$= -x_1F_1 - x_2F_2 - x_3F_3 < 0$$

where

$$F_0 = 0; \quad -F_1 = A^{\mathrm{T}}P_1 + P_1A; \quad -F_2 = A^{\mathrm{T}}P_2 + P_2A; \quad -F_3 = A^{\mathrm{T}}P_3 + P_3A$$

Consequently,

$$x_1F_1 + x_2F_2 + x_3F_3 > 0 \tag{A.41}$$

This shows that a Lyapunov equation can be written in the form of an LMI.

The LMI (A.41) with $A = \begin{bmatrix} a_{11} & a_{12} \\ a_{21} & a_{22} \end{bmatrix}$ can be written as

$$-F_1 = A^{\mathrm{T}}P_1 + P_1A = \begin{bmatrix} a_{11} & a_{21} \\ a_{12} & a_{22} \end{bmatrix}\begin{bmatrix} 1 & 0 \\ 0 & 0 \end{bmatrix} + \begin{bmatrix} 1 & 0 \\ 0 & 0 \end{bmatrix}\begin{bmatrix} a_{11} & a_{12} \\ a_{21} & a_{22} \end{bmatrix}$$

$$= \begin{bmatrix} 2a_{11} & a_{12} \\ a_{12} & 0 \end{bmatrix}$$

$$-F_2 = A^{\mathrm{T}}P_2 + P_2A = \begin{bmatrix} a_{11} & a_{21} \\ a_{12} & a_{22} \end{bmatrix}\begin{bmatrix} 0 & 1 \\ 1 & 0 \end{bmatrix} + \begin{bmatrix} 0 & 1 \\ 1 & 0 \end{bmatrix}\begin{bmatrix} a_{11} & a_{12} \\ a_{21} & a_{22} \end{bmatrix}$$

$$= \begin{bmatrix} 2a_{21} & a_{11} + a_{22} \\ a_{11} + a_{22} & 2a_{12} \end{bmatrix}$$

Similarly, $-F_3 = A^{\mathrm{T}}P_3 + P_3A = \begin{bmatrix} 0 & a_{21} \\ a_{21} & 2a_{22} \end{bmatrix}$

Therefore, $-x_1F_1 - x_2F_2 - x_3F_3 < 0$

gives $\begin{bmatrix} a_{11} & a_{21} & 0 \\ a_{12} & a_{11} + a_{22} & a_{21} \\ 0 & a_{12} & a_{22} \end{bmatrix}\begin{bmatrix} x_1 \\ x_2 \\ x_3 \end{bmatrix} < \begin{bmatrix} b_1 \\ b_2 \\ b_3 \end{bmatrix}$

which is in the form of semi-definite programming problem

$$\widetilde{A}\widetilde{x} < \widetilde{b}$$

A.4.2 Interior-Point method

For the LMI

$$F(x) = F_0 + \sum_{i=1}^{m} x_iF_i > 0 \tag{A.42}$$

where $x \in R^m$ is the variable, and the symmetric positive definite matrices $F_i = F_i^T \in R^{n \times n}$, $i = 0, 1, \ldots, m$ are given, the function

$$\varphi(x) = \begin{cases} \log \det \; F^{-1}(x) & F(x) > 0, \\ \infty & \text{otherwise;} \end{cases}$$

is finite if and only if $F(x) > 0$ and becomes infinite if x approaches the boundary of the feasible set: $\{x | F(x) > 0\}$. It can be shown that φ is strictly convex on the feasible set so that it has a unique minimizer denoted by x^*:

$$\begin{aligned} x^* &= \arg \; \min_x \varphi(x) \\ &= \arg \; \max_{F(x) > 0} \det F(x) \end{aligned}$$

We define here x^* as the analytic center of the LMI, $F(x) > 0$. $F(x^*)$ has the maximum determinant among all positive definite matrices of the form $F(x)$.

Newton's method, with appropriate step length selection, can be used to efficiently compute x^*, starting from a feasible initial point. The algorithm to compute x^* is:

$$x^{k+1} := x^{(k)} - \alpha^{(k)} H\left(x^{(k)}\right)^{-1} g\left(x^{(k)}\right) \tag{A.43}$$

where $\alpha^{(k)}$ is the damping factor of the k-th iteration, $g(x)$ and $H(x)$ denote the gradient Hessian matrix of $\varphi(x)$, respectively, at $x^{(k)}$.

The damping factor

$$\varphi(x) = \begin{cases} 1 & \text{if } \delta\left(x^{(k)}\right) \leq 1/4, \\ 1/\left(1 + \delta\left(x^{(k)}\right)\right) & \text{otherwise,} \end{cases}$$

where

$$\delta(x) \triangleq \sqrt{g(x)^T H(x)^{-1} g(x)}$$

is called the *Newton decrement* of φ at x.

EXAMPLE: 1

Find the analytic center of

$$F(x) = \begin{bmatrix} 1 & x \\ x & 1 \end{bmatrix} > 0$$

We have

$$\varphi(x) = \log \det F^{-1}(x) = -\log\left(1 - x^2\right)$$

$$\frac{d\varphi}{dx} = \frac{2x}{1 - x^2}$$

$$\frac{d^2\varphi}{dx^2} = \frac{2(1 + x^2)}{(1 - x^2)^2}$$

The feasible set in which $\varphi(x)$ is defined is: $(-1 \; 1)$, and the minimum, which occurs at $x = 0$, is also $\varphi(x^*) = 0$.

EXAMPLE: 2

Find the analytic center of

$$F(x) = \begin{bmatrix} 1-x_1 & x_2 \\ x_2 & 1+x_1 \end{bmatrix} > 0$$

We have

$$\varphi(x) = \log \det F^{-1}(x) = -\log\left(1 - x_1^2 - x_2^2\right)$$

$$\frac{d\varphi}{dx_1} = \frac{2x_1}{1 - x_1^2 - x_2^2}, \quad \frac{d^2\varphi}{dx_1^2} = \frac{2\left(1 + x_1^2 - x_2^2\right)}{\left(1 - x_1^2 - x_2^2\right)^2};$$

$$\frac{d\varphi}{dx_2} = \frac{2x_2}{1 - x_1^2 - x_2^2}, \quad \frac{d^2\varphi}{dx_2^2} = \frac{2\left(1 - x_1^2 + x_2^2\right)}{\left(1 - x_1^2 - x_2^2\right)^2};$$

$$\frac{d^2\varphi}{dx_1 dx_2} = \frac{4x_1 x_2}{\left(1 - x_1^2 - x_2^2\right)^2}$$

$$\begin{vmatrix} \varphi_{x_1 x_1} & \varphi_{x_1 x_2} \\ \varphi_{x_2 x_1} & \varphi_{x_2 x_2} \end{vmatrix} = 4\frac{1 - \left(x_1^2 + x_2^2\right)^2}{\left(1 - x_1^2 - x_2^2\right)^4}$$

The feasible set in which $\varphi(x)$ is defined is: $x_1^2 + x_2^2 < 1$, and the minimizer is $x^* = [0 \ 0]^T$, which is an analytical center of the LMI $L(x)$.

Data Used for Relevant Power System Components

B.1 SMIB SYSTEM

$H=2.37$ s, $D=0.0$, $K_A=400$, $R_s=0.0$ pu, $R_e=0.02$ pu, $T_d=5.90$ s, $T_A=0.2$ s, $\omega_s=314$ rad./s, $X_d=1.70$ pu, $X'_d=0.245$ pu, $X_e=0.7$ pu, $X_q=1.64$ pu, $V_{inf}=1.00\angle0°$ pu, $V_t=1.72\angle19.31°$ pu.

B.1.1 SVC and TCSC parameters for SMIB system

Table B.1 SVC and TCSC Parameters

Controller Type	Controller Parameter	X_L (pu)	X_C (pu)	α (°)	Internal Delay (ms)
SVC	$K_{SVC}=5.0$ $K_P=0$ $K_I=1.0$ $T_1=0.5$ s $T_2=0.15$ s	0.225	0.3708	150	$T_{SVC}=20$ s
TCSC	$K_{TCSC}=4.0$ $T_1=0.5$ s, $T_2=0.1$ s	0.0049	0.0189	145	$T_{TCSC}=25$ s

B.2 WSCC TYPE 3 MACHINE, 9 BUS SYSTEM

Table B.2 Load-Flow Data

Bus #	Bus Type	Voltage (pu) $V \angle \theta°$	P_G (pu)	Q_G (pu)	$-P_L$ (pu)	$-Q_L$ (pu)
1	Swing	1.04	0.719	0.546	–	–
2	PV	$1.025 \angle 9.48°$	1.63	0.304	–	–
3	$P-V$	$1.025 \angle 4.77°$	0.85	0.142	–	–

Continued

Table B.2 Load-Flow Data—cont'd

Bus #	Bus Type	Voltage (pu) $V \angle \theta°$	P_G (pu)	Q_G (pu)	$-P_L$ (pu)	$-Q_L$ (pu)
4	$P-Q$	$1.010 \angle -2.26°$	–	–	–	–
5	$P-Q$	$0.972 \angle -4.06°$	–	–	1.25	0.5
6	$P-Q$	$0.989 \angle -3.7°$	–	–	0.9	0.3
7	$P-Q$	$1.011 \angle 3.84°$	–	–	–	–
8	$P-Q$	$0.997 \angle 0.78°$	–	–	1.00	0.35
9	$P-Q$	$1.018 \angle 2.03°$	–	–	–	–

Table B.3 Machine Data

Parameters	Machine #1	Machine #2	Machine #3
R_s (pu)	0.089	0.089	0.089
H (s)	23.64	6.4	3.01
D (pu)	0.2	0.2	0.2
X_d (pu)	0.269	0.8958	1.998
X'_d (pu)	0.0608	0.1198	0.1813
X_q (pu)	0.0969	0.8645	1.2578
X'_q (pu)	0.0969	0.8645	1.2578
T'_{do} (s)	8.96	6.0	5.89
T'_{qo} (s)	0.31	0.535	0.6

Table B.4 Exciter Data

Parameters	Exciter #1	Exciter #2	Exciter #3
K_A	20	20	20
T_A (s)	0.2	0.2	0.2
K_E	1.0	1.0	1.0
T_E (s)	0.314	0.314	0.314
K_F	0.063	0.063	0.063
T_F (s)	0.35	0.35	0.35

$$S_E(E_{fdi}) = 0.0039\exp(1.555E_{fdi}), \quad \text{for } i = 1,2,3.$$

B.2.1 PSS and SVC parameters of 3-machine 9-bus system

$K_{PSS} = 10$, $\tau_1|_{PSS} = 0.4$, $\tau_2|_{PSS} = 0.15$, $K_{svc} = 10.0$, $K_P = 0$, $K_I = 1.0$, $T_1|_{svc} = 0.5\,\text{s}$, $T_2|_{svc} = 0.15\,\text{s}$, $X_L = 0.4925$ pu, $X_C = 1.1708$ pu, $\alpha = 136°$, $T_{svc} = 20$ ms.

B.2.2 **TCSC parameters of 3-machine 9-bus system**

$K_{TCSC}=10.0$, $T_1|_{TCSC}=0.5$ s, $T_2|_{TCSC}=0.15$ s, $X_L=0.0049$ pu, $X_C=0.0284$ pu, $\alpha=145.6°$, $T_{TCSC}=17$ ms.

B.3 **TWO-AREA SYSTEM**

Table B.5 Transmission Line Data on 100 MVA Base

From Bus Number	To Bus Number	Series Resistance (R_s) pu	Series Reactance (X_s) pu	Shunt Susceptance (B) pu
1	5	0.001	0.012	0
2	6	0.001	0.012	0
9	10	0.022	0.22	0.33
9	10	0.022	0.22	0.33
9	10	0.022	0.22	0.33
9	6	0.002	0.02	0.03
9	6	0.002	0.02	0.03
3	7	0.001	0.012	0
4	8	0.001	0.012	0
10	8	0.002	0.02	0.03
10	8	0.002	0.02	0.03
5	6	0.005	0.05	0.075
5	6	0.005	0.05	0.075
7	8	0.005	0.05	0.075
7	8	0.005	0.05	0.075

Table B.6 Machine Data

Variable	Machine at Bus #1	Machine at Bus #2	Machine at Bus #3	Machine at Bus #4
X_1 (pu)	0.022	0.022	0.022	0.022
R_s (pu)	0.00028	0.00028	0.00028	0.00028
X_d (pu)	0.2	0.2	0.2	0.2
X'_d (pu)	0.033	0.033	0.033	0.033
T'_{do} (s)	8	8	8	8
X_q (pu)	0.19	0.19	0.19	0.19
X'_q (pu)	0.061	0.061	0.061	0.061
T'_{qo} (pu)	0.4	0.4	0.4	0.4
H (s)	54	54	63	63
D (pu)	0	0	0	0

Table B.7 Excitation System Data

Variable	Machine at Bus #1	Machine at Bus #2	Machine at Bus #3	Machine at Bus #4
K_A (pu)	200	200	200	200
T_A (pu)	0.0001	0.0001	0.0001	0.0001
K_E	1	1	1	1
T_E (s)	0.314	0.314	0.314	0.314
K_F	0.063	0.063	0.063	0.063
T_F (s)	0.35	0.35	0.35	0.35

Table B.8 Load-Flow Results for the Two-Area System

Bus No	Bus Type	Voltage Mag. (pu)	Angle (°)	Real Power Gen. (pu)	Reactive Power Gen. (pu)	Real Power Load (pu)	Reactive Power Load (pu)
1	PV	1.03	8.2154	7	1.3386	0	0
2	Pv	1.01	−1.504	7	1.592	0	0
3	Swing	1.03	0	7.217	1.4466	0	0
4	PV	1.01	−10.2051	7	1.8083	0	0
5	PQ	1.0108	3.6615	0	0	0	0
6	PQ	0.9875	−6.2433	0	0	0	0
7	PQ	1.0095	−4.6977	0	0	0	0
8	PQ	0.985	−14.9443	0	0	0	0
9	PQ	0.9761	−14.4194	0	0	11.59	2.12
10	PQ	0.9716	−23.2922	0	0	15.75	2.88

B.4 IEEE TYPE 14-BUS TEST SYSTEM

Table B.9 Machine Data

Parameters	Machine-1	Machine-2	Machine-3	Machine-4	Machine-5
MVA	615	60	60	25	25
R_s (pu)	0.0031	0.0031	0.0031	0.0014	0.0014
H (s)	5.148	6.54	6.54	5.06	5.06
D (pu)	0.046	0.046	0.046	0.046	0.046
X_d (pu)	0.8979	1.05	1.05	1.25	1.25
X'_d (pu)	0.2995	0.1850	0.1850	0.232	0.232
X_q (pu)	0.646	0.98	0.98	1.22	1.22
X'_q (pu)	0.646	0.36	0.36	0.715	0.715
T'_{do} (s)	7.4	6.1	6.1	4.75	4.75
T'_{qo} (s)	0.3	0.3	0.3	1.5	1.5

Table B.10 Exciter Data

Parameters	Exciter-1	Exciter-2	Exciter-3	Exciter-4	Exciter-5
K_A	20	20	20	20	20
T_A (s)	0.02	0.02	0.02	0.02	0.02
K_E	1.0	1.0	1.0	1.0	1.0
T_E (s)	0.19	1.98	1.98	0.70	0.70
K_F	0.0012	0.001	0.001	0.001	0.001
T_F (s)	1.0	1.0	1.0	1.0	1.0

Table B.11 Base Case Load Flow Data of 14-Bus Test System

Bus #	Bus Voltage V (pu)	Angle θ (°)	Injection		Generation		Load	
			P_i (MW)	Q_i (MVAr)	P_g (MW)	Q_g MVAr	P_L (MW)	Q_L MVAr
1	1.06	0	−203.69	107.99	−203.69	107.99	0	0
2	1.045	−4.95	−4.264	15.92	−25.96	3.226	−21.7	−12.7
3	1.01	−12.66	84.07	−46.25	−10.12	−65.25	−94.2	−19
4	1.02	−14.39	31.32	59.73	20.12	52.23	−11.2	−7.5
5	1.08	−13.51	0.00	23.61	0.00	23.61	0	0
6	1.021	−10.31	30.97	−48.00	−16.825	−44.10	−47.8	3.9
7	1.032	−8.921	17.44	−26.70	9.8403	−28.30	−7.6	−1.6
8	1.041	−13.51	0.00	0.00	0.00	0.00	0	0
9	1.017	−15.23	42.76	−8.354	13.261	−24.95	−29.5	−16.6
10	1.01	−15.40	2.793	−10.522	−6.206	−16.32	−9	−5.8
11	1.011	−15.04	0.639	−4.056	−2.86	−5.856	−3.5	−1.8
12	1.005	−15.32	3.721	−5.805	−2.378	−7.405	−6.1	−1.6
13	1.001	−15.43	2.078	−14.33	−11.42	−20.13	−13.5	−5.8
14	0.991	−16.41	5.516	−14.76	−9.3839	−19.763	−14.9	−5

Table B.12 Transmission Line Data of 14-Bus System on 100 MVA Base

From Bus Number	To Bus Number	Series Resistance (R_s) (pu)	Series Reactance (X_s) (pu)	Shunt Susceptance ($B/2$) (pu)	Actual Tap Ratio
1	2	0.01938	0.05917	0.0264	1
1	7	0.05403	0.22304	0.0246	1
2	3	0.04699	0.19797	0.0219	1

Continued

Table B.12 Transmission Line Data of 14-Bus System on 100 MVA Base—cont'd

From Bus Number	To Bus Number	Series Resistance (R_s) (pu)	Series Reactance (X_s) (pu)	Shunt Susceptance (B/2) (pu)	Actual Tap Ratio
2	6	0.05811	0.17632	0.0187	1
2	7	0.05695	0.17388	0.017	1
3	6	0.06701	0.17103	0.0173	1
4	7	0	0.25202	0	0.932
4	11	0.09498	0.1989	0	1
4	12	0.12291	0.25581	0	1
4	13	0.06615	0.13027	0	1
5	8	0	0.17615	0	1
6	7	0.01335	0.04211	0.0064	1
6	8	0	0.20912	0	0.978
6	9	0	0.55618	0	0.969
8	9	0	0.11001	0	1
9	10	0.03181	0.0845	0	1
9	14	0.12711	0.27038	0	1
10	11	0.08205	0.19207	0	1
12	13	0.22092	0.19988	0	1
13	14	0.17093	0.34802	0	1

B.4.1 SVC and TCSC for 14-bus system

Table B.13 SVC and TCSC Parameter

Controller	X_L (pu)	X_C (pu)	α_{min} (°)	α_{max} (°)	Internal Delay (ms)
SVC	0.4925	1.1708	120	160	$T_{svc} = 20$
TCSC	0.0049	0.0284	145	160	$T_{TCSC} = 15$

B.5 14-AREA, 24-MACHINE, 203-BUS SYSTEM

Table B.14 Machine Data

Gen.#	P_G	Q_G	R_s	X_d	X_q	X'_d	X'_q	T'_d	T'_q	H	D
Slack bus	25.58	1.5113	0.0981	0.6845	0.646	0.2995	0.646	7.4	0.06	5.148	0.2
2	5.4	0.538	0.0981	0.3845	0.646	0.2995	0.646	7.4	0.06	5.148	0.2
3	5.4	-0.785	0.0981	0.5845	0.646	0.2995	0.646	7.4	0.06	5.148	0.2
4	3.8	-0.690	0.031	0.5979	0.646	0.2995	0.646	7.4	0.06	5.148	0.2
5	5.4	-1.026	0.0981	0.3845	0.646	0.2995	0.646	7.4	0.06	5.148	0.2
6	3.8	0.3662	0.031	0.4979	0.646	0.2995	0.646	7.4	0.06	5.148	0.2
7	2.4	-0.617	0.031	0.5979	0.646	0.2995	0.646	7.4	0.06	5.148	0.2
8	0.9	-0.309	0.031	1.05	0.98	0.185	0.36	6.1	0.3	6.54	0.2
9	0.46	0.026	0.041	1.25	1.22	0.232	0.715	4.75	1.5	5.06	0.2
10	1.8	0.2623	0.031	1.05	0.98	0.185	0.36	6.1	0.3	6.54	0.2
11	2	0.3795	0.031	1.05	0.98	0.185	0.36	6.1	0.3	6.54	0.2
12	1.8	0.1241	0.031	1.05	0.98	0.185	0.36	6.1	0.3	6.54	0.2
13	2.7	-1.576	0.041	0.6979	0.646	0.2995	0.646	7.4	0.06	5.148	0.2
14	0.243	0.0109	0.041	1.25	1.22	0.232	0.715	4.75	1.5	5.06	0.2
15	0.108	0.014	0.041	1.25	1.22	0.232	0.715	4.75	1.5	5.06	0.2
16	0.54	0.015	0.041	1.25	1.22	0.232	0.715	4.75	1.5	5.06	0.2
17	6.0	1.84	0.0981	0.4845	0.646	0.2995	0.646	7.4	0.06	5.148	0.2
18	0.5	0.3894	0.041	1.25	1.22	0.232	0.715	4.75	1.5	5.06	0.2
19	1.5	0.6947	0.031	1.05	0.98	0.185	0.36	6.1	0.3	6.54	0.2
20	6.0	-4.20	0.0981	0.5845	0.646	0.2995	0.646	7.4	0.06	5.148	0.2
21	0.204	0.0297	0.041	1.25	1.22	0.232	0.715	4.75	1.5	5.06	0.2
22	0.225	0.0232	0.041	1.25	1.22	0.232	0.715	4.75	1.5	5.06	0.2
23	0.225	0.02	0.041	1.25	1.22	0.232	0.715	4.75	1.5	5.06	0.2
24	0.4	0.0292	0.041	1.25	1.22	0.232	0.715	4.75	1.5	5.06	0.2

Table B.15 Exciter Data

Exciter #	K_A	T_A	K_E	T_E	K_F	T_F
1	20	0.02	1	0.19	0.0012	1
2	20	0.02	1	0.19	0.0012	1
3	20	0.02	1	0.19	0.0012	1
4	20	0.02	1	0.19	0.0012	1
5	20	0.02	1	0.19	0.0012	1
6	20	0.02	1	0.19	0.0012	1
7	20	0.02	1	0.19	0.0012	1
8	20	0.02	1	1.98	0.001	1
9	20	0.02	1	0.7	0.001	1
10	20	0.02	1	1.98	0.001	1
11	20	0.02	1	1.98	0.001	1
12	20	0.02	1	1.98	0.001	1
13	20	0.02	1	0.19	0.0012	1
14	20	0.02	1	0.7	0.001	1
15	20	0.02	1	0.7	0.001	1
16	20	0.02	1	0.7	0.001	1
17	20	0.02	1	0.19	0.0012	1
18	20	0.02	1	0.7	0.001	1
19	20	0.02	1	1.98	0.001	1
20	200	0.02	1	0.19	0.0012	1
21	20	0.02	1	0.7	0.001	1
22	20	0.02	1	0.7	0.001	1
23	20	0.02	1	0.7	0.001	1
24	20	0.02	1	0.7	0.001	1

Table B.16 Transmission Line Data of 24-Machine, 203-Bus System on 100 MVA Base

From Bus Number	To Bus Number	Series Resistance (R_s) (pu)	Series Reactance (X_s) (pu)	Shunt Susceptance $(B/2)$ (pu)	Actual Tap Ratio
44	96	0.0036	0.0071	0.0108	1
44	91	0.0095	0.0189	0.0287	1
45	53	0.0385	0.077	0.0292	1
27	82	0.0116	0.0232	0.0352	1
27	91	0.0071	0.0143	0.0217	1
46	57	0.031	0.0621	0.0235	1
46	26	0.0053	0.0107	0.0162	1
36	97	0.0178	0.0357	0.0135	1

78	97	0.0185	0.0371	0.0141	1
47	109	0.0407	0.0813	0.0309	1
48	54	0.0134	0.0268	0.0407	1
49	50	0.0125	0.025	0.0379	1
49	26	0.0078	0.0157	0.0238	1
52	32	0.0064	0.0128	0.0195	1
53	42	0.0006	0.0011	0.0004	1
53	76	0.0132	0.0264	0.0401	1
55	30	0.0481	0.0961	0.0365	1
55	85	0.026	0.0519	0.0197	1
56	72	0.0189	0.0377	0.0573	1
96	62	0.0073	0.0146	0.0222	1
96	62	0.0188	0.0377	0.0143	1
96	70	0.0285	0.0571	0.0217	1
96	68	0.0107	0.0214	0.0081	1
96	73	0.0328	0.0656	0.0249	1
96	39	0.0111	0.0221	0.0336	1
96	40	0.0378	0.0756	0.0287	1
96	108	0.0189	0.0378	0.0143	1
58	36	0.0357	0.0713	0.0271	1
58	54	0.0321	0.0642	0.0244	1
59	81	0.0182	0.0364	0.0552	1
59	100	0.0388	0.0776	0.1178	1
60	77	0.0467	0.0934	0.0355	1
60	101	0.0146	0.0292	0.0111	1
61	69	0.009	0.018	0.0274	1
62	26	0.005	0.01	0.0152	1
62	68	0.0093	0.0185	0.007	1
62	84	0.0132	0.0264	0.0401	1
62	92	0.0103	0.0207	0.0314	1
63	74	0.0541	0.1081	0.041	1
63	32	0.0499	0.0999	0.0379	1
31	94	0.0096	0.0193	0.0292	1
30	41	0.0036	0.0071	0.0108	1
30	75	0.0139	0.0278	0.0422	1
30	85	0.0421	0.0842	0.0319	1
30	93	0.0065	0.013	0.0197	1
30	54	0.0185	0.0371	0.0563	1
64	67	0.0435	0.087	0.033	1
65	51	0.0206	0.0412	0.0156	1
65	35	0.0121	0.0243	0.0368	1
51	89	0.0128	0.0257	0.0097	1

Continued

Table B.16 Transmission Line Data of 24-Machine, 203-Bus System on 100 MVA Base—cont'd

From Bus Number	To Bus Number	Series Resistance (R_s) (pu)	Series Reactance (X_s) (pu)	Shunt Susceptance ($B/2$) (pu)	Actual Tap Ratio
32	69	0.0203	0.0407	0.0617	1
32	80	0.0378	0.0756	0.0287	1
32	83	0.0414	0.0827	0.0314	1
32	85	0.0385	0.077	0.0292	1
66	90	0.0143	0.0285	0.0433	1
102	103	0.0043	0.0086	0.0032	1
103	73	0.0043	0.0086	0.0032	1
102	39	0.0078	0.0157	0.006	1
67	36	0.0053	0.0107	0.0162	1
33	73	0.0043	0.0086	0.013	1
33	73	0.0046	0.0093	0.0141	1
26	56	0.0126	0.0251	0.0381	1
51	34	0.0208	0.0417	0.0158	1
51	35	0.0272	0.0545	0.0207	1
34	86	0.009	0.0179	0.0272	1
34	89	0.0087	0.0174	0.0066	1
69	40	0.0155	0.031	0.0471	1
70	40	0.01	0.02	0.0076	1
28	71	0.0012	0.0025	0.0038	1
28	90	0.0089	0.0178	0.0271	1
28	94	0.0104	0.0208	0.0315	1
73	39	0.0136	0.0271	0.0103	1
74	29	0.0021	0.0041	0.0063	1
74	81	0.0556	0.1113	0.0422	1
74	87	0.0349	0.0699	0.0265	1
36	78	0.0285	0.0571	0.0217	1
36	54	0.0678	0.1355	0.0514	1
37	76	0.02	0.0399	0.0606	1
77	101	0.0492	0.0984	0.0374	1
77	43	0.0071	0.0143	0.0054	1
77	98	0.0142	0.0284	0.0108	1
37	98	0.0143	0.0285	0.0108	1
37	88	0.0046	0.0093	0.0141	1
95	79	0.0125	0.025	0.0379	1
87	81	0.0506	0.1013	0.0384	1
85	83	0.0314	0.0628	0.0238	1
40	75	0.0257	0.0514	0.078	1

98	99	0.0056	0.0113	0.0043	1
98	100	0.0178	0.0357	0.0135	1
99	100	0.0121	0.0243	0.0092	1
77	37	0.0071	0.0143	0.0054	1
43	104	0.0317	0.0635	0.0964	1
104	101	0.0193	0.0385	0.0146	1
38	66	0.0014	0.0029	0.0011	1
38	105	0.0007	0.0014	0.0005	1
66	105	0.0014	0.0029	0.0011	1
39	108	0.0039	0.0078	0.003	1
31	107	0.0061	0.0121	0.0184	1
66	106	0.0018	0.0036	0.0054	1
81	109	0.0407	0.0813	0.0309	1
27	113	0.0082	0.0164	0.0249	1
104	110	0.0521	0.1041	0.0395	1
104	111	0.0214	0.0428	0.0162	1
110	111	0.0428	0.0856	0.0325	1
111	43	0.0642	0.1284	0.0487	1
104	43	0.0663	0.1327	0.0503	1
114	45	0.0143	0.0285	0.0108	1
114	53	0.0514	0.1027	0.039	1
116	122	0.0166	0.058	0.2425	1
122	119	0.0014	0.0049	0.0207	1
116	123	0.0037	0.013	0.2169	1
116	128	0.0046	0.016	0.2673	1
116	131	0.0094	0.0328	0.1372	1
116	140	0.0125	0.0438	0.7335	1
118	122	0.0025	0.0088	0.1477	1
118	124	0.0053	0.0184	0.3083	1
118	132	0.009	0.0315	0.5263	1
123	125	0.0012	0.0041	0.0684	1
122	133	0.0006	0.0022	0.0376	1
122	140	0.0064	0.0225	0.3759	1
125	117	0.0046	0.016	0.2669	1
115	126	0.0038	0.0133	0.2233	1
115	126	0.0076	0.0267	0.1118	1
115	132	0.0051	0.0177	0.297	1
126	127	0.0082	0.0287	0.1201	1
117	130	0.0036	0.0126	0.2109	1
119	134	0.001	0.0036	0.0602	1
122	134	0.0011	0.0038	0.0639	1
136	135	0.0055	0.0192	0.7218	1

Continued

Table B.16 Transmission Line Data of 24-Machine, 203-Bus System on 100 MVA Base—cont'd

From Bus Number	To Bus Number	Series Resistance (R_s) (pu)	Series Reactance (X_s) (pu)	Shunt Susceptance ($B/2$) (pu)	Actual Tap Ratio
138	120	0.0069	0.0243	0.406	1
138	137	0.0076	0.0265	0.4436	1
121	136	0.0053	0.0186	0.312	1
121	137	0.0001	0.0004	0.0075	1
141	134	0.0022	0.0076	0.1278	1
132	144	0.003	0.0106	0.1767	1
143	115	0.0091	0.0319	0.1335	1
127	143	0.0055	0.0193	0.0808	1
146	148	0.004	0.0151	0.9204	1
145	148	0.005	0.0189	1.154	1
146	147	0.002	0.0075	0.4545	1
145	147	0.0042	0.0156	0.9516	1
147	153	0.0056	0.021	1.2783	1
145	149	0.0073	0.0273	1.6618	1
149	150	0.0023	0.0087	2.1163	1
149	151	0.0006	0.0022	0.5255	1
151	154	0.0026	0.0099	2.4145	1
152	154	0.0025	0.0093	2.2725	1
150	157	0.0061	0.0228	1.3919	1
146	157	0.0032	0.0122	2.9685	1
155	149	0.0095	0.0355	2.166	1
155	145	0.0022	0.0083	0.5042	1
152	156	0.0034	0.0128	3.1247	1
158	159	0.0114	0.0128	0.0009	1
159	160	0.1027	0.1155	0.0081	1
160	161	0.1912	0.215	0.0038	1
161	162	0.1096	0.1233	0.0022	1
60	163	0.131	0.1723	0.0065	1
164	163	0.0571	0.0642	0.0011	1
145	115	0	0.0132	0	0.95
146	116	0	0.0132	0	0.95
147	117	0	0.0198	0	1
148	118	0	0.0198	0	1
150	119	0	0.0198	0	1.05
151	120	0	0.0198	0	1
152	121	0	0.0198	0	1
154	139	0	0.0198	0	1
149	142	0	0.0397	0	1

155	143	0	0.0198	0	1.05
115	26	0	0.0208	0	1
116	27	0	0.0313	0	0.95
117	28	0	0.0625	0	1
120	29	0	0.0667	0	1
122	30	0	0.0208	0	1
123	31	0	0.0313	0	1
124	32	0	0.0313	0	1
125	33	0	0.0625	0	1
125	33	0	0.0333	0	1
126	34	0	0.0333	0	1
126	34	0	0.0625	0	1
127	35	0	0.0313	0	1
128	36	0	0.0313	0	0.95
129	37	0	0.0313	0	1
130	38	0	0.0313	0	1
131	39	0	0.0313	0	1
132	40	0	0.0313	0	1
133	41	0	0.1	0	1
136	42	0	0.1	0	1
137	43	0	0.1	0	1
140	95	0	0.1	0	1
117	28	0	0.0333	0	1
136	42	0	0.2	0	1
134	112	0	0.0313	0	1
144	72	0	0.0313	0	1
111	164	0	0.2	0	1.05
10	96	0	0.0324	0	1
11	96	0	0.02	0	1
7	140	0	0.0179	0	1
8	133	0	0.0769	0	1
12	41	0	0.0314	0	1
24	41	0	0.1143	0	1
2	117	0	0.0135	0	1
3	147	0	0.0169	0	1
4	148	0	0.0253	0	1
1	149	0	0.0112	0	0.95
5	149	0	0.0177	0	0.95
13	135	0	0.0268	0	0.95
9	101	0	0.1333	0	1
21	98	0	0.25	0	1
22	99	0	0.25	0	1

Continued

Table B.16 Transmission Line Data of 24-Machine, 203-Bus System on 100 MVA Base—cont'd

From Bus Number	To Bus Number	Series Resistance (R_s) (pu)	Series Reactance (X_s) (pu)	Shunt Susceptance ($B/2$) (pu)	Actual Tap Ratio
23	100	0	0.25	0	1
14	158	0	0.1571	0	1
15	159	0	0.3125	0	1
16	104	0	0.0889	0	1
53	162	0	0.1333	0	1.05
17	141	0	0.0101	0	1
6	118	0	0.0405	0	1
18	112	0	0.0485	0	1
19	112	0	0.028	0	1
25	157	0	0.0112	0	0.9875
39	194	0	0.0533	0	1.05
36	201	0	0.127	0	1.05
54	195	0	0.08	0	1.05
64	196	0	0.254	0	1.05
64	196	0	0.4	0	1.05
77	197	0	0.127	0	1.05
86	198	0	0.0533	0	1.05
55	199	0	0.16	0	1.05
55	199	0	0.2	0	1.05
92	200	0	0.08	0	1.05
20	156	0	0.0156	0	0.9875
145	165	0	0.3333	0	1
146	166	0	0.3333	0	1
147	167	0	0.5	0	1
148	170	0	0.5	0	1
150	169	0	0.5	0	1
151	171	0	0.5	0	1
152	172	0	0.5	0	1
154	173	0	0.5	0	1
149	168	0	0.5	0	1
155	202	0	0.5	0	1
115	174	0	0.0741	0	1
116	175	0	0.1111	0	1
117	176	0	0.2222	0	1
120	177	0	0.2222	0	1
122	178	0	0.0741	0	1
123	179	0	0.1111	0	1
124	180	0	0.1111	0	1

125	181	0	0.2222	0	1
125	181	0	0.1111	0	1
126	182	0	0.1111	0	1
126	182	0	0.2222	0	1
127	183	0	0.1111	0	1
128	184	0	0.1111	0	1
129	185	0	0.1111	0	1
130	186	0	0.1111	0	1
131	187	0	0.1111	0	1
132	188	0	0.1111	0	1
133	189	0	0.3333	0	1
136	190	0	0.3333	0	1
137	191	0	0.3333	0	1
140	192	0	0.3333	0	1
117	176	0	0.1111	0	1
136	190	0	0.6667	0	1
134	193	0	0.1111	0	1
144	203	0	0.1111	0	1

Note: Total line = 159 + 35 (Line transformer—3 windings) + 37 (Load transformer—2 windings) + 35 (Load transformer—3 windings) = 266. But here total 265 numbers of lines are considered during analysis, because line 47 has been removed due to its very small impedance. Lines are arranged according to bus voltage level.

Table B.17 Base Case Load Flow Data of 14-Area, 24-Machine, 203-Bus System

Bus No.	Bus Voltage V (pu)	Angle θ (rad.)	P_{cal} (MW)	Q_{cal} (MVAr)	P_g (MW)	P_L (MVAR)	Q_g (MW)	Q_L (MVAr)
1	1.05	0	25.5802	1.5113	25.5895	0	1.5113	0
2	1.04	-0.3067	5.4	0.538	5.4	0	0.538	0
3	1.04	-0.2726	5.4	-0.7857	5.4	0	-0.7857	0
4	1.04	-0.2553	3.8	-0.6909	3.8	0	-0.6909	0
5	1.05	-0.1725	5.4	-1.0262	5.4	0	-1.0262	0
6	1.0433	-0.2162	3.8	0.3662	3.8	0	0.3662	0
7	1.02	-0.3236	2.4	-0.6179	2.4	0	-0.6179	0
8	1	-0.2883	0.9	-0.3094	0.9	0	-0.3094	0
9	1.02	-0.3444	0.46	0.026	0.46	0	0.026	0
10	1	-0.4078	1.8	0.2623	1.8	0	0.2623	0
11	1	-0.4263	2	0.3795	2	0	0.3795	0
12	1	-0.3514	1.8	0.1241	1.8	0	0.1241	0
13	1.01	-0.1747	2.7	-1.576	2.7	0	-1.576	0
14	1.02	-0.3172	0.243	0.0109	0.243	0	0.0109	0
15	1.02	-0.3243	0.108	0.014	0.108	0	0.014	0
16	1.02	-0.3586	0.54	0.015	0.54	0	0.015	0
17	1.04	-0.2758	6	1.84	6	0	1.84	0
18	1.03	-0.3291	0.5	0.3894	0.5	0	0.3894	0
19	1.03	-0.312	1.5	0.6947	1.5	0	0.6947	0
20	1.04	-0.1445	6	-4.2011	6	0	-4.2011	0
21	1.02	-0.3675	0.204	0.0297	0.204	0	0.0297	0
22	1.02	-0.3621	0.225	0.0232	0.225	0	0.0232	0
23	1.02	-0.3663	0.225	0.02	0.225	0	0.02	0

			0.4	0.0292	0.4	0.68	0.0292
24	1	-0.3622	0.4	0.0292	0	0	0.0292
25	1.0964	-0.3579	0	0	0	0	0
26	0.9998	-0.4807	0	0	0	0	0
27	0.9792	-0.4699	-0.68	-0.4214	0	0.68	0.4214
28	1.0107	-0.4248	0	0	0	0	0
29	1.02	-0.4275	0	0	0	0	0
30	1.0017	-0.4166	0	0	0	0	0
31	1.001	-0.4468	-0.6	-0.4	0	0.6	0.4
32	0.9818	-0.4671	-0.51	-0.3161	0	0.51	0.3161
33	0.9831	-0.4802	-2.13	-1.32	0	2.13	1.32
34	0.9825	-0.5076	-2.2	-1.3634	0	2.2	1.3634
35	0.9925	-0.4895	-0.51	-0.3161	0	0.51	0.3161
36	0.9889	-0.4701	0	0	0	0	0
37	1.0025	-0.4217	-0.4	-0.25	0	0.4	0.25
38	1.0071	-0.4233	-0.4	-0.26	0	0.4	0.26
39	0.9922	-0.4751	0	0	0	0	0
40	0.9952	-0.4675	-0.85	-0.5268	0	0.85	0.5268
41	0.9977	-0.4081	-2	-1.2395	0	2	1.2395
42	1.0035	-0.4061	0	0	0	0	0
43	1.025	-0.4065	0	0	0	0	0
44	0.9849	-0.4713	-0.72	-0.4462	0	0.72	0.4462
45	0.982	-0.4237	-0.15	-0.093	0	0.15	0.093
46	0.988	-0.4886	-0.61	-0.378	0	0.61	0.378
47	0.9701	-0.507	-0.22	-0.0983	0	0.22	0.0983
48	0.9586	-0.4771	-0.4	-0.18	0	0.4	0.18
49	0.9828	-0.4947	-0.77	-0.4072	0	0.77	0.4072

Continued

Table B.17 Base Case Load Flow Data of 14-Area, 24-Machine, 203-Bus System—cont'd

Bus No.	Bus Voltage V (pu)	Angle θ (rad.)	P_{cal} (MW)	Q_{cal} (MVAr)	P_g (MW)	P_L (MVAr)	Q_g (MW)	Q_L (MVAr)
50	0.9736	-0.5015	-0.35	-0.2169	0	0.35	0	0.2169
51	0.9708	-0.5114	-0.8	-0.4958	0	0.8	0	0.4958
52	0.9697	-0.4754	-0.86	-0.533	0	0.86	0	0.533
53	1.002	-0.4074	-0.5	-0.3569	0	0.5	0	0.3569
54	0.9674	-0.4672	-0.8	-0.4958	0	0.8	0	0.4958
55	0.9943	-0.4532	0	0	0	0	0	0
56	1.0024	-0.4779	-0.33	-0.2045	0	0.33	0	0.2045
57	0.9603	-0.5074	-0.4	-0.2479	0	0.4	0	0.2479
58	0.961	-0.484	-0.53	-0.3085	0	0.53	0	0.3085
59	1.0051	-0.4536	-0.14	-0.0868	0	0.14	0	0.0868
60	1.0102	-0.4113	-0.32	-0.21	0	0.32	0	0.21
61	0.9682	-0.4881	-0.31	-0.1921	0	0.31	0	0.1921
62	0.9879	-0.4806	-0.57	-0.3533	0	0.57	0	0.3533
63	0.9867	-0.4616	-0.25	-0.1549	0	0.25	0	0.1549
64	0.9995	-0.483	0	0	0	0	0	0
65	0.9786	-0.5031	-0.43	-0.2665	0	0.43	0	0.2665
66	1.0063	-0.4241	-0.39	-0.2417	0	0.39	0	0.2417
67	0.9892	-0.4726	-0.124	-0.0744	0	0.124	0	0.0744
68	0.9834	-0.4789	-0.66	-0.409	0	0.66	0	0.409
69	0.973	-0.4832	-0.74	-0.4586	0	0.74	0	0.4586
70	0.9882	-0.4718	-0.41	-0.2541	0	0.41	0	0.2541
71	1.0083	-0.4264	-0.92	-0.5702	0	0.92	0	0.5702
72	1.0167	-0.4635	-0.52	-0.3223	0	0.52	0	0.3223
73	0.9793	-0.4829	-2.15	-1.3325	0	2.15	0	1.3325

74	1.0146	-0.4331	-0.68	0	-0.4214	0	0.68	0	0.4214
75	0.9945	-0.4396	-0.33	0	-0.2045	0	0.33	0	0.2045
76	0.9966	-0.4183	-0.4	0	-0.2479	0	0.4	0	0.2479
77	1.014	-0.4147	0	0	0	0	0	0	0
78	0.9666	-0.4857	-0.55	0	-0.3409	0	0.55	0	0.3409
79	0.9661	-0.4532	-0.75	0	-0.4648	0	0.75	0	0.4648
80	0.952	-0.4987	-0.46	0	-0.1951	0	0.46	0	0.1951
81	0.9994	-0.4628	-0.23	0	-0.1425	0	0.23	0	0.1425
82	0.9649	-0.4798	-0.56	0	-0.3471	0	0.56	0	0.3471
83	0.9651	-0.4756	-0.4	0	-0.2479	0	0.4	0	0.2479
84	0.9579	-0.5004	-1	0	-0.6197	0	1	0	0.6197
85	0.9783	-0.462	-0.5	0	-0.3099	0	0.5	0	0.3099
86	0.9871	-0.5099	0	0	0	0	0	0	0
87	0.9974	-0.4562	-0.32	0	-0.1983	0	0.32	0	0.1983
88	0.9975	-0.4245	-0.45	0	-0.3203	0	0.45	0	0.3203
89	0.9701	-0.5134	-0.6	0	-0.4348	0	0.6	0	0.4348
90	1.0059	-0.4269	-0.29	0	-0.2097	0	0.29	0	0.2097
91	0.977	-0.4739	-0.54	0	-0.3347	0	0.54	0	0.3347
92	0.9907	-0.4841	0	0	0	0	0	0	0
93	0.9946	-0.4213	-0.5	0	-0.3099	0	0.5	0	0.3099
94	0.9995	-0.4402	-0.57	0	-0.3833	0	0.57	0	0.3833
95	0.987	-0.4392	0	0	0	0	0	0	0
96	0.9932	-0.4665	0	0	0	0	0	0	0
97	0.9745	-0.4804	-0.19	0	-0.1178	0	0.19	0	0.1178
98	1.0139	-0.4168	0	0	0	0	0	0	0
99	1.0158	-0.4164	0	0	0	0	0	0	0
100	1.0166	-0.4206	0	0	0	0	0	0	0

Continued

Table B.17 Base Case Load Flow Data of 14-Area, 24-Machine, 203-Bus System—cont'd

Bus No.	Bus Voltage V (pu)	Angle θ (rad.)	P_{cal} (MW)	Q_{cal} (MVAr)	P_g (MW)	P_L (MVAr)	Q_g (MW)	Q_L (MVAr)
101	1.0184	−0.4035	0	0	0	0	0	0
102	0.9847	−0.4798	−0.13	−0.0806	0	0.13	0	0.0806
103	0.9817	−0.4815	−0.05	−0.031	0	0.05	0	0.031
104	1.0198	−0.4048	0	0	0	0	0	0
105	1.0067	−0.4236	−0.12	−0.082	0	0.12	0	0.082
106	1.0063	−0.4241	−0.016	−0.0087	0	0.016	0	0.0087
107	0.9968	−0.4501	−0.35	−0.1825	0	0.35	0	0.1825
108	0.9902	−0.4751	−0.31	−0.1921	0	0.31	0	0.1921
109	0.9853	−0.4912	−0.12	−0.0744	0	0.12	0	0.0744
110	0.9963	−0.423	−0.35	−0.2269	0	0.35	0	0.2269
111	1.0056	−0.415	−0.28	−0.1935	0	0.28	0	0.1935
112	1.0119	−0.3523	−2	−1.39	0	2	0	1.39
113	0.9724	−0.4746	−0.38	−0.2355	0	0.38	0	0.2355
114	0.9785	−0.4263	−0.35	−0.2169	0	0.35	0	0.2169
115	1.046	−0.4236	0	0	0	0	0	0
116	1.0614	−0.4118	0	0	0	0	0	0
117	1.0354	−0.3744	0	0	0	0	0	0
118	1.0396	−0.3585	0	0	0	0	0	0
119	1.027	−0.3491	0	0	0	0	0	0
120	1.0706	−0.3114	0	0	0	0	0	0
121	1.0787	−0.3091	0	0	0	0	0	0
122	1.0273	−0.3569	0	0	0	0	0	0
123	1.0289	−0.4177	0	0	0	0	0	0
124	1.0129	−0.3948	0	0	0	0	0	0

125	1.0223	-0.4165	0	0	0	0	0	0
126	1.0228	-0.443	0	0	0	0	0	0
127	1.015	-0.4369	0	0	0	0	0	0
128	1.0531	-0.4309	0	0	0	0	0	0
129	1.0025	-0.4217	0	0	0	0	0	0
130	1.026	-0.3872	0	0	0	0	0	0
131	1.0225	-0.4393	0	0	0	0	0	0
132	1.0316	-0.4245	0	0	0	0	0	0
133	1.0261	-0.3558	0	0	0	0	0	0
134	1.0261	-0.349	0	0	0	0	0	0
135	1.0543	-0.2427	-0.3	-0.1859	0	0.3	0	0.1859
136	1.0603	-0.2942	0	0	0	0	0	0
137	1.0785	-0.3096	-0.6	-0.42	0	0.6	0	0.42
138	1.079	-0.3189	0	0	0	0	0	0
139	1.0586	-0.3255	-1	-0.6197	0	1	0	0.6197
140	1.0317	-0.3645	-4	-2.479	0	4	0	2.479
141	1.0238	-0.3327	-0.45	-0.2789	0	0.45	0	0.2789
142	1.0606	-0.2733	0	0	0	0	0	0
143	1.0303	-0.4045	0	0	0	0	0	0
144	1.027	-0.4339	0	0	0	0	0	0
145	1.0549	-0.3603	0	0	0	0	0	0
146	1.0638	-0.3667	0	0	0	0	0	0
147	1.0564	-0.3557	0	0	0	0	0	0
148	1.0608	-0.3425	-7	-4.479	0	7	0	4.479
149	1.0712	-0.2576	-4	-2.8592	0	4	0	2.8592
150	1.0717	-0.3189	0	0	0	0	0	0
151	1.0717	-0.2747	0	0	0	0	0	0
152	1.0942	-0.2945	0	0	0	0	0	0

Continued

Table B.17 Base Case Load Flow Data of 14-Area, 24-Machine, 203-Bus System—cont'd

Bus No.	Bus Voltage V (pu)	Angle θ (rad.)	P_cal (MW)	Q_cal (MVAr)	P_g (MW)	P_L (MVAr)	Q_g (MW)	Q_L (MVAr)
153	1.0208	−0.3992	−2.5	−1.8	0	2.5	0	1.8
154	1.0586	−0.3255	−10	−6.8592	0	10	0	6.8592
155	1.0719	−0.361	0	0	0	0	0	0
156	1.0928	−0.226	0	0	0	0	0	0
157	1.0985	−0.3579	0	0	0	0	0	0
158	1.019	−0.3539	0	0	0	0	0	0
159	1.0163	−0.3569	0	0	0	0	0	0
160	0.9791	−0.3953	−0.15	−0.09	0	0.15	0	0.09
161	0.969	−0.4186	−0.16	−0.07	0	0.16	0	0.07
162	0.9897	−0.4185	0	0	0	0	0	0
163	0.972	−0.4203	−0.15	−0.08	0	0.15	0	0.08
164	0.9822	−0.4241	0	0	0	0	0	0
165	0.9883	−0.466	−0.33	−0.18	0	0.33	0	0.18
166	0.9972	−0.4768	−0.35	−0.18	0	0.35	0	0.18
167	0.9894	−0.466	−0.23	−0.12	0	0.23	0	0.12
168	0.954	−0.4098	−0.31	−0.2	0	0.31	0	0.2
169	0.9656	−0.4643	−0.3	−0.183	0	0.3	0	0.183
170	0.9958	−0.4373	−0.2	−0.12	0	0.2	0	0.12
171	0.9846	−0.4174	−0.3	−0.15	0	0.3	0	0.15
172	0.959	−0.462	−0.35	−0.23	0	0.35	0	0.23
173	0.9653	−0.4976	−0.35	−0.15	0	0.35	0	0.15
174	0.9858	−0.5093	−0.38	−0.181	0	0.38	0	0.181
175	0.9734	−0.4851	−0.13	−0.05	0	0.13	0	0.05
176	0.9891	−0.4604	−0.48	−0.28	0	0.48	0	0.28

177	0.9836	−0.4829	−0.25	−0.154	0	0.25	0	0.154
178	0.9979	−0.4241	−0.1	−0.05	0	0.1	0	0.05
179	0.9832	−0.475	−0.25	−0.154	0	0.25	0	0.154
180	0.9818	−0.4671	0	0	0	0	0	0
181	0.9831	−0.4802	0	0	0	0	0	0
182	0.9787	−0.5128	−0.068	−0.05	0	0.068	0	0.05
183	0.9913	−0.4951	−0.05	−0.01	0	0.05	0	0.01
184	0.9889	−0.4701	0	0	0	0	0	0
185	0.9933	−0.4418	−0.18	−0.08	0	0.18	0	0.08
186	0.998	−0.4432	−0.18	−0.08	0	0.18	0	0.08
187	0.9922	−0.4751	0	0	0	0	0	0
188	0.9952	−0.4675	0	0	0	0	0	0
189	0.9977	−0.4081	0	0	0	0	0	0
190	0.967	−0.4703	−0.28	−0.15	0	0.28	0	0.15
191	0.9935	−0.4557	−0.15	−0.09	0	0.15	0	0.09
192	0.987	−0.4392	0	0	0	0	0	0
193	1.0052	−0.3644	−0.11	−0.06	0	0.11	0	0.06
194	0.9514	−0.4828	−0.13	−0.02	0	0.13	0	0.02
195	0.9671	−0.4681	−0.01	0	0	0.01	0	0
196	0.9698	−0.4909	−0.047	−0.001	0	0.047	0	0.001
197	0.9676	−0.431	−0.12	−0.0161	0	0.12	0	0.0161
198	0.9532	−0.5099	0	0	0	0	0	0
199	0.9675	−0.4605	−0.075	−0.01	0	0.075	0	0.01
200	0.9556	−0.4912	−0.08	−0.029	0	0.08	0	0.029
201	0.9649	−0.484	−0.1	−0.01	0	0.1	0	0.01
202	0.9841	−0.5085	−0.31	−0.15	0	0.31	0	0.15
203	1.0167	−0.4635	0	0	0	0	0	0

B.5.1 PSS and TCSC for 203-bus system

$K_{PSS}=10$, $\tau_1|_{PSS}=0.4\,\text{s}$, $\tau_2|_{PSS}=0.15\,\text{s}$, $K_{TCSC}=2.245$, $T_1|_{TCSC}=1.5\,\text{s}$, $T_2|_{TCSC}=0.11\,\text{s}$, $X_L=0.000526\,\text{pu}$, $X_C=0.00526\,\text{pu}$, $\alpha_{min}\,(°)=145°$, $\alpha_{max}\,(°)=160°$, $T_{TCSC}=17\,\text{ms}$.

Index

Note: Page numbers followed by *b* indicate boxes, *f* indicate figures and *t* indicate tables.

Printed and bound by CPI Group (UK) Ltd, Croydon, CR0 4YY

03/10/2024

01040323-0005